FPST-Vol. 6

FLUID POWER SYSTEMS AND TECHNOLOGY — 1999 —

presented at
THE 1999 ASME INTERNATIONAL MECHANICAL ENGINEERING CONGRESS AND EXPOSITION
NOVEMBER 14–19, 1998
NASHVILLE, TENNESSEE

sponsored by
THE FLUID POWER SYSTEMS AND TECHNOLOGY DIVISION, ASME

edited by
SANJAY I. MISTRY
JOHN DEERE PRODUCT ENGINEERING CENTER

TIM McLAIN
BRIGHAM YOUNG UNIVERSITY

THE AMERICAN SOCIETY OF MECHANICAL ENGINEERS
Three Park Avenue / New York, N.Y. 10016

ISBN No. 0-7918-1647-8

ISSN No. 1096-6455

FOREWORD

The Fluid Power Systems and Technology Division session for the 1999 ASME International Mechanical Engineering Congress and Exposition (IMECE) pertain to the following topics:

- Hydraulic and Electro-Hydraulic Systems
- Research Issues in Fluid Power Systems
- Applications of Advanced Control Techniques
- Hydraulic Valve Controls

In addition to the normal six sessions, this year we were successful in getting one industry/highlight session as well as one industry track session. These types of sessions will enhance interaction between the industry and the academia. These sessions have been used for panel discussion on the following topics:

- Productivity Enhancing Innovations in Fluid Power Systems
- Research Directions — Industry Perspectives

The Division, which was founded in 1989, is still growing and its goals are to increase the interaction between the industry and the academia, and to promote the teaching of fluid power systems in academia.

The Division currently has two Associate Editors for the ASME Journal of Dynamic Systems, Measurements, and Control: Prof. Satish S. Nair of University of Missouri, Columbia and Prof. Finn Conrad of Technical University of Denmark.

On behalf of the FPST Division, the Editors would like to sincerely thank the reviewers for their help.

Sanjay I. Mistry
John Deere Product Engineering Center
Waterloo, IA

Tim McLain
Brigham Young University
Provo, UT

FLUID POWER SYSTEMS AND TECHNOLOGY DIVISION EXECUTIVE COMMITTEE

CONTENTS

FPST-Vol. 6, Fluid Power Systems and Technology — 1999
ASME 1999

MATHEMATICAL MODELING OF A VARIABLE DISPLACEMENT AXIAL PISTON PUMP

Jeff W. DOBCHUK, Richard T. BURTON, Peter N. NIKIFORUK, and Paul R. UKRAINETZ

Department of Mechanical Engineering, University of Saskatchewan
57 Campus Drive
Saskatoon, SK, Canada S7N 5A9
(E-mail:jwd370@mail.usask.ca)

ABSTRACT

The variable displacement axial piston pump has been the subject of much research, having been studied from the controls, noise reduction, and design perspectives. The resulting body of research is large and very diverse in content. A review of the available publications was conducted for this paper in order to identify those works that would be most helpful in developing a complete and accurate mathematical model of an axial piston pump.

Most of the available publications can be classified into one of two general groups; those describing a small group of components to understand specific phenomena or those describing the entire pump for control or design purposes. The significant mathematical developments in various publications regarding specific phenomena, particularly those works involving nonlinear friction or pressure transients, were identified by the authors in this paper. When the mathematical developments of the phenomena specific effects are combined with the widely accepted kinematics equations for the pump, an accurate numerical model can be developed.

Works on linearized lumped parameter models and parameter sensitivity were examined for this paper and the limitations of these types of models were addressed. While linearized models offer mathematical simplicity, they suffer from poor accuracy over a wide range of operating conditions and do not reflect instantaneous swashplate dynamics.

This paper offers insight into the required complexity of a mathematical model that is necessary to achieve a desired accuracy as well as providing the appropriate references to develop that model.

KEYWORDS

Axial Piston Pump, Hydraulics, Modeling

INTRODUCTION

Axial piston pumps are widely used in industry because of their versatility. They offer a good combination of high operating pressure and easily controlled variable flow. The resulting efficiency and circuit control possibilities make pumps of this type very popular. The benefits realized from these pumps are only possible through the application of effective controllers. The design of these controllers depends on the ability to model accurately the axial piston pump. This paper contains a review of many years of research in this field and while it does not claim to be comprehensive, it should give a good starting point for people working in this area.

Three topics were examined in detail for this paper. The first of these was the pump kinematics equations. At any operating point many moving parts are interacting in a semi-constrained manner within the pump. It is essential that the kinematics and dynamics in the pump are properly understood and modeled if accuracy is to be achieved in any mathematical simulation.

The second area of interest was the effect of friction on the operation of the pump. An axial piston pump has many moving parts with metal-to-metal interfaces. The friction relationships are nonlinear and their contribution to the accuracy of the pump model can be significant when simulating small magnitude high frequency effects at extreme operating conditions.

The third topic examined for this paper was the pressure/flow relationships for the pumping and control pistons. Pressure transition refers to the region in a piston/slipper assembly rotation where it is not fully exposed to either the discharge or suction ports, but is somewhere in between. The varying geometry of the orifice formed by the relative motion between the valve plate and the barrel makes this a very difficult phenomenon to model. The equations describing the pressure in the piston chamber are nonlinear and can be complicated by the flow regime into the cylinder, which may be either laminar or turbulent, depending on orifice geometry and operating conditions. The pressure transition region has the largest effect on the torque imbalance on the swashplate. The pressure/flow characteristics of the control piston must also be examined as it is not only the control device, but also receives feedback from the swashplate, which gives information regarding the dynamics of the pump.

BRIEF DESCRIPTION OF PUMP OPERATION

A variable displacement axial piston pump contains five key components within its housing. These important components are the barrel, piston/slipper assembly, swashplate, valve plate, and swashplate control device, which is usually a control piston. These components and their relative positions are illustrated in the exploded view shown in Figure 1.

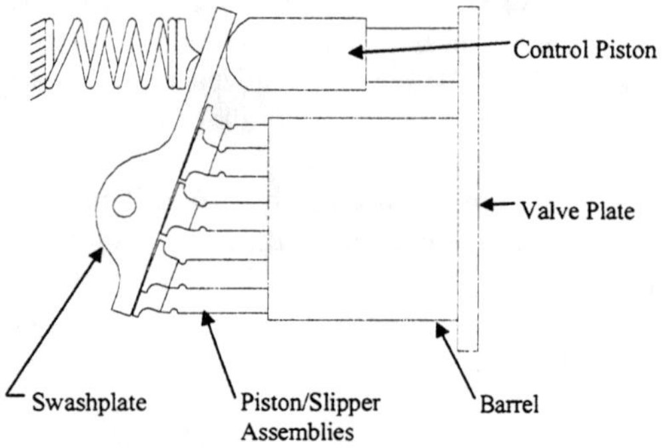

Figure 1. Key Components of the Variable Displacement Axial Piston Pump.

The barrel is attached to the pump's input shaft. Torque applied to this shaft causes the barrel to rotate about its long axis. The piston/slipper assemblies are arrayed around the same axis. As a result, the pistons rotate about the long axis of the barrel.

The piston/slipper assembly consists of two joined components. The pistons are fitted to the slippers by means of a ball joint. This ball joint allows the slippers to slide along the surface of the swashplate regardless of any angle imparted on the swashplate itself. The angle of the swashplate is controlled by an externally applied torque, which is usually supplied by a control piston. This piston must counteract the complex exciting torque profile produced by the pistons if a near constant swashplate angle is to be maintained at steady state.

If the swashplate is rotated about the axis of its supports, a periodic linear translation will occur in the pistons as the barrel rotates about its long axis. The periodic linear translation of the pistons creates the pumping motion. The pistons are exposed to the discharge port of the valve plate as they move upward, and this upward motion causes fluid to be expelled. As the pistons travel downward, they are exposed to the suction port of the valve plate and draw fluid in.

EARLIER RESEARCH; PRIOR TO 1985

Prior to the 1980's large scale computing power was generally not available to run numerical pump simulations. Mathematical analysis was carried out on piston pumps using a transfer function approach and linearization at a given operating condition.

Merritt (1967) typified this type of analysis in that frequency dependent input/output relationships were developed which facilitated the development of electrohydraulic linear control systems. This approach allowed a quick approximation of linearized system parameters, such as damping, stiffness, and undamped hydraulic natural frequency, as well as general system performance characteristics, such as response time, to be inferred. The obvious limitations were in the simplification of the operation of the pump. The pump was assumed to have a flow rate based on shaft velocity only and was assumed to be independent of all other system parameters. Internal leakage was meant to be determined experimentally and implemented in the model as a constant coefficient.

Concurrent research was being performed on many individual components as well. In an effort to understand output noise characteristics, Helgestad et al (1974) examined the pressure transients which occurred in the pumping piston chambers during the transition regions between the suction and delivery ports. Flow equations for the piston were developed which related the flow out of the pumping pistons to the instantaneous pressure in the pistons. These equations are vital to the determination of the forcing function in the equations of motion for the pistons.

Considerable work during this period was also dedicated to the wear and startup performance characteristics of the pump. This was necessary since the linearized models, such as those previously discussed, did not describe the internal operation of the pump such as slipper/swashplate behaviour. Understanding the behaviour of these interactions was prerequisite for understanding low speed and low pressure operations of a pump. Ikeya and Kato (1985) examined the problem of changes in the friction at low speeds where hydrodynamic lubrication tends to break down. This included a discussion of the slipper leakage at various conditions and the piston/barrel friction characteristics under metal-on-metal contact conditions. The results lent credibility to the earlier work of Iboshi and Yamaguchi (1982) who showed that the pressure and the shaft angular velocity had an effect on the relationship of the slipper/swashplate and pumping piston/slipper interfaces. The slipper/swashplate relationship was examined by Iboshi and Yamaguchi (1990) from the perspective of power loss. The effects of viscosity, shaft angular velocity, and swash angle on the possibility of the slipper contacting the swashplate were acknowledged. The condition of slipper contact was confirmed by Tienong et al (1989) as the slipper film thickness was shown to be unstable at the transitions between suction and discharge.

The reduced order model produced by Zeiger and Akers (1985) demonstrated that reasonable dynamic response characteristics could be achieved by a fourth order linear model. Perhaps more importantly, the instantaneous torque equations provided a blueprint for later numerical solving of the pump dynamics. The leakage, friction, and pressure/flow relationships were simplified to facilitate the solving of the describing equations. The analysis, however, was also a steady state study which eliminated many terms from the swashplate and piston describing equations.

LATER RESEARCH; POST 1985

Later, much work was done on the validity of a reduced order model including that by Zeiger and Akers (1985), Schoenau et al (1990), Palmberg and Krus (1985), Zhiwu and Souquan (1989), and the important sensitivity analysis of Kim et al (1987). These models were concerned with the average signal response of the pump and do not reflect the interactions of the swashplate and pistons with the controller. To examine the effect of each component a numerical solution for the swashplate dynamics must be solved, taking into account the components' instantaneous environment. This allows each nonlinear phenomenon to be examined in turn and full advantage can be taken of the mathematical modeling of these phenomena discussed hereafter.

Kinematics

The change in modeling philosophy from a lumped parameter approach to a numerical solution is reflected in the increased attention paid to the detailed consideration given to the pump kinematics and dynamics.

Zeiger and Akers (1985) proposed a model based on the torque applied to the swashplate. They concluded that the average response could be determined from a fourth order model. Kim et al (1987) and Schoenau et al (1990) described similar approaches and results. Although these simplified models showed decreased accuracy in various operating ranges, their limitations do not devalue the describing equations developed therein.

The basic kinematics equations are shown in many publications with similar results. The differences lie only in the detail to which a pump modeler wished to resolve a pump model. Since many similar approaches are available, the authors of this paper developed a summary of the equations with respect to the various analyses discussed earlier and the coordinates shown in Figure 2.

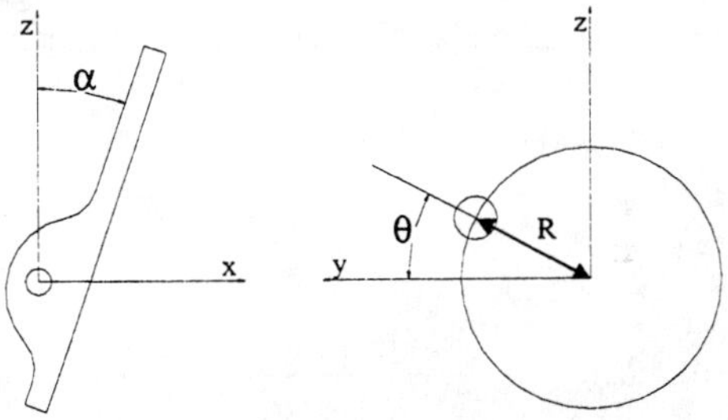

Figure 2. Coordinate system used for the development of the dynamic equations.

The equation for the motion of the swashplate, which is ultimately the device which is to be controlled, is developed from a torque balance including the actions due to the control piston, pumping pistons, as well as the inertial and friction effects of the swashplate itself. The general form for the dynamic equation for the swashplate is

$$\ddot{\alpha} = \frac{1}{J}\left(M_C(t) - \sum_{n=1}^{N} M_n(t) - C_S(t) - K_S \cdot \alpha \right) \tag{1}$$

where K_s is the spring constant of the return spring, $C_s(t)$ is the time varying friction term, $M_c(t)$ is the time varying control moment, and the summation of $M_n(t)$ represents the net exciting torque applied to the swashplate from all the pumping pistons. Approaches to modeling the time varying friction term are addressed in the friction section of this paper. The time varying control moment is normally generated by the application of a control pressure to a control piston. The control moment may be described by

$$M_c = R_c(t) \cdot P_{ch} \cdot A_o \qquad (2)$$

where $R_c(t)$ is the moment arm relative to the swashplate axis of rotation, P_{ch} is the pressure in the control piston chamber, and A_o is the area of the control piston. It was shown by Dobchuk (1998) that the control piston moment arm varies widely over the stroke of the pump and should be calculated at the appropriate swash angle. The control piston chamber pressure is not the control pressure, but is the pressure in the trapped volume created by the controller orifice. This will be discussed in more detail in the pressure/flow relationship section of this paper.

The second term of equation (1), which represents the total torque contribution of all the pumping pistons, must be solved in a piecewise manner. The derivation of the piston moment was performed in this paper relative to a reference angle as shown in Figure 2. The moment can be described in terms of a pumping piston moment arm and a pumping piston forcing function

$$M_n(t) = L_n(t) \cdot F_n(t) \qquad (3)$$

where

$$L_n(t) = \frac{R \cdot \sin(\theta_n) + a \cdot \sin(\alpha) + e}{\cos^2(\alpha)} \qquad (4)$$

and α represents the in-plane vertical offset of the point of action of the pistons from the swashplate axis of rotation, e represents the eccentricity of the swashplate axis of rotation relative to the swashplate centroid, and R represents the radial distance of any piston from the center of the barrel. The $F_n(t)$ term of equation (3) must be examined in depth. To examine the piston forcing function, the external forces must be considered as shown in Figure 3.

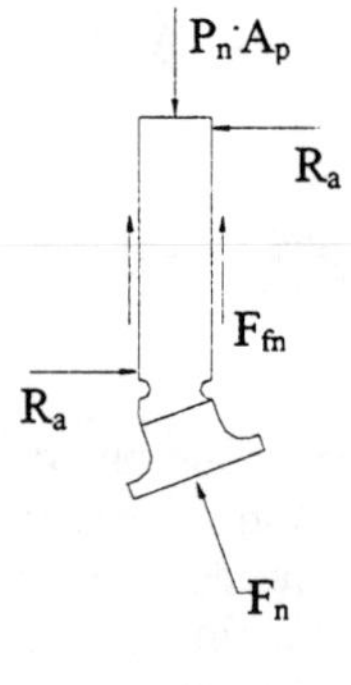

Figure 3. Piston Free Body Diagram.

The dynamic equation of a single piston has been well documented in many papers on the subject of axial piston pumps and is given by

$$F_n(t) = \frac{1}{\cos(\alpha)} \cdot \left(P_n \cdot A_p - m_p \ddot{x}_n - F_{fn}(t)\right) \qquad (5)$$

where the P_n denotes the instantaneous pressure in the pumping piston, A_p and m_p are the area and mass of that piston, respectively, and $F_{fn}(t)$ is the piston specific time variant friction term. This final term will be addressed in more detail in the friction section and the value of P_n will be discussed in the pressure/flow characteristics section of this paper.

A majority of early work in this area was concerned with the steady state accuracy and therefore assumed a static swashplate angle. The displacement of a pumping piston is straightforward and the equation with respect to Figure 2 is given by

$$x_n = R \cdot \sin(\theta_n) \cdot \tan(\alpha) + R \cdot \sin\left(\frac{\pi}{2}\right) \cdot \tan(\alpha_{max}) \qquad (6)$$

where α_{max} is the maximum swashplate angle and the second term sets zero displacement at bottom dead center of the piston stroke when the swashplate is at α_{max}. Given the displacement as shown in equation (6), the velocity and acceleration were determined

$$\dot{x}_n = R \cdot \omega \cdot \cos(\theta_n) \cdot \tan(\alpha) + \frac{R \cdot \dot{\alpha} \cdot \sin(\theta_n)}{\cos^2(\alpha)} \qquad (7)$$

and

$$\ddot{x}_n = \frac{2 \cdot R \cdot \dot{\alpha}^2 \cdot \sin(\theta_n) \cdot \sin(\alpha)}{\cos^3(\alpha)} + \frac{R \cdot \ddot{\alpha} \cdot \sin(\theta_n)}{\cos^2(\alpha)} + \frac{R \cdot \omega \cdot \cos(\theta_n)}{\cos^2(\alpha)} - R \cdot \omega^2 \cdot \sin(\theta_n) \cdot \tan(\alpha) \qquad (8)$$

It is clear from equations (7) and (8) why it would be expedient to perform a static analysis as all terms associated with the swashplate angular velocity and acceleration would be eliminated. The introduction of midrange operating values into equation (8) shows that, relative to the pressure force term in equation (5), the inertial effect of the swashplate may be neglected in many cases. This was shown numerically in the sensitivity analysis performed by Kim et al (1987).

The friction and pressure/flow terms must be addressed before the numerical integration can be performed on equation (1) to reach a solution for the dynamic behavior of the swashplate.

Friction

Phenomena due to friction are seldom included in overall pump modeling. This stems from the inherently nonlinear nature of friction. Friction interfaces are present between the cylinder and bore, slipper and piston, slipper and swashplate, swashplate and pintle, and the barrel and valve plate. There are losses associated with the bearings, but this review considers only those cases which depend on hydrostatic or hydrodynamic lubrication.

The first situation considered was that of the friction present between the pumping piston and the cylinder bore. A viscous friction term was developed and included by Kavanagh (1987). The development assumes an operating condition in which a Poiseuille type flow exists between the piston and bore. It is clear that this type of flow has the potential to break down at high lateral loads on the piston or low operating speeds. This calls into question the validity of a Poiseuille type model over a wide range of operating conditions.

Similarly, Manring and Johnson (1994) included a viscous friction term, C, in their development of the dynamic equation of a piston; however, no development of the coefficient was given. Inoue and Nakasato (1994) discounted friction between the piston and the bore outright as an assumption. This is the same assumption made by Zeiger and Akers (1985).

The works of Yi and Shirakaski (1995) and Ikeya and Kato (1985) showed that it may not be accurate to discount piston bore friction in the development of piston dynamic equations. Rather, there appear to be two separate friction regimes present, one in which shear layer lubrication is present and one in which metal-to-metal contact has occurred. The switch between these two situations is not clear as it is dependent on operating conditions. The two regime scenario is stated by Harris et al (1993) as including a low speed, high pressure boundary lubrication situation and a higher speed lower pressure viscous friction situation.

Viscous friction acting on the swashplate is also regularly neglected in the development of dynamic equations for the swashplate. From a parameter sensitivity analysis, Kim et al (1987) concluded that this does not have a large impact on the steady state response of the pump model. Friction on the swashplate, both viscous friction due to the fluid in the pump case and Coulomb friction between the plate and mounts, does affect the natural frequency of the system and the dynamic response as illustrated by Dobchuk (1998).

In the event that a precise model of the piston/slipper relationship is to be developed, it would be irresponsible to neglect the work of Yi et al (1990). The friction moment present at the piston/slipper interface results in a change in the equations of motion governing the piston/slipper couple if they are to be tracked separately for the entire rotation of the piston. This friction moment is affected by the lubrication regime present at the piston/slipper interface. Harris et al (1993) showed that the behaviour of the piston/slipper lubrication is affected by the operating conditions of the pump, including the rotational velocity and the operating pressure.

Some general conclusions can be drawn from the aforementioned research regarding the breakdown of slipper films. A film may breakdown allowing metal-on-metal contact during a transition region, in a low speed and high pressure condition, and shows a very high sensitivity to viscosity. This sensitivity makes modeling of film thickness very difficult.

If the model is to be extended to include the input torque, two further effects must be examined. The first is the interface between the slipper and the swashplate. This interface accounts for much of the loss associated with axial piston pumps according to Iboshi and Yamaguchi (1990). This interface is also a problematic area in terms of wear as shown by Szafron (1986). The second effect is the oil film present between the barrel and the valve plate. Rui et al (1989) showed the dependence of this oil film thickness on the supply pressure which acts to tilt the barrel, on the viscosity, and on the rotational velocity which acts to strengthen the effect of the hydrodynamic lubrication. Yamaguchi et al (1985) investigated the affect of different fluids and valve plate geometries on the fluid film characteristics between the barrel and the valve plate.

Pressure/Flow Relationships

The torque balance on the swashplate is the most important factor in the control of the output of a variable displacement axial piston pump. The relationship which has the highest potential to destabilize the system is the imbalance caused by the transition zones of the valve plate. Imbalance here is also the cause of noise in the output of the pump. The pressure transition region has been analyzed very closely with the result that an accurate description of the pressure force term in equation (5) is obtainable.

This effect has been examined as a noise problem for many years and was analytically considered by Helgestad et al (1974). Zeiger and Akers (1985) used a development of a similar type and solved it numerically to determine the instantaneous swashplate torque. Manring and Johnson (1994) took a simplified approach by assuming the transition happened linearly and then included a pressure transition angle to account for the torque imbalance. This helped explain torque fluctuations and gave the correct net average torque while avoiding the complications associated with modeling exact orifice geometry. Kavanagh (1987) included a detailed area calculation in his Master's thesis, but did not give any pressure transition numerical results.

The standard model of piston transition pressure was improved by Lin et al (1987) by including a pressure entrapment region. This region occurs when the piston is vented to neither the suction or discharge port of the valve plate and translation of the piston upward causes compression of the fluid and a subsequent pressure rise. Edge and Darling (1986) took into account the fluid inertia in the pressure/flow relationship and thereby obtained an approximation of the overpressure transient that occurs in the transition from the suction to the discharge ports.

Lu et al (1989) attempted to improve the model further by including a correction factor to the discharge coefficient. This was discussed as accounting for bulk modulus changes in the fluid due to large changes in fluid pressure. It may also be required to account for discrepancies caused by a transition from turbulent to laminar flow conditions. Depending on operating conditions, either flow regime could be present at the pumping piston orifice. An operating condition parameter adjustment could acknowledge this without including two separate flow equations. Alternatively, the orifice equation of Ellman and Piche (1996) can by used to account for the laminar/turbulent transition with a function which is differentiable over all pressure differences.

The pressure/flow relationship was solved numerically by Wicke et al (1998) and showed an overpressure due to compressibility and fluid inertia effects. They also showed that the pressure transients at both the suction/discharge and discharge/suction transitions are effected significantly by swashplate angle.

The effects of pressure/flow relationships must also be considered with respect to the control piston. Since the control torque is not applied to the plate through a rigid applicator, but through a trapped volume of compressible fluid, some consideration by the authors was given to this aspect. If the control piston is assumed to be a volume trapped by some control orifice, A_c, then the control piston chamber pressure in equation (2) can be determined and is given by

$$P_{ch} = P_{ch}\big|_{t=0} + \int_{t=0}^{t} \frac{\beta}{V_c} \cdot \left(\frac{dV_c}{dt} - Q_c - Q_{cl} \right) \qquad (9)$$

where V_c is the control piston trapped volume, β is the equivalent fluid bulk modulus, Q_c is the flow through the control orifice, and Q_{cl} is the leakage past the control piston. It is clear from equation (9) that for low control flow the rate of change of the control volume will dominate the equation and the pressure will fluctuate greatly as β is large and V_c tends to be small. This is the case when a compensator is attached to the pump. As the control orifice becomes large, the control flow term and leakage will tend toward the value of the rate of change in control volume. In this situation the standard view, that the control force is simply the control pressure multiplied by the piston area, applies. Kim et al (1987) showed that the steady state response was insensitive to the compressibility of the control piston volume. Exclusion of this parameter will lower the overall order of the pump model and may make it difficult to model the high frequency dynamics of the swashplate.

CONCLUDING COMMENTS

Upon review of the available publications, several general statements are made by the authors regarding the mathematical modeling of axial piston pumps. Reasonable prediction of the of the swashplate input/output characteristics may be obtained from a fourth order model subject to midrange pressure and rotational velocity constraints. This type of model will not resolve the high frequency swashplate response which reflects the direct interaction of the pumping pistons and the swashplate. If the interaction of the swashplate and pistons is to be included in a model, the swashplate dynamic equation must be solved numerically and the effect of friction must be addressed. In most cases the friction characteristics are dependent on the behaviour of a fluid film which is notoriously difficult to model. Two friction regimes are possible, one in which adjacent parts are in boundary lubrication and one in which the parts are separated by a hydrostatic or hydrodynamic layer, which cause a change in the behaviour of the pump when the operating conditions are varied. The pressure/flow relationships involving the pumping pistons and the orifice created by the valve plate are also critical to the correct determination of the equations describing the motion of the swashplate. These relationships are also important from the perspective of noise generation, as the pressure transition region is known to introduce large amounts of system noise.

REFERENCES

P. J. Chapple, *Friction in Hydrualic Motors*, International Conference on Fluid Power Transmission and Control, 1989, pp. 229.

J. Dobchuk, *Mathematical Modeling of an Axial Piston Pump*, University of Saskatchewan Masters Thesis in Progress, 1998.

K. A. Edge and J. Darling, *Cylinder Pressure Transients in Oil Hydraulic Pumps With Sliding Valve Plates*, Proceedings of the Institution of Mechanical Engineers Vol. 200 N B1, 1986, pp. 45.

A. Ellman and R. Piche, *A Modified Orifice Flow Formula for the Numerical Simulation of Fluid Power Systems*, ASME Fluid Power Systems and Technology, 1996, pp.

R. M. Harris, K. A. Edge and D. G. Tilley, *The Spin Motion of Pistons In a Swashplate-Type Axial Piston Pump*, The Third Scandinavian International Conference on Fluid Power, 1993, pp. 95.

B. Helgestad, K. Fosterand and F. Bannister, *Pressure Transients in an Axial Piston Hydraulic Pump*, Proceedings of the Institution of Mechanical Engineers Vol. 188 N 17, 1974, pp. 188.

Naohiro Iboshi and Atsushi Yamaguchi, *Characteristics of a Slipper Bearing for Swash Plate Type Axial Piston Pumps and Motors (1st Report, Theoretical Analysis)*, Bulletin of the JSME, 1982, pp. 1921.

Naohiro Iboshi and Atsushi Yamaguchi, *Effects of Swash Plate Angle, Fluid Viscosity and Operating Conditions on Characteristics of a Slipper Bearing For Swash Plate Type Axial Piston Pumps and Motors*, Fluid Power Control and Robotics - Proceedings of Chengdu 90 ICFPCR, 1990, pp. 274.

Mitsuei Ikeya and Hideki Kato, *Leakage Characteristics of the Hydraulic Slipper Bearing in Swashplate Type Axial Piston Motor at Starting and Low Speed*, International Conference on Fluid Power Transmission and Control, 1985, pp. 695.

Kiyoshi Inoue and Masakasu Nakasato, *Study of the Operating Moment of a Swash Plate Type Axial Piston Pump First Report*, Journal of Fluid Control, Vol 22. N 1, 1994, pp. 7.

Kiyoshi Inoue and Masakasu Nakasato, *Study of the Operating Moment of a Swash Plate Type Axial Piston Pump Second Report*, Journal of Fluid Control, Vol 22. N 1, 1994, pp. 30.

G. P. Kavanagh, *Modeling of an Axial Piston Pump*, University of Saskatchewan Master's Thesis, 1987.

S. D. Kim, H. S. Cho and C. O. Lee, *A Parameter Sensitivity Analysis for the Dynamic Model of a Variable Displacement Axial Piston Pump*, Proceedings of the Institution of Mechanical Engineers Vol. 201 N C4, 1987, pp. 235.

S. J. Lin, A. Aker and G. Zeiger, *Oil Entrapment in an Axial Piston Pump and its Effect Upon Pressures and Swashplate Torques*, National Conference on Fluid Power - 42nd, 1987, pp. 113.

Lu Yuan-Zhang, Chen Zhao-Nen, Qiu Ze-Lin and Xu Zhen-Son, *Measurement and Simulation Model Study of Cylinder Pressure Transient in an Axial Piston Pump*, International Conference on Fluid Power Transmission and Control, 1989, pp. 471.

N. D. Manring and R. E. Johnson, *Swivel Torque Within a Variable-Displacement Pump*, National Conference on Fluid Power - 46th, 1994, pp. 13.

H. E. Merritt, *Hydraulic Control Systems*, John Wiley & Sons, 1967.

J. O. Palmberg and P. Krus, *Dynamic Response Characteristics of Pressure Control Pumps*, International Conference on Fluid Power Transmission and Control, 1985, pp. 110.

Huang Rui, Katstoshi Higuchi, Syunichi Kobayashi, Mitsuei Ikeya and Jyun Hatsuse, *The Study of the Gap Characteristics Between the Valve Plate and Cylinder in Swashplate-Type Axial Piston Motor*, International Conference on Fluid Power Transmission and Control, 1989, pp. 773.

G. J. Schoenau, R. T. Burton and G. P. Kavanagh, *Dynamic Analysis of a Variable Displacement Pump*, Transactions of the ASME Journal of Dynamic Systems, Measurement, and Control Vol 112, 1990, pp. 122.

K. Szafron, *Performance of Components with F.R.F.*, University of Saskatchewan Masters Thesis, 1986.

Zhou Tienong, Han Decai and Pan Yongge, *Study on Dynamic Film Characteristics of the Slipper in Working Conditions*, International Conference on Fluid Power Transmission and Control, 1989, pp. 284.

Volker Wicke, K. A. Edge and N. D. Vaughan, *Investigation of the Effects of Swashplate Angle and Suction Timing on the Noise Generation Potential of an Axial Piston Pump*, ASME Fluid Power Systems and Technology, 1998, pp. 77.

Atsushi Yamaguchi, Hiroyoshi Sekine, Seiji Shimizu and Shuichi Ishida, *Bearing/Seal Characteristics of the Fluid Film Between a Valve Plate and a Cylinder block of Axial Piston Pumps and Motors*, International Conference on Fluid Power Transmission and Control, 1985, pp. 678.

Xu Yaoming and Son Jun, *Hydrostatic Bearing with Intermittent Oil Supply for the Distributor of Axial Plunger Pump*, International Conference on Fluid Power Transmission and Control, 1985, pp. 667.

Fang Yi and Masataka Shirakashi, *Mixed Lubrication Characteristics Between the Piston and Cylinder in Hydraulic Piston Pump-Motor*, Transactions of the ASME Journal of Tribology Vol. 117, 1995, pp. 80.

Fang Yi, Shunichi Kobayashi, Kasuyuki Matsumoto and Mitsuie Ikeya, *Friction Characteristics of a Ball Joint in the Axial Piston Motor*, Fluid Power Control and Robotics - Proceedings of Chengdu 90 ICFPCR, 1990, pp. 262.

G. Zeiger and A. Akers, *Torque on the Swashplate of an Axial Piston Pump*, Transactions of the ASME Journal of Dynamic Systems, Measurement, and Control Vol 107, 1985, pp. 220.

Zhang Zhiwu and Hou Souquan, *Experiment and Research of Model XBPQC-F75FF Pressure-Flow Compensated Load-Sensing Axial Piston Pump*, International Conference on Fluid Power Transmission and Control, 1989, pp. 317.

DESIGN AND ANALYSIS OF A NEW HIGH PERFORMANCE ELECTROHYDRAULIC ACTUATOR

Saeid Habibi
Department of Mechanical Engineering
University of Saskatchewan, 57 Campus drive
Saskatoon, Saskatchewan, Canada S7N 5A9

Andrew Goldenberg
Dept. of Mechanical and Industrial Engineering
University of Toronto, 5 King's College Road,
Toronto, Ontario, Canada, M5S 3G8

ABSTRACT

This paper describes a new high performance robotic actuation system that presents a powerful alternative to existing commercial systems. This actuation system is self-contained and combines the benefits of conventional hydraulic and direct drive electrical actuators, namely high torque/mass ratio and modularity. It is referred to as the ElectroHydraulic Actuator (EHA) and results from the fusion of the above mentioned technologies. EHA is a unique device with its own characteristics and requires hydraulic components that are specifically tailored to its needs. To fulfill these requirements, a new symmetrical Linear Actuator (SLA) was designed as a sub-component of EHA. Experimental data from a prototype of EHA are reported.

INTRODUCTION

A new actuation system referred to as the ElectroHydraulic Actuator (EHA) has been designed and prototyped. The EHA prototype demonstrated an extremely high level of performance and an ability of moving a load of 20 Kg with critical damping at an accuracy of 10 Micron, a bandwidth of 22 Hz, and a rise time of 0.2 seconds. Experimental data from this prototype indicate the possibility of designing a commercially viable system that could combine the benefits so far uniquely associated with either direct drive electrical actuators or hydraulic systems.

EHA is a hydrostatic actuation system that has a bi-directional pump directly connected to a symmetrical actuator. As a result, the actuator pressure varies with the pump speed and there are no requirements for oil reservoirs or electrohydraulic servo-valves. Most hydrostatic circuits use a constant speed piston pump with variable displacement. The flow and pressure in the hydraulic actuator is regulated by moving a swash plate that changes the magnitude and direction of flow, Watton (1989), Korn (1972). These circuits are energy inefficient as the pump would be continuously running irrespective of the motion of the actuator. Furthermore, use of piston pumps also results in ripple effects that could degrade high precision motion. An alternative strategy is to vary the pressure and flow by changing the speed and direction of a fixed displacement pump, Hunter (1985), Desai and Bobrow (1989), Arnautovic (1993).The progress in such hydrostatic

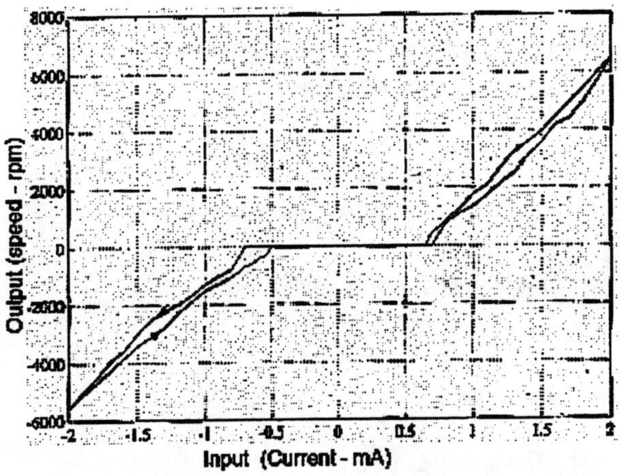

Figure 1: Dead-band in Hydrostatic Actuation, Arnautovic (1993)

actuation has been impeded by the presence of a dead-band (Figure 1) with an effect similar to backlash, thus limiting the positional accuracy and stability of this type of systems, Arnautovic (1993). In the design of EHA, a fixed displacement pump is used and the dead-band is alleviated by:

◊ implementation of a high gain cascaded control strategy involving motor speed feedback, and
◊ appropriate component selection.

While the performance of EHA exceeds that of conventional hydraulic systems as demonstrated by its prototype, lack of an electrohydraulic servo-valve (normally used for precision control) makes it more cost effective.

In this paper, the characteristics of conventional hydraulic and electrical actuation systems are reviewed. EHA and a symmetrical linear actuator that is specifically designed for EHA are described. Experimental results from a prototype of EHA are presented.

NOMENCLATURE

Symbol	Definitions
E	Current multiplier.
I_v, V_v	Control input and voltage.
K_v	Volumetric flow term coefficient.
ψ	Leakage flow term coefficient.
θ	Joint coordinate in rads.
τ	Output torque.
β_e	Effective bulk modulus of hydraulic oil.
K_{sv}	Transfer function relating the differential current input of servo-valve to spool position.
K_c	Motor gain.
K_{valve}	Valve gain.
$K_{fp}, K_{P_{visc}}$	Coefficient of friction and viscosity.
K_ω	Back emf.
R_c, L_c	Motor resistance and inductance.
J_{pm}	Motor/pump inertia.
K_p, K_I	Proportional and integral gains.
ξ	Pump cross-port leakage coefficient.
Q_a, Q_b, Q_{cas}	Pump/actuator flows and pump case drain.
Q_{ac1}, Q_{ac2}, Q_{ac}	Accumulator flow components and total flow.
V_a, V_b	Actuator chamber volumes.
V_o	Pipe plus mean actuator chamber volume.
$V_{acc}, V_{acc_{initial}}$	Accumulator volume.
x_o, x	The mean position and, the displacement from the mean position of linear actuators.
A_1, A_2	Pressure areas in asymmetrical actuators.
A	Pressure area in symmetrical actuators.
l_o, l_1, l_2	Moment arm, linear actuators.
P_s, P_r, P_1, P_2	The supply, the return and the two actuator chamber pressures.
$P_{acc}, P_{acc_{initial}}$	Accumulator pressure.
$P_{valve_{break}}$	Differential crack pressure needed to allow directional flow in check valve.
P_a, P_b, P_{cas}	Pump port and case pressures.
ΔP	Pressure difference.
D_p	Pump volumetric displacement.
T_c	Torque loss due to coulomb friction
T_{load}	Load torque.
t_m	Motor time constant.
V_m	Mean chamber volume, rotary actuators.
ω_p	Pump angular velocity.
$L, L_e,$	Leakage coefficient.

CHARACTERISTICS OF CONVENTIONAL HYDRAULIC ACTUATION SYSTEMS NTRODUCTION

Hydraulics for transmission is well established in some sectors of industry such as aeronautics. Nonetheless, the demand for hydraulic actuation is decreasing and has significantly fallen due to the progress in the field of electric motors. Direct drive electrical actuation systems are becoming increasingly common as they are modular and do not suffer from leakage. However, they in turn have their limitation of producing a comparatively low torque/mass ratio. Heavy-duty systems such as industrial robots and excavators therefore still use hydraulic actuation that offers large torque for a comparatively small final actuator size.

In conventional hydraulic systems, a centralized constant pressure hydraulic power supply is an integral part of the actuation system, with flexible hoses connecting this power supply to the valves, from where solid tubing directs the flow to the actuator. Supply systems are generally composed of a hydraulic pump, an accumulator, a relief valve, a cooling sub-circuit, an accumulator loading/ unloading sub-circuit, check valves and oil filters.

The advantages of conventional hydraulic circuits can be summarized as follows:

◊ Hydraulic actuators have a significantly higher torque/mass ratio in comparison to electrical actuators (between 10-30 times higher). Hydraulic manipulators can therefore handle large loads for a comparatively smaller overall size.

◊ In extensive studies conducted in Habibi etal (1994), Habibi and Richards (1994), it is indicated that as a result of volumetric flow within hydraulic actuators, hydraulic manipulators are more decoupled than their electrical counter-part (in mathematical terms, if a linearized transfer function is obtained for any configuration of the robot, this transfer function is likely to be diagonally dominant and, the volumetric flow reduces the effects of centrifugal and coriolis cross-coupled forces). This is a unique feature that results in a significant improvement in the performance of hydraulic manipulators by reducing the effects of cross-coupling torque. The complexity of controller design for such robots is thus significantly reduced.

Conventional hydraulic systems have a number of disadvantages that can be summarized as:

◊ Requirement for a centralized power system.

◊ Fluctuations in the supposedly constant supply pressure, Habibi and Richards (1994).

◊ Degree of compliance in comparison to electrical actuators.

◊ Leakage.

◊ Large size of the supply unit due to the requirement for a reservoir.

◊ Noise.

◊ Cost: In addition to components of the hydraulic supply unit, each actuator requires an electrohydraulic servo-valve that is expensive.
◊ Low energy efficiency: the hydraulic pump continuously runs at a constant speed irrespective of the operating conditions in order to maintain a constant supply pressure.

In the past two decades, these deficiencies have seriously affected the market share of hydraulic systems.

CHARACTERISTICS OF DIRECT DRIVE ELECTRICAL ACTUATION SYSTEMS

In certain segments of industry such as Robotics, there is a trend away from geared electromagnetic actuators and towards direct drive harmonic actuators, as gearing results in added friction, backlash and flexibility. In turn, the disadvantages of harmonic drives can be listed as:

◊ Low torque/mass ratio.
◊ Relatively high cost.
◊ Increased joint interaction in comparison to hydraulic robots.

Their main advantage, if low torque is required, is the absence of gear which makes them more flexible in their application and less bulky.

THE ELECTROHYDRAULIC ACTUATOR (EHA)

The Electrohydraulic Actuator is based on the principle of closed-circuit hydrostatic transmission. In this system, as opposed to the conventional open-circuit hydraulics, the main component is a bi-directional pump that would rotate with a variable speed in the direction of joint movement. As a result, the supply pressure is variable, and there are no requirements for relief valves, reservoirs or electrohydraulic servo-valves. The main constituents of EHA are:

◊ An electrical motor.
◊ A bi-directional pump.
◊ A symmetrical actuator.
◊ Pressure and position sensors.
◊ Accumulator and related sub-circuit.
◊ Filtering sub-circuits.

A schematic of EHA is depicted in Figure 2. The motor is coupled to the bi-directional pump that would cause oil flow and pressure changes in the hydraulic actuator. The pressure difference in the actuator chambers in turn results in exertion of force on the external load and the movement of this load according to the speed and direction of the pump.

Special Design considerations that have affected the circuit diagram of EHA include:

◊ High-gain inner-loop velocity-controller: this control strategy is used to alleviate the effect of a dead-band in the input current/output torque characteristics of the device. This dead-band is by pump/motor static friction, cross port leakage and pressure drop across hydraulic components such as filters.
◊ Service life: EHA can include a filtering mechanism as depicted in Figure 2 that would extend its service life for commercial applications.
◊ System safety: the check valves depicted in Figure 2 are used to prevent excessive pressure build up in the system.
◊ Symmetrical Actuator: in order to connect the actuator directly to the pump, EHA requires a symmetrical actuator with equal in and out flows.

A symmetrical linear actuator is specifically developed for EHA as described in the following section. A prototype of EHA with the new Symmetrical Linear actuator has been produced following extensive modeling and simulation. The other components that are used in this working prototype and their models are described in the following sections.

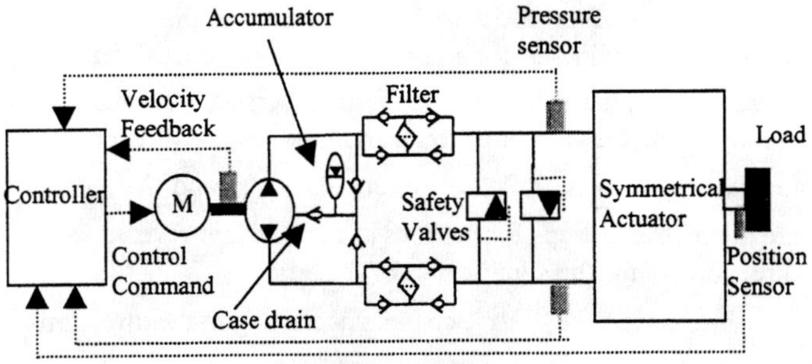

Figure 2: ElectroHydraulic Actuator

DESIGN OF A SYMMETRICAL LINEAR ACTUATOR FOR EHA

Conventional linear hydraulic actuators are composed of two pressure chambers, which are separated by a solid surface. The solid surface is, conventionally, connected to a single rod or to two rods as depicted in Figure 3. In the former, the active areas in the two chambers of the cylinder are not equal (asymmetrical actuator). This results in different characteristics for the forward and reverse stroke directions in the following:

- Maximum output torque,
- Maximum speed,
- Quantity of hydraulic fluid displaced by the two chambers, and
- dynamic response and characteristics of the actuator.

In the symmetrical double rod piston (Figure 3a), the active areas are made equal by the provision of two rods. In applications where the second rod is not utilized, a dead-space needs to be provided on the fixed side of the piston as depicted in Figure 3a. The provision of the dead-space can be restrictive for applications where the effective length of the cylinder needs to be short or is constrained by the fixtures surrounding the piston.

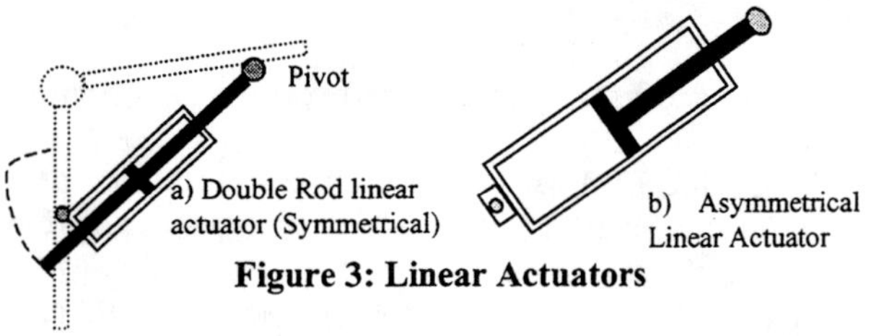

Figure 3: Linear Actuators

A mathematical model of asymmetrical actuators is presented in Habibi etal (1994), Habibi and Richards (1994), as follows:

$$\dot{\tau} = EI_v - K_v\dot{\theta} - \psi\tau - L\beta_e l_o \Delta P x_o / x_o^2 - x^2$$

The asymmetry of pistons results in unequal flow rates for the two directions of stroke. This means that the input/speed transfer function of asymmetrical actuators and their dynamic characteristics in the forward motion is different from that of the reverse motion. This observation is confirmed by an inspection of the term $-L\beta_e l_o \Delta P x_o / x_o^2 - x^2$. Due to different sizes of the two active pressure areas in the asymmetrical cylinder, for a same speed of motion and output torque, the pressure drop across the chambers ΔP is much greater in the forward direction than in the reverse direction i.e. $\Delta P_{forward} \neq \Delta P_{reverse} > 1$ where $\Delta P_{forward}$ and $\Delta P_{reverse}$ are the pressure drops for motions in forward and reverse directions. Thus the significance of the term $-L\beta_e l_o \Delta P x_o / x_o^2 - x^2$ changes according to the direction of movement, resulting in two different dynamic characteristics. The degradation in performance resulting from this asymmetry can be observed in a comparison of performance between symmetrical and asymmetrical actuator reported in Habibi and Richards (1994) and shown in Figure 4. The linear asymmetrical actuator exhibits a lower positional accuracy as well as a changing performance according to the direction of stroke.

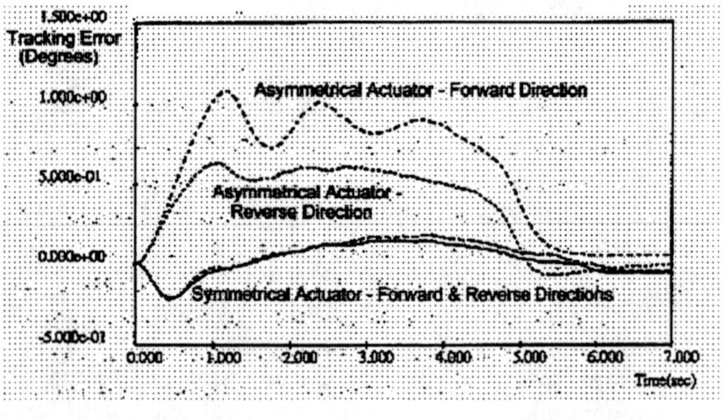

Figure 4: Performance of Symmetrical Vs Asymmetrical Actuators

A new design for linear actuators is proposed for applications requiring symmetry such as hydrostatic

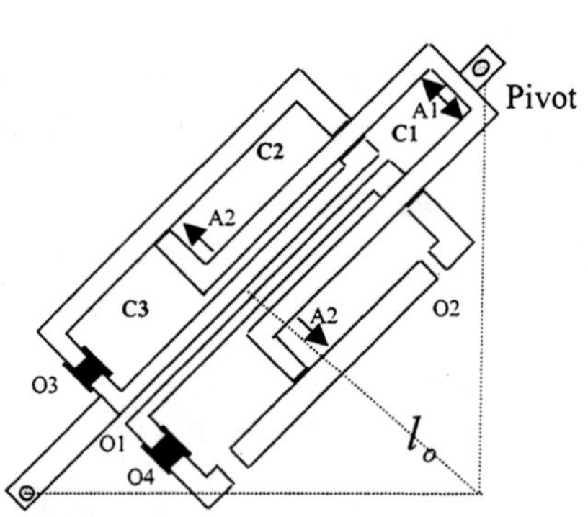

Figure 5: Linear Actuator

transmission. In this design, there are two working chambers C1 and C2 as illustrated in Figure 5. The solid rod of the conventional piston is replaced by a hollow cylinder that is closed at one end and has a circular disc around the opening at the other end. The surface area of the closed end of the cylinder, A1, is made equal to the area A2 of the disc. A1 and A2 are the active areas of the two pressure chambers and are therefore made equal. Chamber C3 can be opened to the atmosphere or could be filled with pressurized gas or fluid in applications where a bias is required to counter balance a dead weight acting under gravity. The contact between the cylinder and the fixed body of the actuator is through low friction seals. Hydraulic fluid enters the two chambers of the actuator through openings O1 and O2. It is recognized that there will be some external leakage across the seal into chamber C3. The openings O3 and O4 allow draining of fluid from this chamber. Opening O5 is provided to allow pressurization of chamber C3 for specialized applications. Following the method proposed in Habibi etal (1994) and referring to the nomenclature, the mathematical model for this actuator is obtained as:

$\dot{\tau} = EI_v - K_v\dot{\theta} - \psi\tau$ where,

$$E = \left(\frac{l_o \beta_e K_{sv}}{x_o - \frac{x^2}{x_o}}\right)\left(\sqrt{P_s - P_1} + \sqrt{P_2 - P_r} - \frac{x}{x_o}\left(\sqrt{P_s - P_1} + \sqrt{P_2 - P_r}\right)\right)$$

$$K_v = 4A^2 l_o \beta_e l_1 l_2 \sin(\theta)\frac{V_m^2 - v^2}{V_m}$$

$$\psi = \frac{2L\beta_e V_m}{V_m^2 - (Ax)^2} - \frac{\dot{l}_o}{l_o}$$

This actuator does not require a dead-space and has equal active pressure areas that results in the following:

- maximum saturation speed and torque are the same for the two directions of stroke,
- the dynamic term $-L\beta_e l_o \Delta P x_o / x_o^2 - x^2$ and its undesirable effects are alleviated, resulting in a simpler dynamic structure,

- the dynamic characteristic of SLA is the same for the two directions of stroke and therefore, controllers for its regulation will have a simpler structure, and
- the quantities of hydraulic fluid displaced by the two primary chambers of the actuator are equal.

The new actuator satisfies the prime requirement of closed circuit hydrostatic transmissions, which is having equal volumetric in and out flow to and from its chambers. This requirement is not satisfied by conventional single rod pistons. The structure and the mathematical model of this actuator is described by the following flow equations:

$$Q_a = A\dot{x} + \frac{A(x_o + x)}{\beta_e}\frac{dP_1}{dt} + LP_1$$

$$Q_b = A\dot{x} - \frac{A(x_o - x)}{\beta_e}\frac{dP_2}{dt} - LP_2$$

The design specification for the actuator used in the prototype of EHA and its dimensions are as follows:

Design Specification	Value
Volumetric displacement	0.505 ltr/m
Maximum stroke length	0.12 m
Maximum fluid chamber pressure:	3000 psi
Maximum gas chamber pressure	550 psi
Maximum burst pressure	5000psi
Surface finish	8 to 20 μm
Active pressure Area (fluid	0.000505 m^2
Maximum Output Force	10.5KN

The actuator has cushioning features that restrict the velocity of the hollow cylinder (moving part) as the stroke reaches its limits in either direction. The cushioning features reduce the area of the in/out flow orifices of the actuator as the limits are reached as encircled in Figure 6.

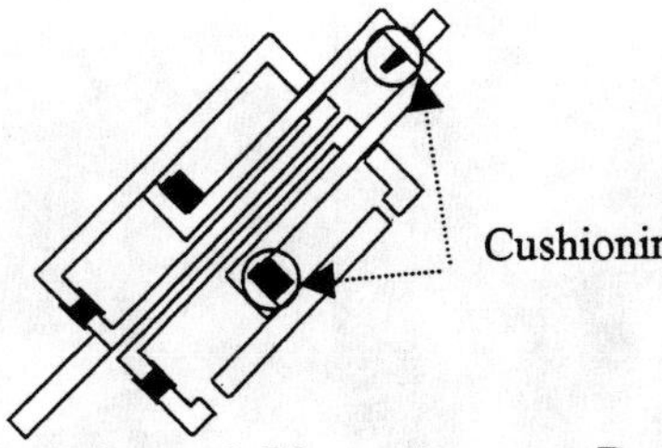

Figure 6: Linear Actuator Prototype

The size and dimension of the cushioning features were derived by using computer simulation as shown in Figure 7. The velocity of the moving part of the actuator is sharply reduced to zero as the moving part reaches the limits and cushioning comes into effect.

HYDRAULIC PUMP

A small bi-directional gear pump with a volumetric displacement of 1.06cc/rev is used in the EHA prototype. The internal seals of the pump were modified to withstand a maximum case pressure of 100 psi. This allowed the attachment of the pump's case drain to EHA's accumulator which is maintained at 45psi. This connection is made via a check valve of 1psi break pressure. The selection of the pump was based on simulation of a nonlinear model of EHA containing the pump. This model calculates the pump in/out flows (Q_a / Q_b), case drain flow (Q_{case}), and Load Torque (T_{load}) by using the following equations:

$$Q_a = D_p\omega_p - \xi(P_a - P_b) - L_e(P_a - P_{case}) + \frac{V_a}{\beta_e}\frac{dP_a}{dt}$$

$$Q_b = D_p\omega_p + \xi(P_a - P_b) - L_e(P_b - P_{case}) + \frac{V_b}{\beta_e}\frac{dP_b}{dt}$$

$$Q_{case} = L_e(P_a + P_b - 2P_{case})$$

$$T_{load} = D_p(P_a - P_b) + K_{fp}\omega_p + T_c$$

where, pipe elasticity is considered negligible due to the use of solid tubing and, $V_a = V_o + Ax$ and $V_b = V_o - Ax$. Various off the shelf pumps were simulated before selection of a model with the following specifications:

Design Specification	Symbol	Value
Pump displacement	D_p	1.065 cc/rev
Coeffient of external leakage (estimated)	L_e	$1.5\text{x}10^{-13}$ $m^3/s/Nm^{-2}$
Pump cross-port leakage coefficient (estimated)	ξ	$1.5\text{x}10^{-13}$ $m^3/s/Nm^{-2}$
Pipe plus mean actuator chamber volume.	V_o	$3.6\text{x}10^{-5}$ m^3

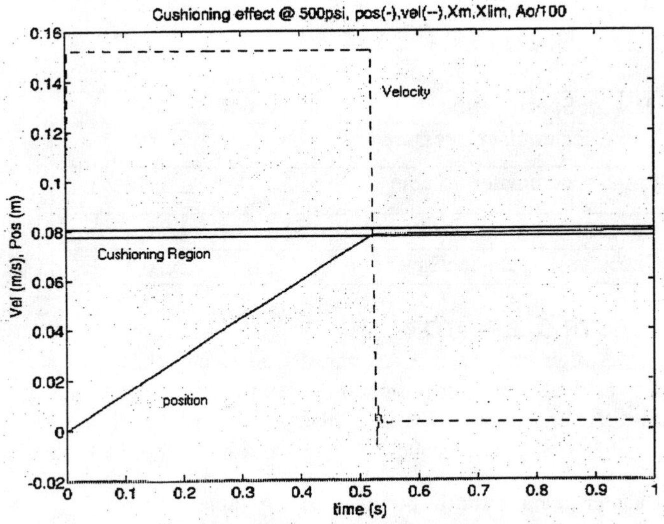

Figure 7: Cushioning Effect

ACCUMULATOR

A small bladder-type accumulator in the circuit as shown in Figure 8 is used to prevent:

- Cavitation, and
- excessive case pressure build up in the pump.

The model used for this component as follows:

If $P_{acc} - P_a > P_{valve_{break}}$, $Q_{acc_1} = K_{valve}(P_{acc} - P_a)$

If $P_{acc} - P_a < P_{valve_{break}}$, $Q_{acc_1} = 0$

If $P_{acc} - P_b > P_{valve_{break}}$, $Q_{acc_2} = K_{valve}(P_{acc} - P_b)$

If $P_{acc} - P_b < P_{valve_{break}}$, $Q_{acc_2} = 0$

If $P_{case} - P_{acc} > P_{valve_{break}}$, $Q_{acc_{in}} = Q_{case}$

If $P_{case} - P_{acc} < P_{valve_{break}}$, $Q_{acc_{in}} = 0$

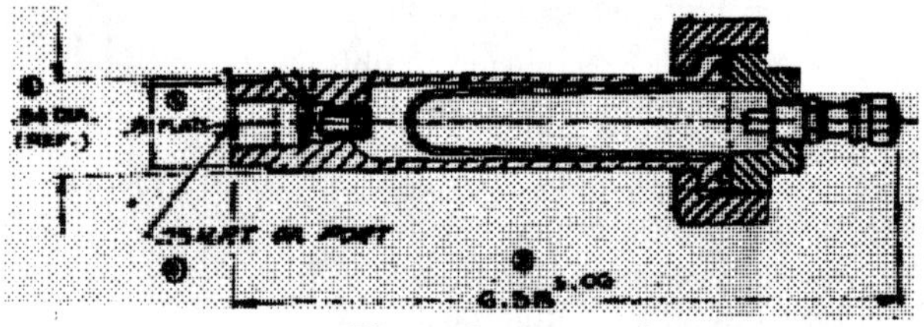

Figure 8: Accumulator

The total in/out flows and the pressure of the accumulator are obtained as:

$$Q_{acc} = Q_{acc_{in}} - Q_{acc_1} - Q_{acc_2}$$

Assuming isothermic expansion and compression of gas:

$$P_{acc} = \frac{P_{acc_{initial}} V_{acc_{initial}}}{V_{acc_{initial}} + \int Q_{acc} dt}$$

The accumulator selected for EHA is shown in Figure 8 and has the following specifications:

Design Specification	Symbol	Value
Initial accumulator pressure	$P_{acc_{initial}}$	50 Psi
Initial accumulator volume	$V_{acc_{initial}}$	2.5x10^{-5} m^3
Check valve break pressure	$P_{valve_{break}}$	1 Psi
Check valve gain coefficient	K_{valve}	5x10^{-9}

ELECTRICAL MOTOR

A high performance AC pan-shaped electrical motor with low as well as high speed-control accuracy was chosen for the bi-directional pump. Given a high-gain proportional-integral inner-loop speed-controller $G_{CS} = K_p + \frac{K_I}{s}$, and the pump model, the transfer function for the pump speed is obtained as:

$$\omega_p = \frac{G_{CS} K_C G_1 G_2}{1 + G_{CS} K_C G_1 G_2} V_v - \frac{G_2}{1 + G_{CS} K_C G_1 G_2} T_c$$

$$- \frac{G_2}{1 + G_{CS} K_C G_1 G_2} D_P (P_a - P_b)$$

where: $G_1 = \frac{I_c}{V_c} = \frac{1/R_c}{(L_c/R_c)s + 1}$ and $G_2 = \frac{1/(K_{P_{visc}} + K_{fric} + K_\omega K_c G_1)}{\frac{J_{pm}}{(K_{P_{visc}} + K_{fric} + K_\omega K_c G_1)} s + 1}$.

A high control gain, G_{CS}, and a careful selection of the motor such that $|K_c G_1| >> |G_2|$ eliminated the deadband in the EHA prototype caused by T_c. Given the chosen electrical motor[1], the transfer functions are obtained as: $G_1 = \frac{2}{0.008s + 1}$,

$G_2 = \frac{0.83}{1.5 \times 10^{-5} s^2 + 0.002s + 1} \approx \frac{0.83}{0.002s + 1}$, and $G_{cs} = 50 + \frac{100}{s}$.

EHA PROTOTYPE DESIGN

A prototype of EHA with the new Symmetrical Linear actuator has been produced following modeling and simulation. This prototype has achieved an extremely high level of performance and is capable of moving large loads with a very high level of accuracy and precision. The step response of this prototype when attached to a horizontal load of 20Kg (as shown in Figure 9) is given in Figure 10. Its performance specifications are as follows:

Maximum working Pressure	210 Bar (3000 psi)
Maximum Output Force	10.5 kN
Maximum Piston Stroke	12 cm
Maximum speed	14 cm/s
Closed loop position accuracy	0.01 mm (limited by encoder accuracy)
System Bandwidth	22 Hz (load of 20kg)
Force to weight ratio	1.1 kN/kg (actuator)
Deadband	Negligible
Rise time	0.2 sec (load of 20 kg)
Settling time	0.6 sec (load of 20 kg)

Figure 9: EHA Prototype

[1] Estimated values used in this simulation, are as follows $J_{pm} = 0.0016 Kg - m^2$, input=0.1sin(2πt) Volts, amplifier voltage saturation at +/-10V with gain of 3350, K_c =1Nm/A, $K_\omega = 62.8V/KRPM$, $K_{fric} = 0.02Nm/KRPM$, $K_{P_{visc}} = 0.2Nm/KRPM$, $L_c = 0.004h, R_c = 0.5\Omega$, $T_{DB} = 2Nm$.

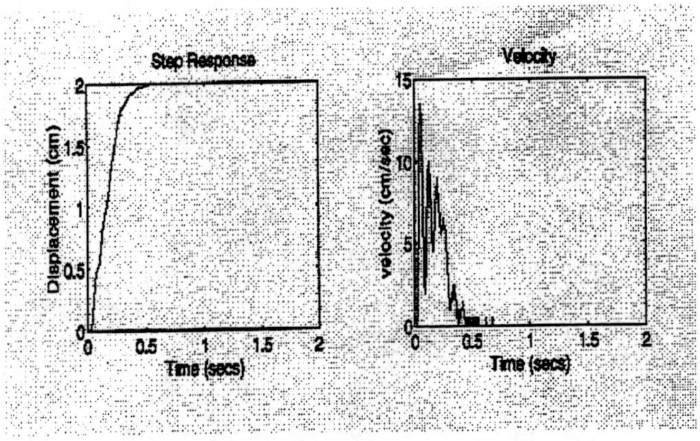

Figure 10: EHA Prototype Step Response

CONCLUSIONS

A new hydrostatic device (EHA) has been produced to combine the two major benefits of hydraulic and direct drive electrical motors. Analytical and experimental studies conducted on EHA indicate that the following characteristics can be expected:

◊ Large torque/mass ratio comparable to conventional hydraulic circuits.
◊ Modularity: EHA is self-contained and does not require a centralized supply system.
◊ A reduction of the effects of cross coupling torque (or external disturbances) due to the volumetric flow within the actuation system.
◊ Cost: the use of expensive components such as electrohydraulic servo-valves or harmonic drives are avoided.

This new class of modular hydraulic actuation system has applications across all fields of industrial hydraulics. Its design removes a number of important drawbacks of the existing technology and, with the exception of special cases, the use of this device can benefit most hydraulic applications.

ACKNOWLEDGMENTS

The authors would like to thank the Manufacturing Research Corporation of Ontario (MRCO) for their financial support.

REFERENCES

Watton J., 1989, “Fluid Power Systems”, Prentice Hall.

Korn J., 1972, “Hydrostatic Transmission Systems”, *Int. Textbook Ltd.*

Hunter I., 1985, “Novel Actuators for Use in Robotics and Telerobotics”, *Internal Report, McGill University.*

Desai J. and Bobrow J.E., 1989, “Modeling and Analysis of a High Torque Hydrostatic Actuator for Robotic Applications”, *Exp. Robotics.*

Arnautovic S., 1993, “Electrohydraulic Actuator”, *Internal Report,Robotics and Automation Laboratory, Univ. Toronto.*

Habibi S., Richards R. and Goldenberg A., 1994, “Hydraulic actuator analysis for industrial robot multivariable control”, Am. Cont. Conf.

Habibi S. and Richards R., 1991, “Computed-torque and variable-structure multivariable control of a hydraulic industrial robot” *Pr. Inst Mech Eng., Part I, Vol 205.*

FPST-Vol. 6, Fluid Power Systems and Technology — 1999
ASME 1999

Dynamic Modeling of an Indexing Valve Plate Pump

J. Cho and **X. Zhang,** Graduate Students
S. S. Nair, Associate Professor
Computer Controlled Systems Laboratory
Mechanical and Aerospace Engineering
University of Missouri, Columbia MO 65211

and

N. D. Manring, Program Manager
Hydraulic Components Research
Technical Center, Building G
Caterpillar, Inc., Peoria, IL 61656-1875

The swash-plate in a variable displacement pump experiences very large forces and moments that try to dislocate its position and therefore a large device is required for adequate control. In this paper, the dynamics of an alternative pump design using an indexing valve plate to position the swash-plate are reported. The indexing valve plate design is aimed at controlling the pressure transition for a piston, which is moving from a high-pressure port to a low-pressure port and back. In this paper, the governing equations for the pump are derived and the detailed open-loop, which is necessary for understanding the overall dynamic characteristics of the pump, is reported. Also, linear and nonlinear modeling approaches for the system are compared.

NOMENCLATURE

A_p	effective pressured area within a single piston
C_s	systemic viscous damping coefficient
H_{z_n}	*z*-location of the force exerted on the *n*th slipper from the swash plate
I'_{yy}	mass moment-of-inertia for the swash plate
I''_{yy}	mass moment-of-inertia for the slipper
k_a	spring rate used to bias the pump into stroke
L	spring force moment-arm about the swash plate pivot
M_p	single piston mass
M_{sw}	the swash plate mass
M_s	the slipper mass
P_d	discharge pressure
P_i	intake pressure
Pl	spring load when $\alpha = 0$
P_n	pressure rise-rate within the *n*th piston chamber
r	piston pitch radius
$T_c^{y'}$	scalar control and containment moment exerted on the swash plate in the y' direction
ω	rotational speed of the pump shaft and the cylinder block

1. Introduction

The axial-piston pump converts rotating mechanical power into fluid power, which is transmitted to other components within the hydraulic circuit. One advantage of using swash-plate type hydrostatic pumps for distributing power within mechanical systems is that the power distribution can be continuously varied without the use of clutches or fixed gear-ratios. This variable power distribution is accomplished by adjusting the output flow of the pump through instantaneously altering the angular position of the swash plate. Since changing the rotational displacement of the swash plate is fundamental for altering the output power of the pump, it is important to understand the moments that are exerted on the swash plate by the reciprocating pistons and any control devices that exist within the pump. The dynamics of these components can often dictate the overall dynamic behavior of the hydraulic system. The moments exerted on the swash-plate by the pumping elements themselves result from the fluid pressure within the piston chambers and the inertia of the reciprocating piston-slipper assemblies. The net moment exerted on the swash plate by these components has a nonlinear and partly discontinuous behavior, and must be reacted against by the swash-plate control actuator. In the past, various authors have developed mathematical models for such pumps by considering the pressure transients within the pistons. Due to the complexity of the pressure transitions between pump ports, some authors have used an approximate pressure-profile schematic to characterize the pressure transitions [1-5]. Zeiger and Akers developed mathematical models and investigated the effects of different design, tolerance variations, and varying operating conditions. Manring proposed an approximate pressure distribution and also developed equations for the torque components acting on the swash-plate. He also introduced the pressure carry-over to explain the pressure transition at dead centers [6]. Other simplifications have been introduced into the analysis of these complex systems by using sensitivity techniques for reducing the order of the dynamic model of the axial-piston pump [7].

In general, the swash-plate experiences very large forces and moments that try to dislocate its position and therefore a correspondingly large device is required for adequate control. The control device, which biases the machine and is used to stroke and de-stroke the swash-plate, is generally comprised of hydrostatic actuators and springs that overwhelm any other forces acting on the swash-plate. These control mechanisms are generally large, inefficient, and costly to manufacture. Moreover, they are massive and flexible and may often add unfavorable dynamics to the system response. Due to the disadvantages of the conventional control mechanism, it is desirable to replace this design altogether with another device that is smaller, less costly, more efficient, and more dynamically responsive.

This paper deals with a mathematical model of the swash-plate using a new control mechanism, which is called an indexing valve-plate. The open-loop system dynamic characteristics of the pump and the indexing valve-plate are considered. The indexing valve-plate uses the pressurized fluid from the discharge port to rotate the ported geometry of the valve plate about the centerline of the shaft and ports. This alteration in porting locations modifies the pressure characteristics within the pump so as to adjust the swash-plate angle, α, itself. Based upon this concept, the governing equation of motion for the swash-plate is derived with an assumed piston pressure distribution. A simplified model is then developed for the overall dynamics.

2. Governing Equations of Motion

2.1 Swash-Plate

The equations that govern the behavior of the swash-plate, the slipper, and the pistons of a swash-plate type hydrostatic axial-piston machine have been derived by Manring [8]. The basic components are shown in Fig. 1. Each machine component is analyzed separately and the major forces are shown in the respective free body diagrams. The coordinate systems used vary depending upon the component considered. Fixed and moving coordinate systems are used whenever necessary and any force analysis is considered only in the x and z directions since the pump rotates about an axis that is perpendicular to the x - direction. A primed coordinate system describes the centerline of the swash-plate pivot and a double primed coordinate system is fixed at the mass center of the slipper which translates and rotates together with the slipper. The free body diagram of the swash plate is shown in Fig. 2a.

There are two types of forces acting on the swash plate: the reaction forces between the swash plate and the slippers ($-R_n'$) and any external forces exerted by control and containment devices (F_c'). The time rate-of change of angular moment of the swash plate about its pivot is given by

$$\dot{H}_{sw}' = \sum_{n=1}^{N} M_n' + M_c' \quad (1)$$

where the two components represent moments generated by the two force components R_n' and F_c' about the swash plate pivot. Since the swash-plate rotates only about the y - axis, its time rate-of-change of angular momentum is given by

$$\dot{H}_{sw}' = I_{yy}'\ddot{\alpha} \quad (2)$$

where I_{yy}' represents the principal moment-of-inertia about the y' - axis. Using Eqns. 1-2, it can be shown that the swash-plate torque due to the control and containment devices is given by

$$T_c^{y'} = I_{yy}'\ddot{\alpha} + \sum_{n=1}^{N} R_n'(z_n \cos(\alpha) + x_n \sin(\alpha) + H_{z_n}) \quad (3)$$

Expressions for R_n', H_{z_n}, x_n and z_n need to be determined, in Eqn. (3).

2.2 Slipper

There are two forces acting on the nth slipper: the reaction force of the nth piston against the nth slipper ($\boldsymbol{F}_{s_n}''$) and the equal and opposite reaction force of the swash-plate against the nth slipper ($\boldsymbol{R}_n''$). Summing these forces and setting them equal to the time rate-of-change of linear momentum of the slipper, it can be seen that

$$M_s\ddot{\boldsymbol{X}}_{s_n}'' = \boldsymbol{F}_{s_n}'' + \boldsymbol{R}_n'' \quad (4)$$

A corresponding moment about the mass center of the nth slipper is generated by each force on the nth slipper. Therefore, the time rate-of-change of angular momentum of the nth slipper about its mass center is given by

$$\dot{\boldsymbol{H}}_{s_n}'' = \boldsymbol{M}_{s_n}'' + \boldsymbol{M}_n'' \quad (5)$$

where $\boldsymbol{M}_{s_n}''$ and $\boldsymbol{M}_n''$ are the respective moments generated by $\boldsymbol{F}_{s_n}''$ and $\boldsymbol{R}_n''$ about the mass center of the slipper.

The position of the nth slipper's mass center relative to the un-primed coordinate system of Fig. 2b is given by

$$\boldsymbol{X}_{s_n} = (x_n - X_s\cos(\alpha))\,\hat{\imath} + (z_n + X_s\sin(\alpha))\,\hat{k} \quad (6)$$

where the position, (x_n, z_n), locates the center of the piston-slipper ball-joint relative to the un-primed coordinate system of Fig. 1. The time rate-of change of angular momentum of the nth slipper about its mass center is

$$\dot{H}_{s_n}'' = I_{yy}''\ddot{\alpha} \quad (7)$$

From the swash-plate and slipper equations, the reaction force between the swash-plate and the nth slipper can be obtained by

$$R_n' = M_s(\ddot{x}_n \sec(\alpha) + X_s\tan(\alpha)\ddot{\alpha} + X_s\dot{\alpha}^2) - F_{s_n}^x \sec(\alpha) \quad (8)$$

In addition, the dimension, H_{z_n}, may be expressed by

$$H_{z_n} = \frac{1}{R_n'}((I_{yy}'' + M_sX_s^2)\ddot{\alpha} + M_sX_s(\ddot{x}_n\sin(\alpha) + \ddot{z}_n\cos(\alpha))) \quad (9)$$

The only unknown quantities now are R_n' and $F_{s_n}^x$. The reaction force, R_n', between the nth slipper and the swash plate can be obtained by calculating the force, $F_{s_n}^x$. This is the reaction force between the nth slipper and the nth piston. Since linear motion of the pistons is of interest, all forces acting on the single piston can be summed and set equal to the time rate-of-change of the linear momentum in the x – direction. This results in

$$M_p \ddot{x}_n = F^x_{s_n} - C_p \dot{x}_n - A_p P_n \tag{10}$$

where C_p is piston viscous drag coefficient. Therefore, the reaction force, R_n', between the nth slipper and the swash plate can be written as

$$R_n' = ((M_p + M_s)\ddot{x}_n + A_p P_n + C_p \dot{x}_n)\sec(\alpha) + M_s(X_s \tan(\alpha)\ddot{\alpha} + X_s \dot{\alpha}^2) \tag{11}$$

Expressions for x_n and z_n need to be determined next.

2.3 Kinematics of the Piston-Slipper Ball-Joint

The piston-slipper ball-joint makes a curvilinear trajectory about the x-axis generated by the rotation of the cylinder block and a rectilinear trajectory along the x-axis generated due to the inclination of the swash plate from the vertical axis. From Fig. 2c, the location of the nth piston-slipper ball-joint relative to the fixed origin in the x and z-directions is given by

$$x_n = r\sin(\theta_n)\tan(\alpha) \tag{12}$$

$$z_n = r\sin(\theta_n) \tag{13}$$

where θ_n is the angular polar-coordinate of the nth piston. Differentiating these with respect to time, the instantaneous velocity and acceleration of the assembly can be obtained.

2.4 Complete Swash-Plate Dynamics

Substituting Eqns. (9), (11), (12), and (13) into (3), the moment exerted on the swash-plate about its pivot is given by

$$T_c^{y'} = \Phi_1 \ddot{\alpha} + \Phi_2 \dot{\alpha}^2 + \Phi_3 \dot{\alpha} - \frac{N(M_p + M_s) r^2 \omega^2 \tan(\alpha)}{2\cos^2(\alpha)} + \frac{\sum_{n=1}^{N} A_p P_n (r\sin(\theta_n))}{\cos^2(\alpha)} \tag{14}$$

where

$$\Phi_1 = I'_{yy} + N(I''_{yy} + M_s X_s^2) + (M_p + M_s)\frac{N r^2}{2\cos^4(\alpha)}$$

$$\Phi_2 = N\left(M_s + M_p\right)\left(\frac{r^2 \sin(\alpha)}{\cos^5(\alpha)}\right), \text{ and}$$

$$\Phi_3 = \left(\frac{N C_p r^2}{2\cos^4(\alpha)}\right).$$

This model is referred to as the full nonlinear model. The pressure summation term in the model can be simplified, and this is discussed in the next section.

2.5 Piston Chamber Pressure and Distribution

The general equation describing the time rate-of-change of the pressure within the nth piston bore is

$$\dot{P}_n = \beta / V_n (Q_n - \dot{V}_n) \tag{15}$$

where β is fluid bulk modulus and V is the fluid volume entrapped in the piston chamber.

The flow rate Q_n is given by using the orifice equation which assumes a high velocity flow in and out of the piston bore. This flow is given by

$$Q_n = [\text{sign}(P_b - P_n)]\; C_d A_o \sqrt{2 \mid P_b - P_n \mid / \rho} \tag{16}$$

where the 'sign' function takes on the value ± 1 depending upon the sign of its argument, P_n is fluid pressure, P_b is the boundary pressure outside of the control volume, C_d is the orifice discharge coefficient for the control volume of a single piston-bore, ρ is fluid density and A_o is the varying discharge-area of the control volume for the piston bore. When the piston passes the top dead center, A_o varies in four stages, passing at the end of the discharge port, at the top dead center, at the slot of the intake port, and completely over the slot. Due to symmetry, the same holds at the bottom dead center. If an indexing valve plate concept is considered, the stages in area variation occur early or late depending on the direction of the valve plate indexing, +δ or -δ, and this will result in changes in the pressure transient at the top and bottom dead centers. Details of the pressure transient effect with the indexing valve plate will be discussed later. The instantaneous volume of the nth piston bore can be determined from geometry and may be mathematically expressed as

$$V = V_o - A_p r\tan(\alpha)\sin(\theta) \tag{17}$$

Using Eqns. 15-17, the pressure rise-rate within the nth piston chamber can be written as

$$\frac{dP_n}{d\theta_n} = \frac{\beta}{V_o - A_p \tan(\alpha)\sin(\theta_n)} * \left\{\frac{sign(P_{b_n} - P_n) C_d A_o}{\omega}\sqrt{\frac{2 \mid P_{b_n} - P_n}{\rho}} + A_p r\tan(\alpha)\cos(\theta_n)\right\} \tag{18}$$

where $dt = d\theta / \omega$.

Fig. 3 shows the piston pressure characteristics and an approximate pressure-profile schematic assumed to simplify the pressure term Eqn. (14). In this schematic, it can be seen that the pressure is constant as the piston passes directly over either port and that it undergoes a transition in pressure as the piston passes over the grooves on the valve-plate. This transition occurs through some average angular-distance, which is noted in Fig. 3b as γ. This angle, introduced in [6], is referred to as the pressure carry-over angle, γ. This pressure distribution is given mathematically by a discontinuous equation as follows,

$$P_n = \begin{cases} P_d & 3\pi/2 + \gamma < \theta_n < \pi/2 \\ P_d - m(\theta_n - \pi/2) & \pi/2 < \theta_n < \pi/2 + \gamma \\ P_i & \pi/2 + \gamma < \theta_n < 3\pi/2 \\ P_i + m(\theta_n - 3\pi/2) & 3\pi/2 < \theta_n < 3\pi/2 + \gamma \end{cases} \tag{19}$$

where $m = (P_d - P_i)/\gamma$. This expression thus assumes that the pressure profile created by the nth piston remains constant as the piston passes over the intake or discharge ports (outside the transition regions) and that the pressure transition between ports occurs linearly over the range of the pressure carry-over angle, γ.

2.6 Variation of Pressure Carry-Over Angle

The pressure carry-over angle, γ, determines the magnitude of the torque and so should be examined in more detail. Since the volumetric changes of the piston chamber are small near the pressure transition area, the instantaneous pressure change in this region can be approximated as

$$\frac{dP}{d\varphi} = \frac{\beta Q}{V\,\omega} \qquad (20)$$

where $d\varphi = \omega dt$. As shown in Fig. 3a, the discharge area of the valve-plate increases with φ, therefore, using the general form of Eqn. 16, the volumetric flow out of the piston chamber may be approximated as

$$Q = -\varphi K_a \sqrt{P_d - P_i} \qquad (21)$$

where K_a contains information regarding the geometry of the discharge area and the density of the fluid. Substituting Q from Eqn. 20 and treating the volume of the piston chamber as a constant V', Eqn. 21 can be rearranged in integral form as

$$\int_{P_d}^{P_i} \frac{1}{\sqrt{P - P_i}} dP = -\int_0^{\gamma_o} \frac{\beta K_a}{V'\omega} \varphi\, d\varphi \qquad (22)$$

Integrating Eqn. 22 and rearranging terms yields the following result for the pressure carry-over angle:

$$\gamma_o = \left(\sqrt{\frac{4V'}{\beta K_a}}\right) \sqrt{\omega}\ (P_d - P_i)^{1/4} \qquad (23)$$

From Eqn. 23, it can be seen that the pressure carry-over angle grows with the angular speed and the pressure difference between the two ports.

3. Simplified Model with Indexing

A simplified model for the swash-plate dynamics is developed using the preceding equations and assuming a fixed non-zero valve plate position, i.e., $\delta \neq 0$. The indexing valve-plate is aimed at controlling the pressure transition for a piston, which is moving from a high pressure port to a low pressure port and back. This is based upon the idea that the pressure transition occurs near the top and bottom dead center locations of the piston trajectory, over an average angular distance. By rotating the valve-plate relative to its center, this angular distance and location may be modified.

Fig. 3a also shows the indexing valve plate geometric features. In this paper, it is assumed that the diameter of the piston chamber is less than the angular distance between the ports. The valve-plate indexing can be in either positive (in the direction of shaft rotation) or negative directions.

Since the swash-plate dynamic equation, Eqn. 14, is non-linear, it is desirable to replace this equation with a linear one using the concept of the approximate pressure distribution and the pressure carry-over angle. Numerical analysis is used to find the relationship between the pressure carry-over angle, γ, and the indexing valve plate angle, δ. Fig. 4a shows a set of numerical results for the pressure transition of a single piston within a pump and a direct relationship between δ and γ. The valve-plate slots are of the constant-area type and are identical at top (θ = 90°) and bottom (θ = 270°) dead centers. The results of Fig. 4 have been generated for a mid-angle displacement of the swash-plate, α = 0.12 rad, with the pump running at a typical operating speed, ω = 210 rad/sec, and at a mid-range discharge pressure of 25 MPa. Three different indexing angles, δ, are compared: δ = –0.1, 0, and 0.1 radians. As noted from Eqn. 23, the pressure carry-over grows as the discharge pressure increases. In addition, Fig. 4a shows that a change in δ is approximately equal to a change in the pressure carry-over angle; therefore, this steady-state relationship can be written as $\Delta\delta \approx \Delta\gamma$, which leads to $\delta \approx \gamma - \gamma_o$, where γ_o is the carry-over angle when $\delta = 0$. The primary reason for using indexing on the valve-plate is to alter the pressure carry-over angle, γ. Indexing in the negative direction causes the torque about the swash-plate pivot axis to increase (become more positive) while indexing in the positive direction causes the torque about the swash-plate pivot axis to decrease. To summarize the importance of this quantity, it should be observed that a non-zero value of the pressure carry-over produces a torque about the swash-plate pivot axis that is proportional to the working pressure of the pump. Fig. 4b describes the behavior of the pressure carry-over angle, γ, as it varies with pressure.

Based on these investigations, it can be seen that the torque acting on the swash-plate is significantly altered by indexing the valve plate, which results in a change of the pressure carry-over angle. To get an average pressure history, the summation sign is replaced by $\frac{N}{2\pi}\int_0^{2\pi}$ to get the average value of the swash-plate torque with respect to the pressure carry-over angle that describes instantaneous quantities and the equation is integrated with respect to θ_n. By such an analysis, the following result is obtained:

$$\frac{\sum_{n=1}^{N} A_p P_n (r\sin(\theta_n))}{\cos^2(\alpha)} = \frac{N A_p}{2}\frac{r}{\pi}\sec^2(\alpha)(P_d - P_i)\gamma \qquad (24)$$

Using this result, Eqn. 14 can be rewritten, with a viscous damping coefficient, C_s, as follows,

$$T_c^{y'} = \Phi_1\,\ddot{\alpha} + \Phi_2\,\dot{\alpha}^2 + \Phi_3\,\dot{\alpha} - \frac{N(M_p + M_s) r^2 \omega^2 \tan(\alpha)}{2\cos^2(\alpha)} + \frac{N A_p}{2}\frac{r}{\pi}\sec^2(\alpha)(P_d - P_i)\gamma \qquad (25)$$

where

$$\Phi_1 = I'_{yy} + N(I'_{yy} + M_s X_s^2) + (M_p + M_s)\frac{N r^2}{2\cos^4(\alpha)}$$

$$\Phi_2 = N\left(M_s + M_p\right)\left(\frac{r^2 \sin(\alpha)}{\cos^5(\alpha)}\right), \text{ and}$$

$$\Phi_3 = \left(\frac{N C_p r^2}{2\cos^4(\alpha)} + C_s\right).$$

When springs are used to bias the pump into stroke (Fig. 1), the spring torque balances the torque in Eqn. 25. Therefore, the torque on the swash-plate is equal to the spring torque, i.e.,

$$T_c^{y'} = P\,l\,L - k_a L^2 \tan(\alpha) \quad (26)$$

where the spring pre-load Pl is designed such that $Pl \geq k_a L\,\alpha_{max}$. Substituting this and linearizing Eqn. 25 about $\alpha = 0$ and $\dot{\alpha} = 0$, a linearized swash-plate equation is obtained as

$$\ddot{\alpha} = -\,\Phi\,\{\left(\frac{N C_p r^2}{2} + C_s\right)\dot{\alpha} - P\,l\,L + k_a L^2\,\alpha$$

$$-\frac{N(M_p + M_s) r^2 \omega^2 \alpha}{2}$$

$$+\frac{N A_p}{2}\frac{r}{\pi}(P_d - P_i)\gamma\,\} \quad (27)$$

where

$$\Phi = \frac{1}{I'_{yy} + N\left(I''_{yy} + M_s X_s^2\right) + \frac{N r^2}{2}\left(M_p + M_s\right)}.$$

Such a linear model is useful for control purposes.

4. Relationship of γ and δ to α, and Different Operating Envelopes

In the previous section, reduced order nonlinear and linear models for the swash-plate dynamics were derived. Comparisons of typical results from the full-nonlinear model and the reduced order models of Eqns. 25 and 27 are shown in Fig. 5, for $P_d = 25\text{MPa}$. The errors are small justifying approximation of the full-nonlinear model of Eqn. 14 by the simpler form of Eqn. 26 and its linearized form of Eqn. 27. At steady-state (the time derivatives of α are zero in Eqn. 25), a direct linear relationship between the swash-plate angle, α, and the pressure carry-over angle, γ, can be obtained as

$$\gamma = -\frac{2}{(P_d - P_i) N A_p}\frac{\pi}{r}$$

$$\{-\frac{N(M_p + M_s) r^2 \omega^2 \alpha}{2} - P\,l\,L + k_a L^2\,\alpha\,\} \quad (28)$$

Based on the linear relationship between α and γ, Eqn. 28 is plotted (labeled 'analytical') in Fig. 6a, using the relationship $\Delta\delta = \Delta\gamma$, for $P_d = 25MPa$. Since there are large deviations between the full-nonlinear and simplified model at the lower and upper boundaries, it is necessary to consider some acceptable deviations at the boundaries. For this case, $\pm 5\%$ deviation is considered, and this is also shown in Fig. 6a. A $\pm 5\%$ deviation for γ results in about 8% steady-state error in the swash-plate angle α, and δ can be replaced by γ within these limits. Similar curves for the indexing valve plate angle, δ, versus the swash-plate angle, α, can be obtained for other discharge pressures also.

Several envelopes have been investigated in this simulation study and the results are shown in Fig. 6b: (i) The max. and min, curves are obtained by complete nonlinear simulation as boundaries when α no longer changes with change in δ. (ii)If the engineer defines acceptable deviations of $\pm$ x% from linearity, an envelope can be obtained where the linear relationship $\delta = \gamma - \gamma_o$ is valid. (iii)Using a $\pm 5\%$ band, an envelope for this case is shown Fig. 6b. If the engineer defines acceptable overshoot limits, another operational envelope can be obtained where overshoot is less than those limits. Using $\pm 30\%$ overshoot as limits, such an envelope is also shown in Fig. 6b for the pump. Thus, this parametric simulation study shows several possible operational envelopes and their relationships. Such studies provide insights into the dynamic characteristics, which are useful for control design.

5. Summary and Conclusion

Issues related to dynamical modeling of a valve plate controlled variable displacement pump are reported. The governing equations of motion for the swash-plate have been derived. Since the mathematical model is highly nonlinear and complex, an approximate pressure distribution is used to simplify it. The concept of the pressure carry over helps to simplify the model and the direct relationship between the valve plate actuation and the pressure carry-over for non-zero δ can be expressed by $\delta = \gamma - \gamma_o$. A reduced order model is developed and validated. Then, several possible operational envelopes have been investigated for the indexing valve plate pump. The insights and models developed will be used for on-going control studies.

References

1. Manring, N., and Johnson, R. 1996, "Modeling and Designing a Variable-Displacement Open-Loop Pump," *ASME Journal of Dynamic Systems, Measurement, and Control*, Vol. 118, pp. 267-271.
2. Zeiger, G., and Akers, A. 1985, "Torque on the Swash Plate of an Axial Piston Pump," *ASME Journal of Dynamic Systems, Measurement, and Control*, Vol. 107, pp. 220-226.
3. Schoenau, G., Burton, R., and Kavanagh, G. 1990, "Dynamic Analysis of a Variable Displacement Pump," *ASME Journal of Dynamic Systems, Measurement, and Control*, Vol. 112, pp. 122-132.
4. Zeiger, G., and Akers, A. 1986, "Dynamic Analysis of an Axial Piston Pump Swash Plate Control," *Proceeding of the Institution of Mechanical Engineeers*, Vol. 200, pp. 49-58.

5. Yamaguchi, A., and Ishikawa, T. 1979, "Characteristics of Displacement Control Mechanisms in Axial Piston Pumps," *JSME*, Vol. 22, No. 165, pp. 356-361.
6. Manring, N. D., 1998, "The Torque on the Input Shaft of an Axial-Piston Swash-Plate Type Hydrostatic Pump," *ASME Journal of Dynamic Systems, Measurement, and Control*, Vol. 120, pp. 57-62.
7. Kim, S., Cho, H., and Lee, C. 1987, "A Parameter Sensitivity Analysis for the Dynamic Model of a Variable Displacement Axial Piston Pump," *Proceedings of the Institution of Mechanical Engineers*, Vol. 201, No C4, pp. 235-243.
8. Manring, N. 1998, "The Analysis, Design, and Control of Axial Piston Swash Plate Type Hydrostatic Machines," *Unpublished Notes*.

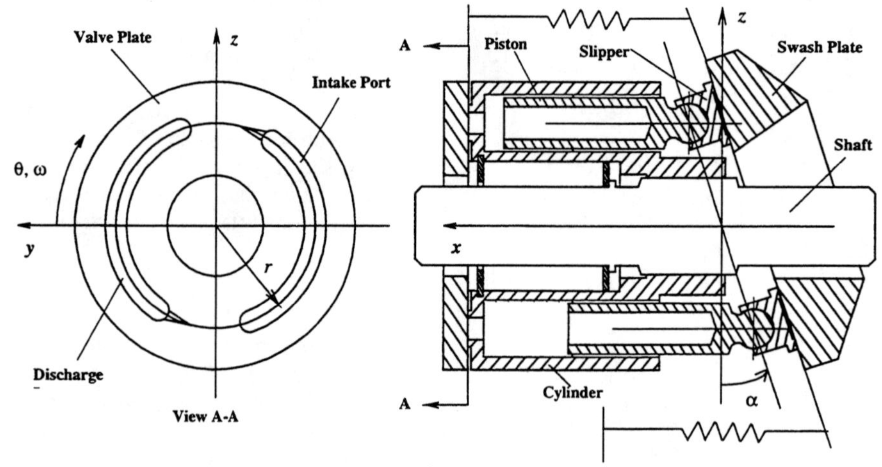

Figure 1. General Configuration of Overall System

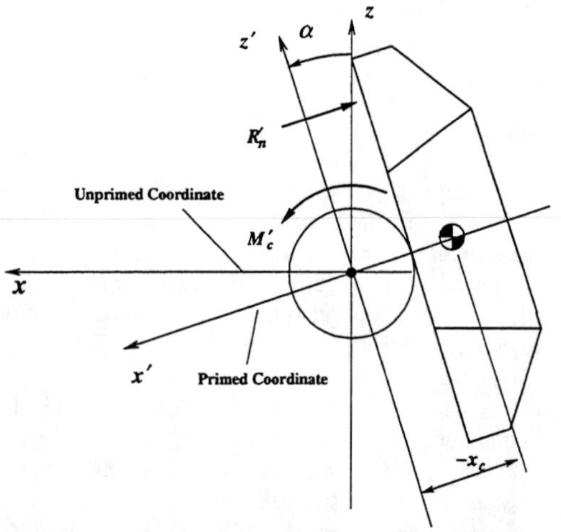

2a. Free-Body Diagram of Swash-Plate

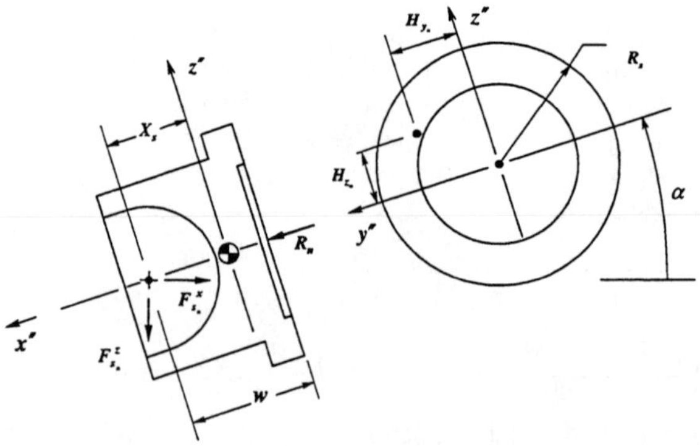

2b. Free-Body Diagram of Slipper

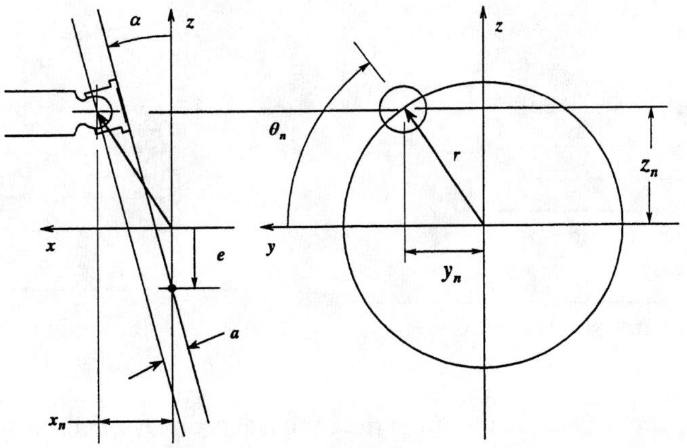

2c. Piston-Slipper Ball Joint Kinematics ($a = e = 0$)

Figure 2. Geometrical parameters and forces for the model

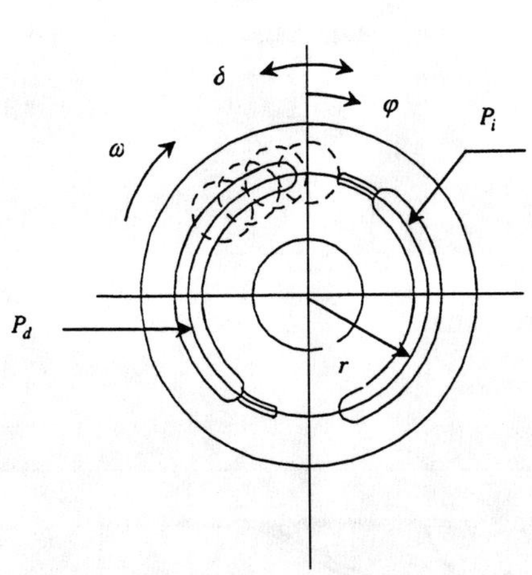

3a. The pressure transition concept

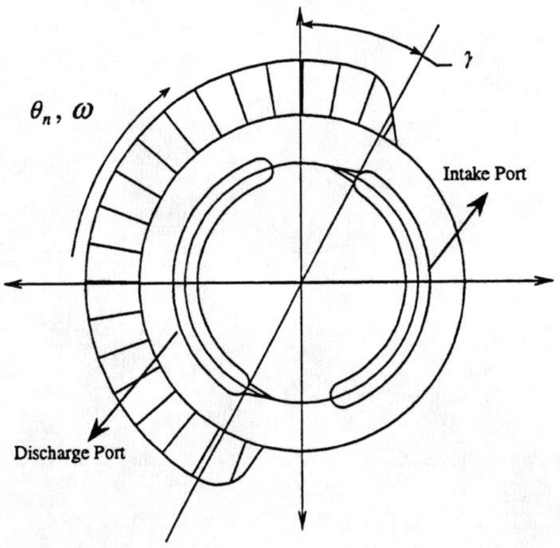

3b. Approximate Pressure Distribution

Figure 3. Piston pressure characteristics

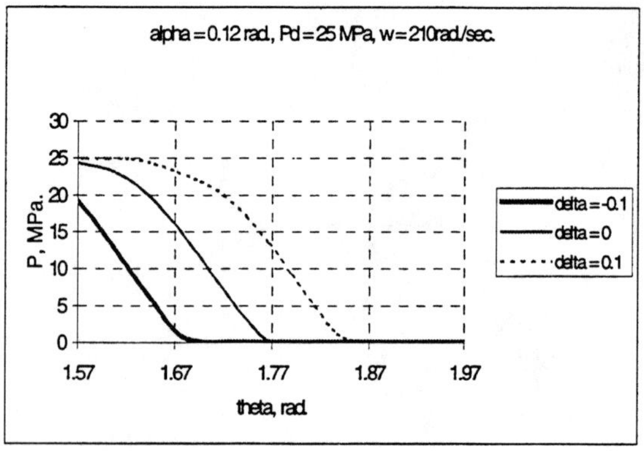

4a. Variation in pressure carry-over

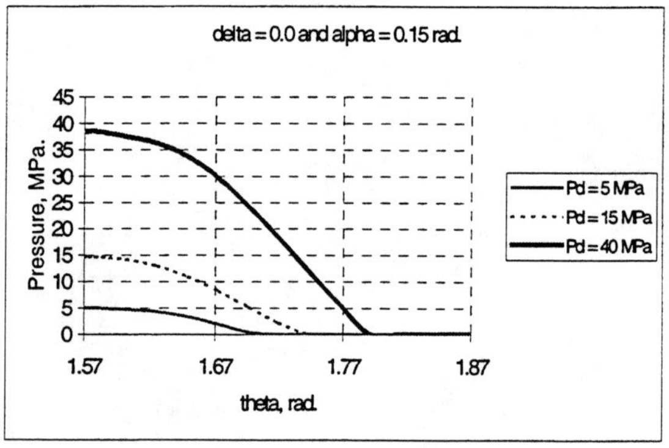

4b. Relationship between carry-over and discharge pressure

Figure 4. Observations from the complete nonlinear model simulations

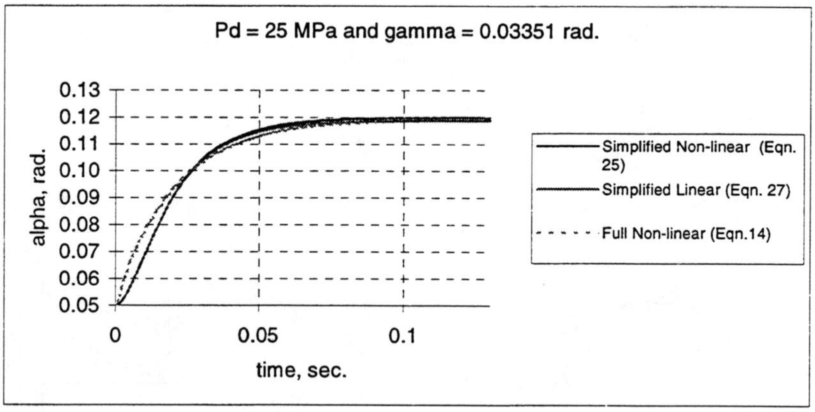

Figure 5. Comparisons of responses

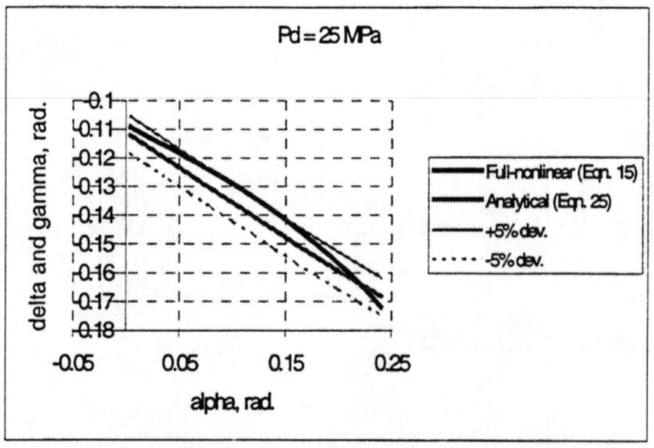

6a. Relationship of γ and δ to α at steady state

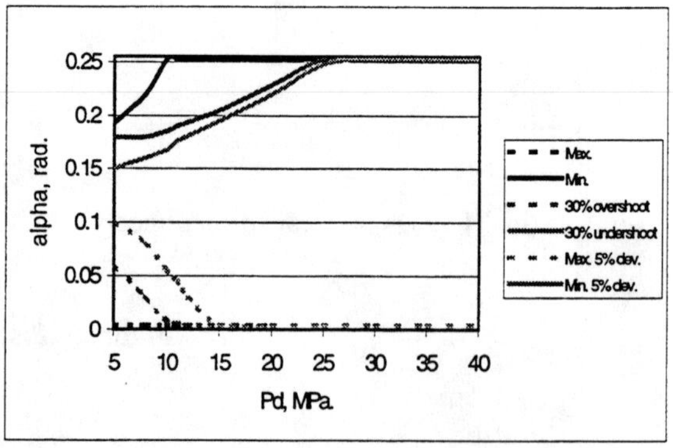

6b. Operating envelopes

Figure 6. Development of the different operating envelopes

FPST-Vol. 6, Fluid Power Systems and Technology — 1999
ASME 1999

FREQUENCY DOMAIN ANALYSIS OF VIBRATIONS IN LIQUID-FILLED PIPING SYSTEMS

Zongxia, Jiao, Professor, Ph.D.

Qing Hua, Ph.D. student

Kai Yu, Lecturer, Ph.D.

Automatic Control Department

Beijing University of Aeronautics and Astronautics

Beijing 100083, P.R. CHINA

ABSTRACT

In the analysis of liquid-filled piping systems there are Poisson-coupled axial stress waves in the pipe and liquid column, which are caused by the dilation of the pipe. In some conditions the influence of viscous friction that is usually frequency-dependent should not be omitted, which in fact is another kind of coupled form. It directly influences the amplitude of vibration of piping systems to some degree. The larger the viscosity of the liquid is, the greater the influence will be. Budny (1991) included the viscous friction influence in time domain analysis of fluid-structure interaction, but did not give frequency domain analysis. Lesmez (1990) gave the model analysis liquid-filled piping systems without considering friction. If the friction is not included in frequency domain analysis, the vibration amplitude will be greater than that when friction is included, especially at harmony points, cause large errors in the simulation of fluid pipe network analysis, although it may have little influence on the frequency of harmony points. The present paper will give detail solutions to the transfer matrix that represents the motion of single pipe section, which is the basis of complex fluid-structure interaction analysis. Combined with point matrices that describe specified boundary conditions, overall transfer matrix for a piping system can be assembled. Corresponding state vectors can then be evaluated to predict the piping and liquid motion. At last, a twice-coordinate transformation method is adopted in joint coupling. Consequently, the vibration analysis of spatial liquid-filled piping systems can be carried out. It is proved to be succinct, valid and versatile. This method can be extended to the simulation of the curved spatial pipeline systems.

Keywords: Fluid-structure interaction, Frequency-dependent friction, Coordinate transformation, Impedance

INTRODUCTION

The development trend of hydraulic technology in aircraft is high pressure, great flowrate and smart control etc. The coupling vibration of fluid-structure and flowrate surge of pump is becoming more important problem. The F14 of USA and F8 of China both have ever been damaged due to the vibration of fluid pipeline network system. The accuracy analysis on the fluid-structure interaction can optimize the system design and improve its performance.

Generally, in the vibration analysis of liquid-filled piping systems, there are both Poisson coupling and joint coupling. The motion of liquid-filled pipe reaches in which longwave assumptions are valid can then be described by Poisson coupled axial stress waves in the pipe and in the fluid column. The joint coupling takes place at piping discontinuities. Skalak (1956) extended Joukouski's analysis to include Poisson coupling. Walker and Philips (1977) developed a six-equation model with three wave families, which included radial equilibrium and continuity in the equations of motion for the fluid and pipe. Wiggert and Otwell et al (1985) simplified the above models, and a four-equation model with Poisson coupling has been developed. Budny (1991) introduced the Zielke's frequency-dependent fluid friction into the coupled equations and gave the characteristic's solution method and demonstrated that the model could satisfactorily predict fluid pressure and structural velocity. It is very important to get the transfer matrix between two points of a pipeline in the frequency analysis of fluid lines. D'Souza and Oldenburger (1964) derived the transfer matrix between two points of fluid line, which has been taking important role in the dynamic analysis of fluid network. Lesmez et al (1990) has developed the transfer matrix of liquid-filled single pipe section, but fluid viscous friction is neglected. In fact fluid viscous friction is another distributed coupling form. The present paper derived the transfer matrix through solving eigenvalues and eigenvectors, in which Zielke's frequency-dependent friction is included. A twice-coordinate transformation method introduced in this paper is proved to be succinct, valid and versatile. This method can be extended to the simulation of the curved spatial pipeline.

NOMENCLATURE

V: liquid velocity amplitude; P: pressure amplitude;
U: pipe velocity amplitude; F: force amplitude;
R: rotational velocity; M: moment; A: cross-sectional area;
r: inner radius of pipe cross-section; D : inner diameter of pipe cross-section; I: moment of inertia; J: polar moment of inertia; G: shear modulus of rigidity; e: thickness of pipe wall;
ρ : density; ν : Poisson's ratio; E: modulus of elasticity;

K: bulk modulus; ν_f : kinematic viscous; S: complex variable;

t: time; J_0, J_1 : Bessel Functions; a: wave speed.

Superscripts: x, y, z: principal directions; L : left; R : right.
Subscripts: p : pipe; f : fluid.

The transfer matrix

Considering the columned liquid-filled pipe element shown in Figure 1.

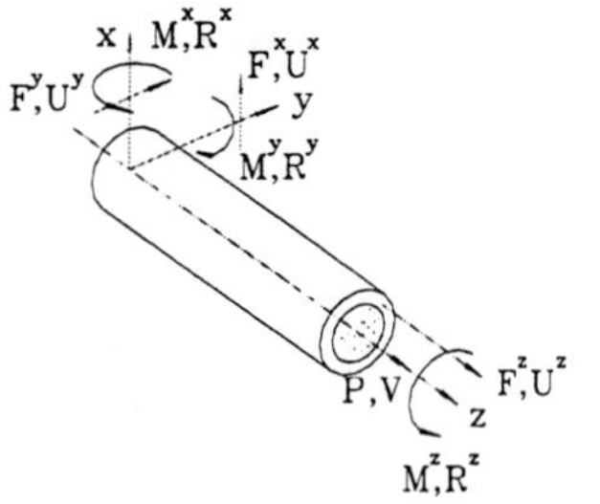

Figure 1 Elementary liquid-filled pipe section

The following equations (see reference [6]) are used to describe the characteristics of fluid pipeline systems. But in the equations of pipe element's axial stress, pressure, and velocities, the viscous friction that is introduced by Budny (1991) and was proved in the time domain analysis is involved. The present paper will infer the transfer matrix in frequency domain by them. The differential equations describing the coupled structural and fluid motion of a single pipe section with assumptions: long waves; low March numbers; linear elastic behavior of piping material; homogeneous, and isotropic, round and thin-walled pipeline; no buckling; negligible cavitation; negligible damping and negligible inertia in the radial direction.

(1) Axial stress, pressure and velocities:

$$\frac{\partial V}{\partial t}+\frac{1}{\rho_f}\frac{\partial p}{\partial z}+\frac{\tau_0 \pi D}{A_f \rho_f}=0 \tag{1}$$

$$\frac{\partial F^z}{\partial z}\frac{1}{A_p}+\frac{\tau_0 \pi D}{A_p}-\rho_p\frac{\partial U^z}{\partial t}=0 \tag{2}$$

$$\frac{\partial p}{\partial t}-2\nu K^*\frac{\partial U^z}{\partial z}+K^*\frac{\partial V}{\partial z}=0 \tag{3}$$

$$\frac{\partial F^z}{\partial t}\frac{1}{A_p}-E\frac{\partial U^z}{\partial z}-\frac{\nu r}{e}\frac{\partial p}{\partial t}=0 \tag{4}$$

(2) Shear and bending in the x-z plane:

$$\frac{\partial M^y}{\partial t}=EI_p\frac{\partial R^y}{\partial z} \tag{5}$$

$$\frac{\partial M^y}{\partial z}+F^x=(\rho_p I_p+\rho_f I_f)\frac{\partial R^y}{\partial t} \tag{6}$$

$$\frac{\partial F^x}{\partial z}=(\rho_p A_p+\rho_f A_f)\frac{\partial U^x}{\partial t} \tag{7}$$

$$\frac{\partial F^x}{\partial t}=A_p G(\frac{\partial U^x}{\partial z}-R^y) \tag{8}$$

(3) Shear and bending in the y-z plane:

$$\frac{\partial M^x}{\partial t}=EI_p\frac{\partial R^x}{\partial z} \tag{9}$$

$$\frac{\partial M^x}{\partial z}-F^y=(\rho_p I_p+\rho_f I_f)\frac{\partial R^x}{\partial t} \tag{10}$$

$$\frac{\partial F^y}{\partial z}=(\rho_p A_p+\rho_f A_f)\frac{\partial U^y}{\partial t} \tag{11}$$

$$\frac{\partial F^y}{\partial t}=A_p G(\frac{\partial U^y}{\partial z}+R^x) \tag{12}$$

(4) Torsion in the pipe wall:

$$\frac{\partial M^z}{\partial z}=\rho_p J\frac{\partial R^z}{\partial t} \tag{13}$$

$$\frac{\partial R^z}{\partial z}=\frac{1}{GJ}\frac{\partial M^z}{\partial t} \tag{14}$$

where

$$K^*=\frac{K}{1+\frac{2r\,K}{eE}(1-\nu^2)} \tag{15}$$

$$\tau_0=\frac{4\rho_f\mu_f}{r}(v-u^z)+\frac{2\rho_f\mu_f}{r}\int_0^t\frac{\partial(v-u^z)}{\partial t}(\beta)W(t-\beta)d\beta \tag{16}$$

(see reference [4])

It's Laplace formulation is

$$\tau_0(S)=\frac{\rho_f rS}{\frac{j\lambda J_0(j\lambda)}{J_1(j\lambda)}-2}(V-U^z)=A(S)(V-U^z) \tag{17}$$

in which $\lambda=\sqrt{\frac{S}{\nu_f}}r$ (18)

Taking the Laplace transformation of above equations results□

$$\dot{X}=TX, \tag{19}$$

$$X=[V\ \ P\ \ U^z\ \ F^z\ \ R^y\ \ M^y\ \ U^x\ \ F^x\ \ R^x\ \ M^x\ \ U^y\ \ F^y\ \ R^z\ \ M^z]^T \tag{20}$$

$$T_{fp}=\begin{bmatrix} 0 & -(\frac{2\nu^2 r}{eE}+\frac{1}{K^*})S & 0 & \frac{2\nu S}{EA_p} \\ -[\rho_f S+\frac{\pi D}{A_f}A(S)] & 0 & \frac{\pi D}{A_f}A(S) & 0 \\ 0 & \frac{\nu r S}{eE} & 0 & \frac{S}{EA_p} \\ -\pi D A(S) & 0 & [\pi D A(S)+\rho_p A_p S] & 0 \end{bmatrix} \tag{21}$$

$$T_{xz}=\begin{bmatrix} 0 & \frac{S}{EI_p} & 0 & 0 \\ (\rho_p I_p+\rho_f I_f)S & 0 & 0 & -1 \\ 1 & 0 & 0 & \frac{S}{A_p G} \\ 0 & 0 & (\rho_p A_p+\rho_f A_f)S & 0 \end{bmatrix} \tag{22}$$

$$T_{yz}=\begin{bmatrix} 0 & \frac{S}{EI_p} & 0 & 0 \\ (\rho_p I_p+\rho_f I_f)S & 0 & 0 & 1 \\ -1 & 0 & 0 & \frac{S}{A_p G} \\ 0 & 0 & (\rho_p A_p+\rho_f A_f)S & 0 \end{bmatrix} \tag{23}$$

$$T_{tz}=\begin{bmatrix} 0 & \frac{S}{GJ} \\ \rho_p JS & 0 \end{bmatrix} \tag{24}$$

the overall matrix is:

$$T=\begin{bmatrix} T_{fp} & 0 & 0 & 0 \\ 0 & T_{xz} & 0 & 0 \\ 0 & 0 & T_{yz} & 0 \\ 0 & 0 & 0 & T_{tz} \end{bmatrix} \quad (25)$$

where (see reference [8])

$$A(S)=\frac{\rho_f rS}{\dfrac{j\lambda J_0(j\lambda)}{J_1(j\lambda)}-2}=\frac{\rho_f rS}{2(\alpha-1)} \quad (26)$$

in which $\alpha=\dfrac{j\lambda J_0(j\lambda)}{2J_1(j\lambda)}$ (27)

Taking $X=X_0$ as the initial conditions, these differential equations can be solved to get:

$$X=GX_0, \quad (28)$$

G is transfer matrix.

$$G=Ee^{\lambda Iz}E^{-1}, \quad (29)$$

λ is the eigenvalue of matrix T; E is the matrix composed of eigenvectors . Obviously, G is related to z and S. The above relation can also be expressed as $X_0=G^{-1}X$, where

$$G^{-1}=Ee^{-\lambda Iz}E^{-1}. \quad (30)$$

To simplify the calculation, the eigenvalues of T are given:

(1) Corresponding to the matrix T_{fp}:

$$\lambda_{1,2,3,4}=\pm\sqrt{\frac{A}{2}\pm\sqrt{\frac{1}{4}A^2-B}}=\lambda'_{1,2,3,4}S \quad (31)$$

where (A_f/A_p is simplified as r/2e):

$$A=\frac{S^2}{a_p^2}\{[\frac{r}{2e}(5-4\nu)+\frac{E}{K}]\frac{\rho_f}{\rho_p}\beta+(\frac{E}{K}+\frac{2r}{e})\frac{\rho_f}{\rho_p}+1\} \quad (32)$$

$$B=\frac{S^4}{a_p^2a_f^2}[1+\beta(1+\frac{r}{2e}\frac{\rho_f}{\rho_p})] \quad (33)$$

in which (when $\lambda > 3$)(see reference[5])

$$\beta=\frac{1}{\alpha-1}=\frac{2}{\lambda}+\frac{3}{\lambda^2}+\frac{15}{4\lambda^3} \quad (34)$$

Eigenvectors:

$$y_i=\begin{bmatrix} \frac{1}{A_p}(2a_f^2\nu\lambda_i'^2-\beta) \\ \frac{1}{A_p}\lambda_i' K*[(1-2\nu)\beta-2\nu] \\ \frac{1}{A_p}(a_f^2\lambda_i'^2-1-\beta) \\ a_f^2\lambda_i'\{\lambda_i'^2E-[\frac{r(2-\nu)}{e}+\frac{E}{K}]\rho_f\beta-(\frac{E}{K}+\frac{2r}{e})\rho_f\} \end{bmatrix} \quad (35)$$

i=1,2,3,4

Then the matrix E_{fp} is $E_{fp}=[y_1y_2y_3y_4]$ (36)

(2) Corresponding to the matrix T_{xz}:

$$\lambda_{1,2,3,4}=\pm\sqrt{\frac{A}{2}\pm\sqrt{\frac{1}{4}A^2-B}} \quad (37)$$

where :

$$A=S^2(\frac{\rho_p+\rho_f\frac{I_f}{I_p}}{E}+\frac{\rho_p+\rho_f\frac{A_f}{A_p}}{G}) \quad (38)$$

$$B=\frac{S^4}{EG}(\rho_p+\rho_f\frac{I_f}{I_p})(\rho_p+\rho_f\frac{A_f}{A_p})+\frac{S^2}{EI_p}(\rho_pA_p+\rho_fA_f) \quad (39)$$

Eigenvectors:

$$y_i=\begin{bmatrix} \lambda_i-\frac{1}{\lambda_i}\frac{S}{G}(\rho_p+\rho_f\frac{A_f}{A_p}) \\ \frac{EI_p}{S}[\lambda_i^2-\frac{S}{G}(\rho_p+\rho_f\frac{A_f}{A_p})] \\ 1.0 \\ (\rho_pA_p+\rho_fA_f)\frac{S}{\lambda_i} \end{bmatrix} \quad (40)$$

i=5,6,7,8

Then the matrix E_{xz} is $E_{xz}=[y_5y_6y_7y_8]$ (41)

(3) Corresponding to the matrix T_{yz}:

$$\lambda_{1,2,3,4}=\pm\sqrt{\frac{A}{2}\pm\sqrt{\frac{1}{4}A^2-B}} \quad (42)$$

where:

$$A=S^2(\frac{\rho_p+\rho_f\frac{I_f}{I_p}}{E}+\frac{\rho_p+\rho_f\frac{A_f}{A_p}}{G}) \quad (43)$$

$$B=\frac{S^4}{EG}(\rho_p+\rho_f\frac{I_f}{I_p})(\rho_p+\rho_f\frac{A_f}{A_p})+\frac{S^2}{EI_p}(\rho_pA_p+\rho_fA_f) \quad (44)$$

Eigenvectors:

$$y_i=\begin{bmatrix} \frac{1}{\lambda_i}\frac{S}{G}(\rho_p+\rho_f\frac{A_f}{A_p})-\lambda_i \\ \frac{EI_p}{S}[\frac{S}{G}(\rho_p+\rho_f\frac{A_f}{A_p})-\lambda_i^2] \\ 1.0 \\ (\rho_pA_p+\rho_fA_f)\frac{S}{\lambda_i} \end{bmatrix} \quad (45)$$

i = 9,10,11,12

Then the matrix E_{yz} is $E_{yz}=[y_9y_{10}y_{11}y_{12}]$ (46)

(4) Corresponding to the matrix T_{tz}:

$$\lambda_{1,2}=\pm\sqrt{\frac{\rho_p}{G}}S \quad (47)$$

Eigenvectors:

$$y_i=\begin{bmatrix} 1 \\ \frac{\lambda iGJ}{S} \end{bmatrix} \quad (48)$$

i=13,14

Then the matrix E_{tz} is $E_{tz}=[y_{13}y_{14}]$ (49)

The overall matrix is

$$E=\begin{bmatrix} E_{fp} & 0 & 0 & 0 \\ 0 & E_{xz} & 0 & 0 \\ 0 & 0 & E_{yz} & 0 \\ 0 & 0 & 0 & E_{tz} \end{bmatrix} \quad (50)$$

And the transfer matrix G can be obtained.

Coordinate transformation in joint coupling

The choice of appropriate local coordinate system is of importance in the joint coupling. In general, the global coordinate system is known, which is hardly used directly to express the mathematics models of fluid-structure interaction in piping systems. The paper presents a new method of determining local coordinate system. Consequently, the joint coupling can be realized through twice plane coordination transformation, in which the coupling of forces and displacements of liquid and pipe can be added. The new method of determining local coordinate system is:

1. Direction of z--axis is correspondent to the direction of the fluid transmission;
2. To the bend pipe, direction of x--axis is correspondent to the direction of the normal line of the plane of the linked two pipe sections;

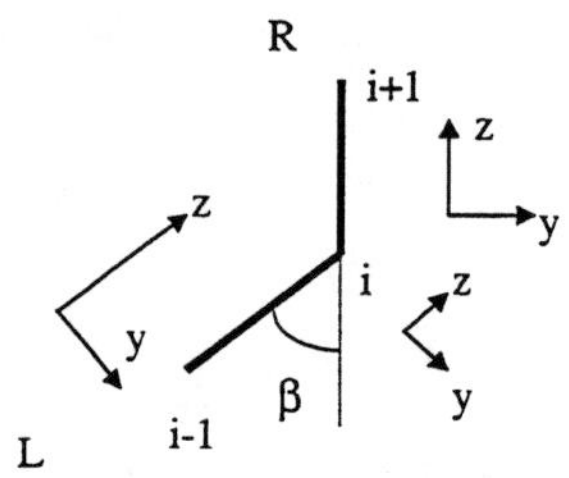

Figure 2 Coordinate transformation of coupling pipe joint

According to these principles, the deduction of the transformation matrix of bend pipe (see Figure 2) is as follows:

(1) Define the coordinate x of point i of the bend pipe, the direction of x--axis is correspondent to the direction of the normal line of the plane i-1_ i _ i+1 with the equation:

$$\vec{x}_i = \vec{R}_{i-1,i} \times \vec{R}_{i,i+1} \tag{51}$$

where $\vec{R}$ is the direction vector of the pipe segment.

(2) The coordinate transformation from point i^L to point i-1: since the z--axis isn't changed, we can only adjust x--axis and y--axis, the adjustment angle α is:

$$\alpha = \arccos \frac{\vec{x}_{i-1} \cdot \vec{x}_i}{|\vec{x}_{i-1}| \cdot |\vec{x}_i|} \tag{52}$$

then the transformation matrix is

$$t_{xy}(\alpha) = \begin{bmatrix} \cos\alpha & -\sin\alpha \\ \sin\alpha & \cos\alpha \end{bmatrix} \tag{53}$$

(3) The coordinate transformation from point i^R to point i^L: the angle of the bend pipe is:

$$\beta = \arccos \frac{\vec{R}_{i-1,i} \cdot \vec{R}_{i,i+1}}{|\vec{R}_{i-1,i}| \cdot |\vec{R}_{i,i+1}|} \cdot \tag{54}$$

Since the x--axis isn't changed, coordinate transformation is only used to y--axis and z--axis, the transformation matrix is

$$t_{yz}(\beta) = \begin{bmatrix} \cos\beta & -\sin\beta \\ \sin\beta & \cos\beta \end{bmatrix}. \tag{55}$$

With joint coupling exists at the joint 'i', all related state variables can be transferred through matrix $t_{xy}(\alpha)$ and $t_{yz}(\beta)$, whereas the following vectors have increments due to the coupling effect of force and displacement (see reference [6] which gave a transfer matrix describing the plane joint coupling):

$$\Delta V_i^R = \sin\beta \cdot U_i^{yL} - (1-\cos\beta) \cdot U_i^{zL} \tag{56}$$

$$\Delta F_i^{yR} = A_f \sin\beta \cdot P_i \tag{57}$$

$$\Delta F_i^{zR} = A_f (1-\cos\beta) \cdot P_i \tag{58}$$

Using above equations can amend the elements in the transformation matrix.

The point transfer matrix can be obtained through the above process. A global transfer matrix that relates the state vector at one end of the piping system to that at the other end is obtained by a systematic product of the interior point and field matrices along a defined path.

The another advantage of this method is that in the computing of joint coupling transfer matrix if the angle $\alpha$and $\beta$are known, the transfer matrix can be gotten more directly. In this paper the above twice-transformation method is adopted to solve joint coupling problem of a spatial piping system. The solution procedure is:

a) Compute the angles (α and β)describing two cross pipe's relation;
b) Get the transfer matrix $t_{xy}(\alpha)$ and $t_{yz}(\beta)$ to do the transformation on the related variables;
c) Add the increments to V_i^R, F_i^{yR}, and F_I^{zR}.

Examples

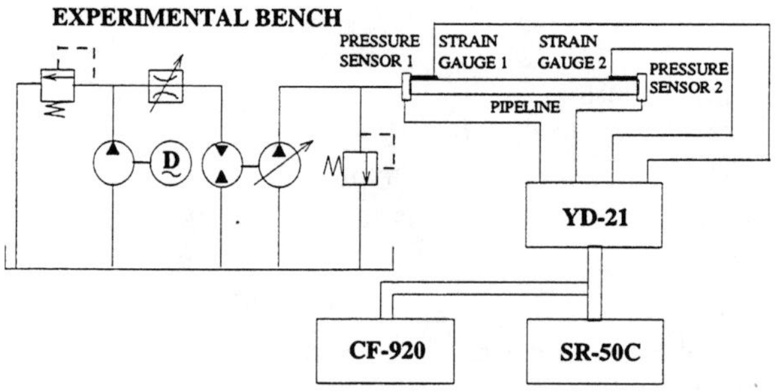

Figure 3 Experimental apparatus

The test rig shown in Figure 3 is a hydraulic motor velocity control system, which is controlled by a handle control orifice valve. The hydraulic motor drives an aerial pump which connects with the pipe network and load. Because pump pumps oil in pulse, it can be used as input signal generator, sweep frequency from 10Hz to 1000Hz by varying pump rotation speed. In high frequency band, the second harmony wave of flowrate pulse is made used of, which is realized through frequency spectrum analysis in dynamic signal analyzer (CF-920). The pressure of the fluid and the axial strain of pipe wall at each terminated end are measured by transducers (YD-21 is a dynamic resistance strain meter), which are recorded by portable data recorder SR-50C and analyzed by mini FFT analysis system CF-920.

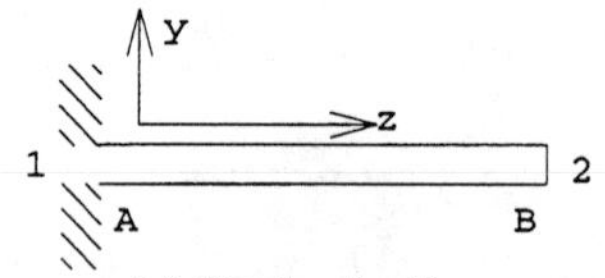

(a) Single pipeline

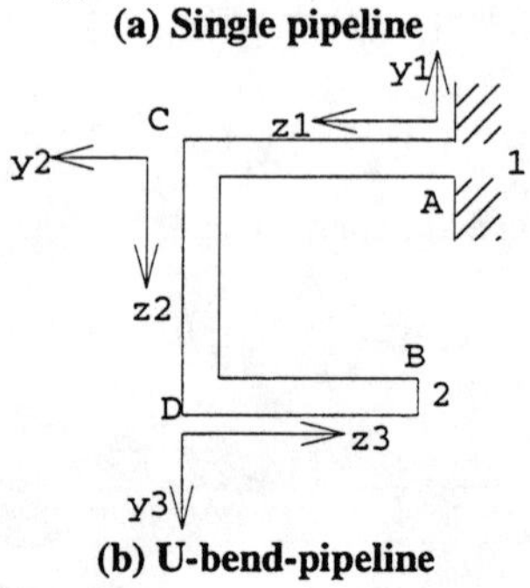

(b) U-bend-pipeline

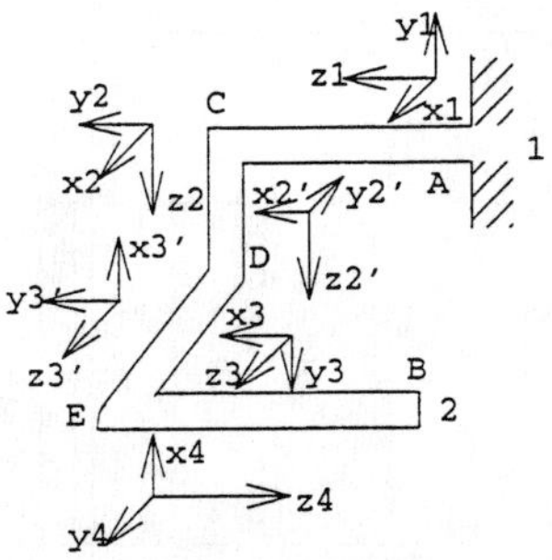

(c) Spatial pipeline
Figure 4 Piping system

Figure 4 (a) (b) (c) are three kinds of pipelines, the thickness of the pipe is 1mm, inner radius is 8mm, oil bulk modulus is 1500MPa, density is $872\,kg/m^3$, pipe modulus of elastic is 210GPa, density is $7800\,kg/m^3$, Poisson's ratio is 0.3. The angle of the bend pipe is 90°.

In each example, the entry of the pipeline is rigid supported (fixed) and the other end is rigid supported or free, but closed. Regard as to the fourteen-equation models of fluid-filled pipe, there are twenty-eight unknown quantities in all, but the number of equations is fourteen. In order to resolve them, the other fourteen conditions are necessary. At the rigid supported entry of the pipe: $\dot{U}^z = \dot{U}^x = \dot{U}^y = 0,\ R^x = R^y = R^z = 0$,
and at the other rigid supported (closed) end: $\dot{U}^z = \dot{U}^x = \dot{U}^y = 0,\ R^x = R^y = R^z = 0, V = 0$.
If the other end is free, the boundary conditions will be: $V = \dot{U}^z, F^z = A_f P,\ F^x = F^y = 0,\ M^x = M^y = M^z = 0$.
Then the impedance of the terminated end can be obtained.

In order to compare the result of simulation and experiments, this paper presents four groups of results.

(1) The influence of the frequency-dependent friction on simulation result:

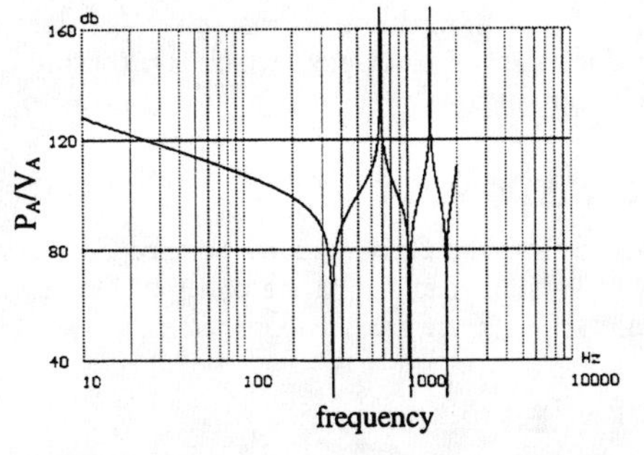

(a) Simulation curve without viscous friction

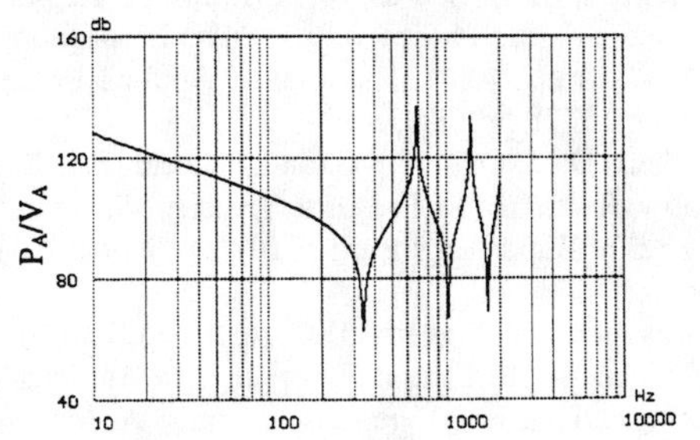

(b) Simulation curve with viscous friction
Figure 5 Comparison of two kinds of simulation curve

Figure 5 shows the simulation result of the impedance P_A/V_A at the entry of the pipe (point A). If the viscous friction is neglected, pressure amplitude at the harmony point will be very large (see (a)), which is far from the reality. The simulation curve with frequency-dependent friction is more closely to reality. From Figure 5 (a) and (b), we can see that the friction has slight influence on the harmony frequency.

(2) Simulation curves compared with experimental curves:

The dotted curve is experimental result and the continuous curve is simulation result.

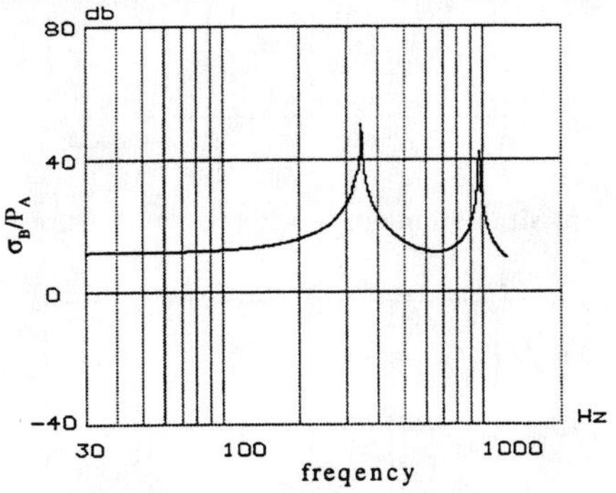

(a) Simulation curve with viscous friction

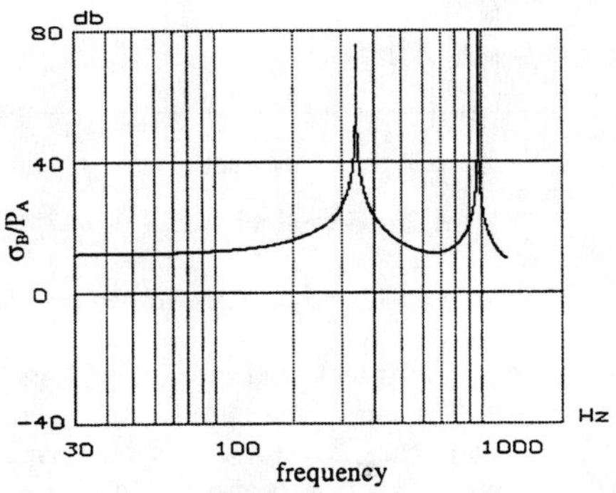

(b) Simulation curve without viscous friction

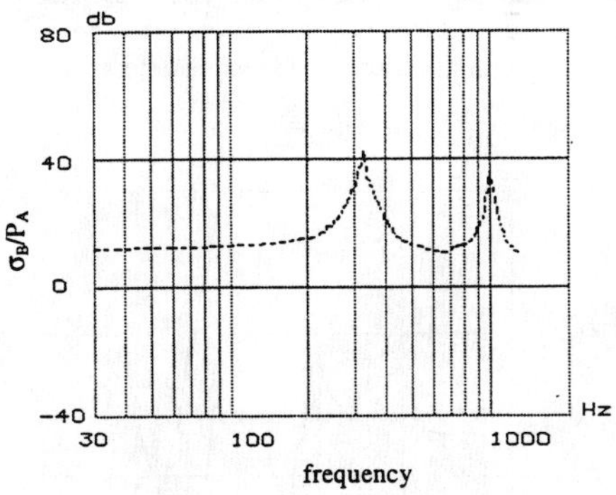

(c) Experimental curve

Figure 6 Free-end single pipeline impedance curve

Figure 6 is related to the single straight pipeline with its end free (see Figure 4 (a)). The curve shows the impedance σ_B/P_A of simulation and experiment. In the figure, the effect of the viscous friction can be seen: the amplitude of the peaks at the harmony frequency in the curve without viscous friction is much higher than that of the curve with viscous friction and the experimental curve. The curve with viscous friction is more close to the experiment, which means that the transfer matrix is correct.

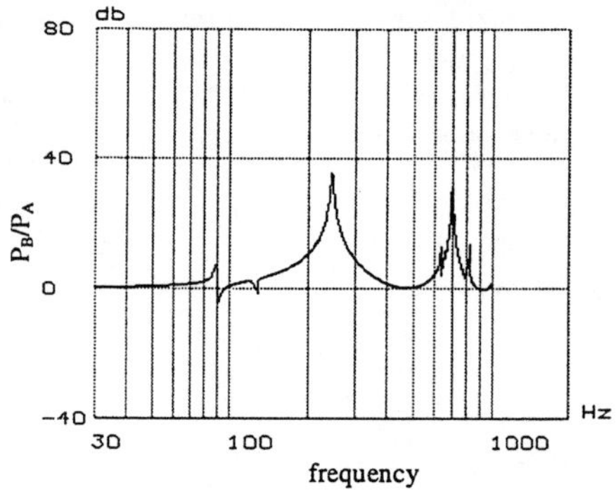

(a) Simulation curve with viscous friction

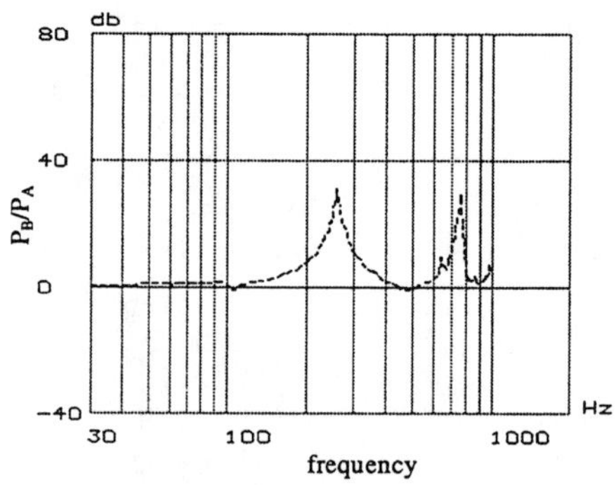

(b) Experimental curve

Figure 7 Fixed-end U-bend pipeline impedance curve

Figure 7 is related to the U-bend pipeline with its end fixed (see Figure 4 (b)). The curve shows the impedance P_B/P_A of simulation and experiment. Because the pipeline is in a plane, only one time coordinate transformation needed. The coupling matrix only related to angle β and the increments. The frequency of the peaks in the simulation curve is about the same as that in the experimental curve, which proves the coupling matrix correct. The viscous friction has the same effects on the curve as above.

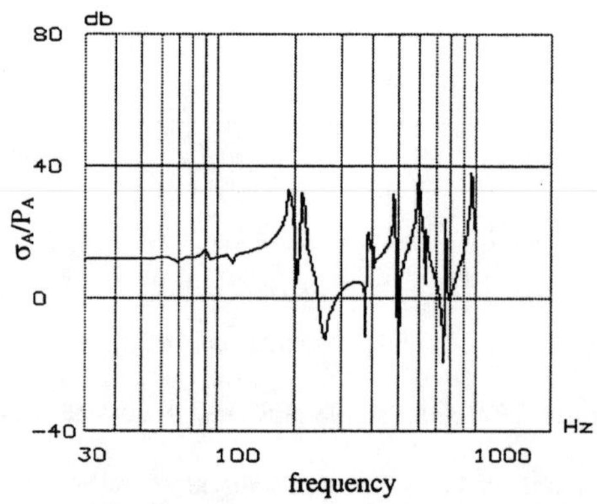

(a) Simulation curve with viscous friction

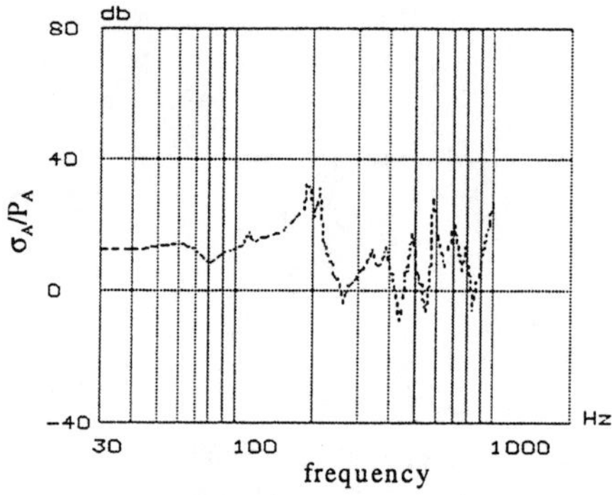

(b) Experimental curve

Figure 8 Free-end spatial pipeline impedance curve

Figure 8 is related to the spatial pipeline with its end free (see Figure 4 (c)). The curve shows the impedance σ_A/P_A of simulation and experiment. In the simulation, the twice-coordinate transformation method introduced in this paper is adopted. The coupling matrix is related to angle α, β and the increments. The main peaks shown in the simulation curve also appear in the experimental curve, which means that the twice-coordinate transformation method is correct. The viscous friction has the same effects as in Figure 6.

This work is the basis of analysis of complex fluid-structure interaction. It can be used in technical field to predict the vibration of piping system, which is helpful to the design of piping system.

Conclusion

The viscous friction is one of coupled form of liquid-filled piping systems. It is of great influence on the amplitude of vibration of the system simulation, but less influence on the harmonic frequency. In the present paper, the Zeilke's frequency-dependent friction is introduced. In the joint coupling, a new coordinate transformation method, twice-coordinate transformation method is adopted, which is proved to be validity and general purpose.

ACKNOWLEDGMENT

Financial aid from the China Natural Science Foundation under Grant No.59305035 and China Aeronautics Foundation under Grant No. 93H51126 are greatly appreciated.

REFERENCES

1. Skalak, R., "An Extension of the Theory of Water Hammer", Trans. ASME, **78**(1) (1956), pp.105-116.
2. Walker, J.S. and Philips, J.W., "Pulse Propagation in Fluid Tubes", Trans. ASME, Journal of Applied Mechanics, 99 (March 1977), pp.31-35.
3. Wiggert, D.C., Otwell, R.S. and Hatfield, F.J., "The Effect Elbow Restraint on Pressure Transients", Trans. ASME, Journal of Fluids Engineering, **107**(3) (Sept. 1985), pp.402-406.
4. Budny, D.D., Wiggert, D.C. and Hatfield, F.J., "The Influence of Structural Damping on Internal Pressure During a Transient Pipe Flow", Trans. ASME, Journal of Fluids Engineering, **113** (Sept. 1991), pp.424-429.
5. D'Sousa, A.F. and Odenburger, R., "Dynamic Response of Fluid Lines, Journal of Basic Engineering", Trans. ASME, Series D, **86**(3) (Sept. 1964), pp.589-598.

6. Lesmez, M.W., Wiggert, D.C. and Hatfield, F.J., "Modal Analysis of Vibrations in Liquid-Filled Piping Systems", Trans. ASME, Journal of Fluids Engineering, **112** (Sept. 1990), pp.311-318.
7. Brown, F.T., "The Transient Response of Fluid Lines", Trans. ASME ser.D, **84**(4) (Dec. 1962), pp.547-553.
8. Zielke, W., "Frequency-dependent Friction in Transient Flow", Journal of Basic Engineering, Trans. ASME, Ser.D, **90** (March 1968), pp.109-115.
9. Wylie, E.B and Streeter, V.L., 1983, "Fluid Transients", FEB Press.

FPST-Vol. 6, Fluid Power Systems and Technology — 1999
ASME 1999

MODELING OF HYDRAULIC SYSTEMS FOR HARDWARE-IN-THE-LOOP SIMULATION: A METHODOLOGY PROPOSAL

Jorge A. Ferreira
Department of Mechanical Engineering
University of Aveiro
3810 Aveiro, Portugal

João E. de Oliveira
Department of Electronic Engineering
University of Aveiro
3810 Aveiro, Portugal

Vítor A. Costa
Department of Mechanical Engineering
University of Aveiro
3810 Aveiro, Portugal

ABSTRACT

The present paper proposes a methodology to organize model libraries of electro-hydraulic components. This methodology holds in the association of an object oriented modeling, equation based, language for model structure description, with a graphical formalism suitable for the dynamic behavior description of reactive hybrid systems. That is, a recent general purpose language for physical modeling, called Modelica, is used to develop object-oriented libraries of models for different physical domains; the hierarchical description of the dynamic behavior of each model is obtained by means of the Statecharts formalism.

With the method proposed, complex models are built by model interconnection schemes; it is possible to organize the models for different complexity levels adapting them for distinct simulations (real-time or off-line), by refining their behavior using the Statecharts graphical formalism.

The developed hydraulic models, once compiled into C-code, are assembled as a whole application, and executed by a digital signal processing card (DSP). The simulation task is based on hardware-in-the-loop techniques in such a way that the "virtual" hydraulic application is controlled by real hardware.

1. INTRODUCTION

Hydraulic systems have been, for a long time, often used in industrial manufacturing and in heavy machinery. Hydraulic hardware suffered a great evolution during the last years, from hydro-mechanical devices to microprocessor controlled electro-hydraulic systems. The use of electronics and microprocessors contributes to improve the dynamic performance and to enhance the traditional systems with new features, as well as with new control possibilities. Due to the complex dynamics and non-linearity of these systems, the control algorithms usually applied in linear systems, such as the traditional PID algorithms, can have a poor performance for sophisticated hydraulic applications. Nowadays, there is an increased interest on strategies for using electro-hydraulics in advanced manufacturing systems, where dynamic performance and precision are very important parameters as, for example, in high-speed machining, injection molding systems or high-speed assembly hydraulic robots. It is important to investigate how advanced control schemes can improve the hydraulic actuation of this type of machinery. According to [Edge, 1997], it is important that, for a given application, the relative merits of different control schemes can be evaluated, being the computer simulation one of the best evaluation tool. Supporting this idea is [Ellman et al., 95] referring that a simulated environment is the cheapest and fastest way to test control algorithms. Modeling and real time simulation of complex systems still is an area to explore [Burrows, 1998] and, with the growing of computating power, more complex systems can be simulated in real time with decreasing costs [Lennevi et al., 1995].

The application of new control schemes on real hydraulic systems is a difficult task due to the cost and/or size of the hardware and its working conditions. In many applications, it is impossible to reproduce, in the laboratory, the hydraulic systems and their operating conditions. Some specific studies have been performed, where the operating environment and the hydraulic machinery hardware were reproduced by simulation and 3D visualization of the simulated model's dynamic behavior. Examples of such studies can be found in [Gonthier and Papadopoulos, 1998], [DiMaio at al., 1998] or [Schothorst, 1997]. However, these works use private modeling methodologies, thus precluding the interchange and the refining of the models.

Why a new methodology proposal?

Although the strong recent evolution in electro-hydraulic hardware, the project of hydraulic systems still is essentially based on tradition and experience. This fact, added to the growing complexity of modern hydraulic systems, can lead to unexpected behavior and errors. One way to predict these situations is the use of computer simulation, in many cases with ad-hoc simulation programs, usually written in FORTRAN, or with physical prototypes [Ellman et al., 1995].

The modeling language to be used must deal with various physical domains involved (hydraulics, mechanics, electrics, etc.) and with hybrid systems, where the involved components have a continuos and/or discrete dynamic behavior.

The representation of models in a systematic and flexible way has been studied in recent years, and there are specific domain and general purpose modeling languages [Otter and Cellier, 1995]; in either case, two main philosophies can be identified: traditional programming techniques and object-oriented methodologies. Nowadays, the tendency is focused toward object-oriented languages mainly due to the simplicity when reusing, expanding or adapting models. Confirming this idea is the work done by [Beater, 1998], referring that the largest time consuming step in system modeling can be speed up by using modern object-oriented simulation languages and component libraries.

The integration of models from different domains is a complex and time consuming task because, although powerful libraries exist, they are generally based on different modeling languages, almost invariably, not compatible. To minimize this situation there is a strong effort of a working group (including simulation tool builders, computer scientists and users from different domains) in order to build a unified object-oriented language for physical systems modeling. This language, named Modelica [Mattsson and Elmqvist, 1997], is intended for modeling virtually any application domain (electrical circuits, multi-body systems, hydraulics, thermodynamic systems, chemical systems, etc.), and it allows the inclusion of several formalisms for behavior description (ODE, DAE, bond graphs, finite state automata, petri nets or statecharts).

It is believed that the association of an object oriented modeling language, for model structure description, with a graphical formalism suitable for the dynamic behavior description of hybrid systems can be an interesting and useful approach for electro-hydraulic systems modeling.

The objective of this paper is to present a methodology where such an association, using Modelica and Statecharts, is embedded in a hardware-in-the-loop simulation environment.

2. HYBRID STATECHARTS AND MODELICA

A brief review on the formalism of Statecharts

The formalism of Statecharts [Harel, 1987] is intended to describe the dynamic behavior of complex reactive systems. It is viewed as an extension of the finite-state-machine (FSM) formalism with the add-ons of *hierarchy*, *parallelism* and *broadcast communication*.

Hierarchy is a well accepted approach for dealing with complexity, helping the human abstraction process. In Statecharts, hierarchy is used to group sets of states together, allowing high level description and step-wise development. The designer can start with an high level description of the model and proceed with the refinement of states by means of *AND/OR* decomposition operations. This hierarchical organization encourage "zoom" capabilities for moving easily back and forth between different levels of abstraction.

Concurrence inside a statechart can be described with AND states, allowing modeling of concurrent activities (*parallelism*) in the same model via *orthogonal* states. All these orthogonal states are activated when an *AND* state is entered and deactivated when it is exited.

Figure 1 shows the high level description of a system with states A and B. These states can be refined through state decomposition. State A will be a compound *OR* state with substates A1, A2 and A3. State B will be a compound *AND* state with substates B1 and B2. This process can proceed until low level description is achieved.

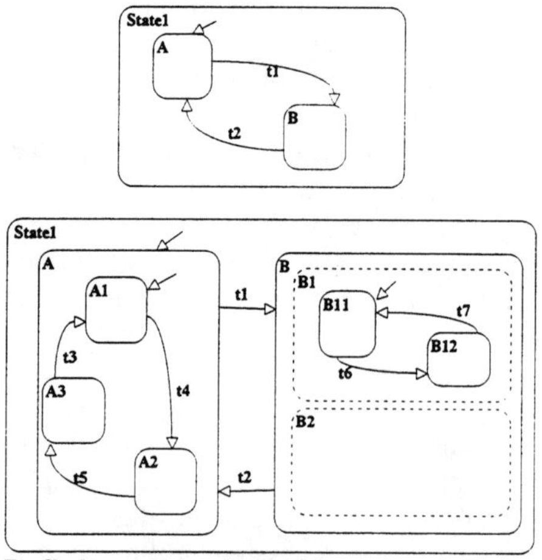

Fig. 1 - Refining states to achieve low level system's description

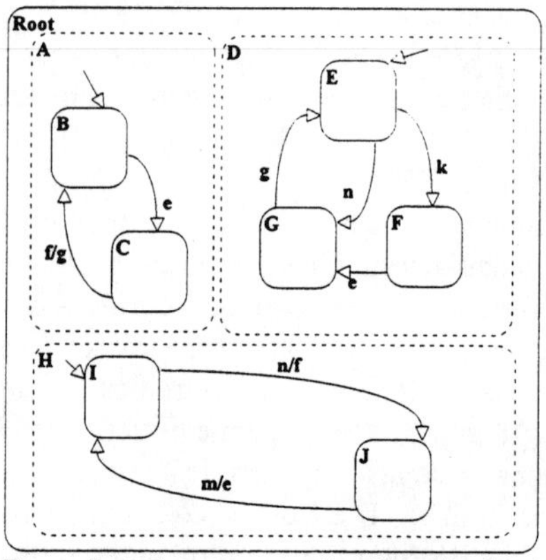

Fig. 2 - Broadcast communication in Statecharts

When orthogonal components are not truly independent, communication between them must be specified by associating an action with a transition. This action is assumed to be broadcasted and so every system component in the statechart will recognize the message. This *broadcast communication* mechanism allows that, when one part generates an event (attaching an action to a transition), all the other parts sense it, acting in response if it is so specified. In the example of figure 2, state *A* generates an event when transition *f* is fired; this event is sensed if state *G* is active, although state *G* belongs to another orthogonal state.

Another enhancement over the FMS formalism is the association of an event *action* with a transition, when it is taken, or with a state, when the state is entered or exited. Continuous *activity* can also be associated with a state for modeling continuous behavior when the state is active. These *action* and *activity* concepts allow the modeling of hybrid systems: *actions* capture the discrete features of the system while *activities* describe the continuous part.

Modelica

Modelica is a new language for physical modeling that is being developed as an international effort with the main objective of making easy the exchange of models and model libraries. The language is built on non-causal modeling with algebraic and differential equations, and uses object-oriented constructs to facilitate model reuse, through hierarchical modeling, encapsulation, and inheritance.

Models and submodels are declared as classes with connection interfaces called *connectors*. This connection capability allows the use of model libraries to compose complex models with the drag and drop, and connection drawing facilities of modern graphical editors.

With Modelica, the modeling of hybrid systems is supported via mixed continuous/discrete systems of equations. Discontinuous models can be handled with *if-then-else* expressions, allowing the modeling of phenomena with different expressions in different operating regions. Models with different complexity levels can be supported by the use of conditional equations, in such a way that changes on behavior are obtained by just setting a parameter. Discrete event and discrete time models are supported by *when* statements. The equations in a *when* clause are conditionally activated at event instants where the *when* condition becomes true.

The Statecharts library in Modelica

A small library to implement the hybrid Statecharts formalism in Modelica language was already developed [Ferreira and Oliveira, 1999]. Two statecharts implementation levels were considered: library models and component models. Statecharts library models are responsible for capturing events related to the firing of transitions and to the activation and deactivation of states that must be performed when transitions are taken. Library models have also to code the activation or deactivation of the substates, in order to implement what is expected with OR states and AND states. Component models create the statechart, with the basic models provided by the Statecharts library, make the state-transition-state connections and define the transition events or describe the continuous activities within states.

The approach followed to generate the code for statechart implementation is to consider a statechart as a model in Modelica. This model will be composed by states and transitions. Modelica models are developed for the basic elements of the statechart. The final model of the statechart is a set of Modelica models connected by state-transition-state connectors.

The equation based modeling of the Modelica language, along with the connector constructions, proves its efficiency by passing activation/deactivation messages instantaneously through nested states. Also, the broadcast communication mechanisms of Statecharts can easily be fulfilled, just by setting the value of a variable; this is automatically transmitted to all the statechart components because, in fact, the statechart is implemented with differential algebraic equations (DAE) that are evaluated concurrently.

Behavior inheritance of the Statecharts library

The main guideline followed by Harel [Harel and Gery, 1997], concerning the Statecharts behavior inheritance, is to base the two statecharts on the same underlying state/transition topology. Thus, B inherits all A states and transitions. Although these cannot be removed, certain changes are allowed. States can be modified in three ways: breaking down a basic state by OR (into substates) or by AND (into orthogonal components) decompositions; adding substates to an OR state; adding orthogonal components to any state. This last way is the most important because it enriches A behavior capabilities.

Transitions can also be added to the statechart, and some modifications are allowed in the inherited ones. For example, if the transition is labelled by *event[guard]action*, changes can be made in the trigger *event*, the *guard* or even in the *action* list. Although, explicitly, it is not possible to remove a transition, it can be done implicitly by making its *guard* false.

All these features were implemented [Ferreira and Oliveira, 1999] in the statecharts Modelica library by means of boolean switch parameters associated with the status of each state or transition. That is, each state or transition can be inhibited, at compile time, by setting its enabling parameter to false.

3. MODELING METHODOLOGY

The main directions to model the dynamic behavior of modern hydraulic systems are: object oriented libraries of models; hierarchical description of the dynamic behavior; model interconnection is used to build more complex models; various levels of model complexity achieve different simulation experiments; ability to refine or redefine behavior; graphical description of dynamic behavior to enhance model understanding.

The concept behind the methodology is to consider a model, of a physical component, as a composition of two complementary perspectives: its structure and its behavior.

The structure characterize the static part of the model, its parameters, its connection terminals, etc. Object oriented techniques and the design of hierarchies of models, connected by inheritance mechanisms, are important issues to develop reusable model's libraries. Mechanisms for model interconnection can also make easier complex system's modeling.

The dynamic part of the model, that is, its time evolution, is described by its behavior. This behavior depends on time, captured events or changing attributes that derive from the model's connection terminals. The behavior of an hydraulic model can be reactive when stimulated through its connection terminals. Thus, it is believed that a graphical formalism (such as Statecharts) is suitable to describe reactive hybrid systems and very useful for the hierarchical description of the model's dynamics.

The proposed methodology holds in the association of an object oriented modeling, equation based, language (Modelica) for model structure description, with a graphical formalism (Statecharts) suitable for the dynamic behavior description of reactive hybrid systems. The implementation of this graphical formalism inside the modeling language allows the methodology to be supported by a single global language.

The association of the object oriented modeling language with the graphical formalism leads to two types of inheritance for the component models. The model structure can be inherited through the usual mechanisms of object oriented languages. For model behavior description, this work purposes the adoption of the hybrid Statecharts formalism to obtain the required behavior inheritance.

The organization of model's knowledge is simple because the model's behavior can be hierarchically detailed by nesting statecharts. In fact, the statechart states can be decomposed until low level behavior description is achieved. An interesting feature of this approach is the possibility of inheriting and refining model's behavior in a very understandable manner, and with well-defined rules; this allows the organization, for the same component, of models whose behaviors can have various complexity stages suitable for different simulation experiments. For example, component models can be organized with a complexity order number; then, by just defining a certain degree of complexity, a complex model can be composed by several connected component models, either for real time simulation or for off line simulation experiments. The user can always inhibit the behavior of models and define a new behavior, for a part or for the whole model, in a very elegant way.

A simple example is presented shortly to illustrate the method.

The relief valve example

Consider the example of a component, relief valve (fig. 3), that limits the pressure in a hydraulic system. It is a closed loop system but it is usually modelled by its (static) input/output characteristic [Beater, 1998]. Consider, as a first approach, that the valve has the behavior depicted on the left side of figure 4.

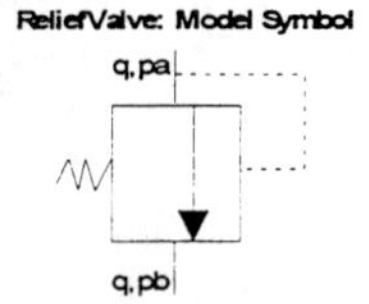

Fig. 3 – Symbol for a hydraulic relief valve

The behavior of the simple relief valve model can be described by considering that the valve has two possible stable states, and also hysteresis when it changes from one state to another. When the pressure difference (*dp*) is bigger than the pressure required to open the valve (*pressureOpen*), the valve will be totally open (*q=(dp–pressureClose)*gOpen*); when the pressure difference is smaller than the close pressure (*pressureClose*) the valve will be totally closed (*q = dp*gLeak*).

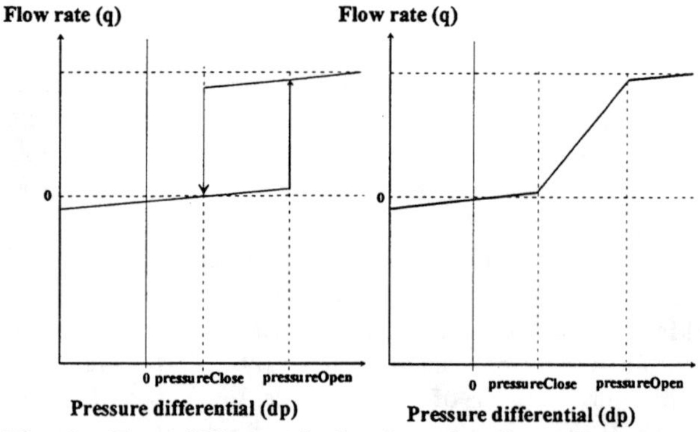

Fig. 4 – Two different behaviors for the relief valve

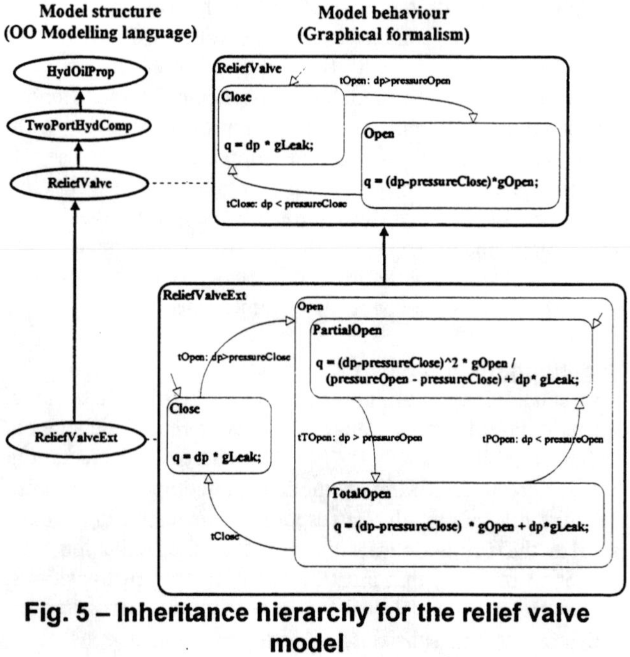

Fig. 5 – Inheritance hierarchy for the relief valve model

Using the formalism of Statecharts, the dynamic behavior of the valve (fig. 5) can be represented by an OR statechart (*ReliefValve*) that contains two substates (*Close* and *Open*) and two state transitions (*tOpen* and *tClose*). Initially the valve is closed (*Close* is the default substate). The continuous behavior of the relief valve is described by its during activity ($q = dp * gLeak$), that is enabled while the valve is in its *Close* state and is not exiting. If the valve is totally closed and its pressure difference (*dp*) exceeds *pressureOpen*, then a transition (*tOpen*) takes place and the state *Open* becomes active. The during activity of state *Open* will be evaluated for all the simulation steps while the state *Open* is active, that is, while transition *tClose* does not occur. This transition is fired when pressure *dp* reaches a value below the pressure difference defined by the parameter *pressureClose*.

Refining the relief valve model

To improve the relief valve model, the characteristics displayed at the right of figure 4 can be used. One possible solution is carried out by the statechart associated with the *ReliefValveExt* model of figure 5. The statechart shows that the states *Close* and *Open* and transition *tOpen* were redefined. The transition *tClose* is the only one that is inherited without changes from the ancestor model (*ReliefValve*).

It should be noted that the state *Open* was refined, being now an OR state with *PartialOpen* and *TotalOpen* substates. The model will inherit all the structure and, concerning the behavior, all the equations that were not redefined. For instance, the equation that defines the trigger event for the transition *tClose* (*tClose.event = dp<pressureClose*) will be inherited from the model *ReliefValve*, while the trigger event equation for transition *tOpen* must be redefined.

A subset of the Modelica code for the relief valve model's hierarchy is presented bellow. The models use the Statecharts Modelica library described in [Ferreira and Oliveira, 1999].

```
model HydOilProp "Oil properties"
  parameter Real KVisc=46e-6;          // m^2/s; kinematic viscosity
  parameter Real rho=865;              // kg/m^3; mass density
  parameter Real bulkModulus=1e9;      // Pa; bulk modulus
end HydOilProp;
```

```
connector HydConnector "Hydraulic connector"
  Real p;       // Pa; fluid pressure
  flow Real q;  // m^3/s; fluid flow
end HydConnector;
```

```
model TwoPortHydComp
  extends BasicOil;
  HydConnector HydA; // hydraulic connector A
  HydConnector HydB; // hydraulic connector B
  Real dp;       // Pa, pressure difference
  Real q;        //m^3/s; flow through component
equation
  dp = HydA.p - HydB.p;
  ...
end TwoPortHydComp;
```

```
model ReliefValve
  extends TwoPortHydComp;
  parameter Real pressureOpen=55e5; //Pa
  ...
  RootStateS Root;
  StateS Close(defaultState = true);
  StateS Open;
  TransitionS tOpen;
  TransitionS tClose;
equation
  tOpen.event = event(dp > pressureOpen);
  tClose.event = event(dp < pressureClose);
  q = if Open.active then (dp-pressureClose)*gOpen
      else dp * gLeak;
end ReliefValve;
```

```
model ReliefValveExt
  extends ReliefValve;
  ...
  StateS PartialOpen(defaultState = true);
  StateS TotalOpen;
  TransitionS tPOpen;
  TransitionS tTOpen;
equation
  ...
  tOpen.event = event(dp > pressureClose);
  tPOpen.event = event(dp > pressureOpen);
  tTOpen.event = event(dp < pressureOpen);
  q=if Closed.active then dp * gLeak
    else if PartialOpen.active then (dp-pressureClosed)^2 *
         gOpen / (pressureOpen-pressureClose) + dp*gLeak
    else (dp-pressureClosed) * gOpen + dp*gLeak;
end ReliefValveExt;
```

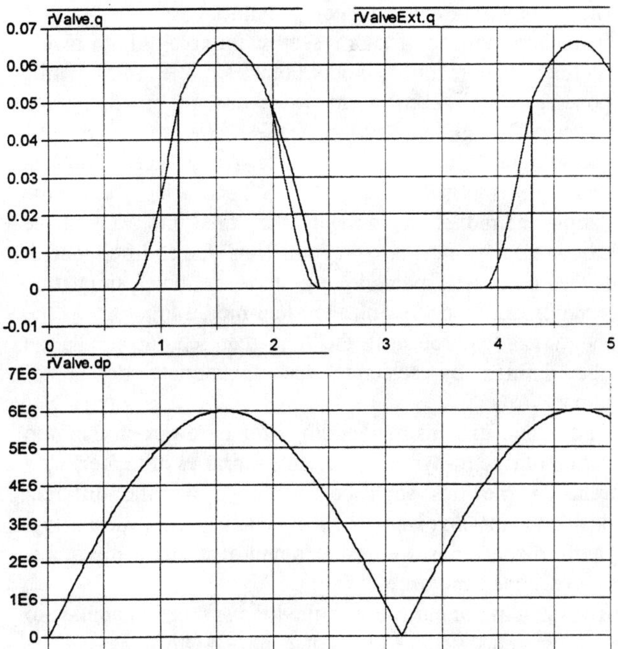

Fig. 6 – Behavior of ReliefValve and ReliefValveExt models

The dynamic behavior of instances of *ReliefValve* and *ReliefValveExt* models (*rValve* and *rValveExt*, respectivelly) is shown in the following graphics; the absolute value of a sinusoidal signal was used as the pressure difference *dp*. This model was tested with the Dymola package [Dymola] that supports the Modelica language.

This example shows how simple can be the refining of behavior of, for example, model libraries of hydraulic components, if the present methodology is used.

4. HARDWARE-IN-THE-LOOP SIMULATION OF HYDRAULIC SYSTEMS

Hardware-in-the-loop simulation refers to a technology where some of the components of a pure simulation are replaced with actual hardware. This type of procedure is useful, for example, to test a controller which, instead of being connected to the real equipment under control, is connected to a real time simulator. The controller must "think" that it is working with the real system and so the accuracy of the simulation and its electrical interfacing to the controller must be adequate [Maclay, 1997]. This technology provides a way for testing control systems over the full range of operating conditions, including failure modes. Testing a control system prior to its use in a real plant can reduce the cost and the development cycle of the overall system. Hardware-in-the-loop simulation has been used, with success, in the aerospace industry and is now emerging as a technique for testing electronic control units. It has been applied to solve some specific problems but is seldom used as a platform to test the real time behavior of hardware components.

The main purpose of the present methodology is to give a well defined support for the model libraries of electro-hydraulic components, to be used in hardware-in-the-loop simulation experiments. The global performance is related to the model's complexity, thus, for diverse type of simulations, different model's behavior shall be used. When using real time simulation, the model's complexity shall take into account the dedicated real time hardware that will "run" the codified model. With the proposed methodology, the real-time simulation experiments can be done with the same model topology as the off-line simulations, because the time required for simulation can be defined by selecting the appropriate degree of complexity.

In order to use this methodology for hardware-in-the-loop simulation of electro-hydraulic systems, libraries of models with different complexities shall be developed for the different physical domains involved in such systems. This means that new basic models must be designed and also that some others must be redefined and/or refined.

The main use of hardware-in-the-loop simulation applied to hydraulic systems is the testing of control algorithms and the development of new control schemes. Usually this implies a real controller, operating over a real time simulated hydraulic plant.

The objective of future work is to be able to modify the complexity of a composed hydraulic model, just by setting parameter values; that way a model can be compiled with the complexity adjusted to the hardware platform.

Figure 7 presents an outline of the general block diagram for the modeling methodology testbench. This platform will be used in a near future to emulate electro-hydraulic systems in order to test real controllers and algorithms.

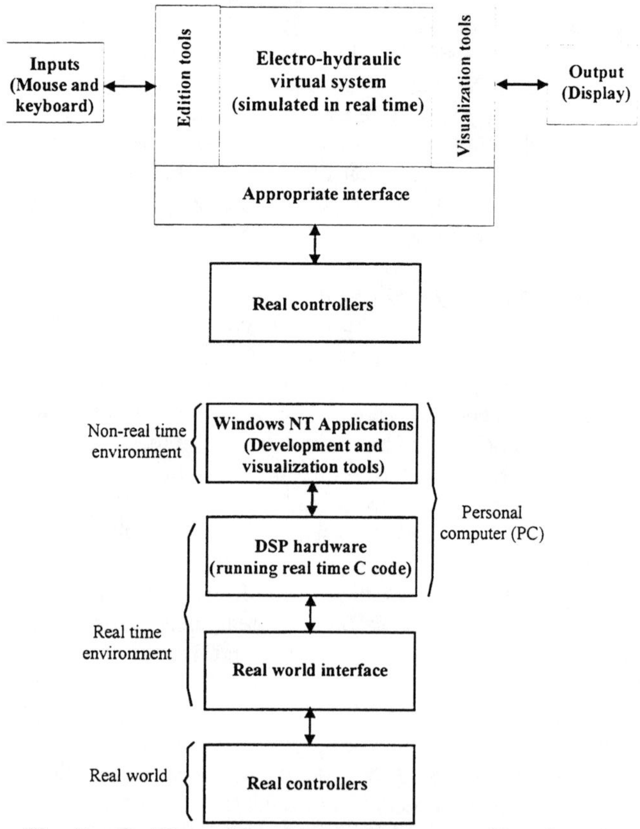

Fig. 7 – Outline of the block diagrams of hardware and software components

5. CONCLUSIONS

The present paper proposes a methodology to organize model libraries of electro-hydraulic components, in order to easily manage complex models for hardware-in-the-loop simulation experiments.

The concept behind the methodology is the following: the model of a physical component is composed by two complementary perspectives, its structure and its behavior. For that, the methodology is based on the association of an object-oriented language, Modelica, to model the structure, with the hybrid Statecharts formalism to describe the dynamic behavior of the model.

By making use of the Statecharts inheritance rules, this work also proposes a solution to refine or redefine the behavior of hydraulic systems. This way, different simulation experiments can be achieved by choosing the appropriate behavior for each model.

REFERENCES

[Beater, 1998] – P. Beater, "Object Oriented Modeling and Simulation of Hydraulic Drives," Simulation News Europe, March, 1998.

[Burrows, 1998] – C. R. Burrows, "Fluid Power Systems – An Academic Perspective," JHPS Journal of Fluid Power Systems, vol 29, n°1, January 1998.

[DiMaio et al., 1998] – S. P. DiMaio et al., "A Virtual Excavator for Controller Development and Evaluation," Proceedings of the 1998 IEEE International Conference on Robotics & Automation, Belgium, May 1998.

[Dymola] – Dymola package: Homepage: «hyperlink http://www.dynasim.se»

[Edge, 1997] - K. A. Edge, "The control of fluid power systems – responding to the challenges," Proceedings of Institute of Mechanical Engineers, Vol 211 part 1, 1997.

[Ellman et al., 1995] - A. Ellman, S. Sanerma, M. Salminen, R. Piché and T. Virvalo, "Tools for Control and Hydraulic Circuit Design of a Hydraulic-Driven Manipulator Mechanism," Proceedings of the 9th European Simulation Multiconference, June, Czech Republic, 1995.

[Ferreira and Oliveira, 1999] – J. A. Ferreira, J. Estima de. Oliveira, "Modeling hybrid systems using Statecharts and Modelica," To be presented in the 7th IEEE International Conference on Emerging Technologies and Factory Automation, October 1999, Spain.

[Gonthier and Papadopoulos, 1998] – Y. Gonthier, E. Papadopoulos, "On the Development of a Real Time Simulator for an Electro-hydraulic Forestry Machine," Proceedings of the 1998 IEEE International Conference on Robotics & Automation, Belgium, May 1998.

[Harel, 1987] - D. Harel, "Statecharts: A visual formalism for complex systems," Science of Computer Programming 8, 231-274, 1987.

[Harel and Gery, 1997] - D. Harel, E. Gery, "Executable Object Modeling with Statecharts," IEEE Computer, July 1997.

[Lennevi et al., 1995] – J. Lennevi, J. Palmberg, A. Jansson, "A Simulation Tool for the Evaluation of Control Concepts for Vehicle Drive Systems," Proceedings of the 4th Scandinavian International Conference on Fluid Power, Tampere, September, 1995.

[Maclay, 1997] – D. Maclay, "Simulation gets into the loop," IEE Review, May 1997.

[Mattsson and Elmqvist, 1997] – S. E. Mattsson, H. Elmqvist, "An International Effort to Design the Next Generation of Modeling Language," Proceedings of the IFAC Symposium on Computer Aided Control Systems Design - CACSD'97, Gent, Belgium, April 1997.

[Otter and Celier, 1995] – M. Otter, and F.E. Cellier, "Software for Modeling and Simulating Control Systems," The Control Handbook (W.S. Levine, ed.), CRC Press, Boca Raton, FL, 1995.

[Schothorst, 1997] – G. Schothorst, "Modelling of Long-Stroke Hydraulic Servo-Systems for Flight Simulator Motion Control and System Design," PhD thesis, Mechanical Engineering Systems and Control Group, Delft University of Technology, The Netherlands, 1997.

Research on A New Type of Flow Control System of Hydraulic Elevators with Car Speed Feedback

Xu Bing(bxu@sfp.zju.edu.cn)
Yang Huayong(yhy@sfp.zju.edu.cn)
(The State Key Laboratory of Fluid Power Transmission and Control of Zhejiang University, Hangzhou, 310027 P.R.China)

1. Abstract

This paper introduces an innovation design of flow control system for hydraulic elevators with car speed feedback. By measuring pulse signal emitted by photoelectric encoder mounted on the elevator car, a single-chip computer controls the electrohydraulic proportional integrated flow valve, then drives the elevator car by a hydraulic cylinder. With this set of single-chip computer control system, good speed performance of the elevator can be achieved. Furthermore, it has strong robustness against the disturbance of oil temperature variation, mechanical friction and other disturbance factors. In order to obtain better ride comfort and soft start of the elevator, PID control algorithm is also proposed. Some mathematical models both for static and dynamic states have been established, consequently simulation results are found to be in close agreement with the experimental results.

[Keywords] control system, hydraulic elevators, speed feedback

2. Nomenclature:

a —Effective area for spool of up pilot valve.

A_h —Piston working area.

A_1, A_x —Effective area for spool of main orifice valve.

A_s —Effective area for spool of check valve.

B_x, B_y, B_s —Flow coefficients of main orifice valve□pilot valve and check valve.

E_β —Effective bulk modulus of the fluid.

e —Error between preset voltage and car speed feedback voltage.

F_{fx}, F_{fy}, F_{fs} —Static flow force of main orifice valve□pilot valve and check valve.

F_k —Friction force between cylinder and piston.

F_{em} —Output force of proportional solenoid.

g — Acceleration due to gravity.

G_1, G_2, G_3 —Conductance of fixed orifice of R1□R2 and R3.

K_p —Proportional amplifier gain of PID regulator.

K_{sx}, K_{ss} —Spring rigidity of main orifice valve and check valve.

m_x, m_y, m_s, m_h —Plungers mass of main orifice valve□pilot valve□check valve and cylinder.

P_1, P_2, P_3, P_4 —Pressure of each chamber.

R —Coil resistance of proportional solenoid.

R_y —Spool radius of up pilot valve.

T_i —Integral constant of PID regulator.

T_d —Differential constant of PID regulator.

u —Input voltage of proportional amplifier.

x, y, s, h —Stroke of main orifice valve□pilot valve□check valve and piston.

x_0, s_0 —Spring preset compression of main orifice and check valve.

v_1, v_2, v_3, v_4 — Volumes of each chamber.

3. Introduction:

Hydraulic elevators have many advantages compared with traction elevators, for example, no top machine room□agility of machine room arrangement, heavy load, high safety and architecture privileges, etc. Three critical factors in measuring elevator performance are: (a) the rapidity and (b) comfort(means start acceleration and stop deceleration) with which it travels form one stop to the next, and (c) the accuracy with which it stops at and maintains floor level under all loads and temperatures regardless of travel direction. There are two kinds of control systems of hydraulic elevators generally.

A. Volume controlled systems: pump control and VVVF (variable voltage and variable frequency) control.

B. Valve controlled systems: on-off or mechanical valve control and proportional valve control .

Proportional valve control systems are the research subject of this paper. At present, proportional control system of hydraulic elevators with flow feedback is widely used and its performance is better than mechanical controlled valve. But many application and researches indicate that flow feedback system can't give a satisfied ride performance, and its ability against diversified disturbances, such as oil temperature variation, the difference of static and dynamic friction between piston and cylinder, and rope's elasticity variation, etc. , are poor. Based on careful research of hydraulic elevator control systems and recently wide used technology, digital electrical control unit and car speed feedback idea are suggested, and the stability and ride comfort of hydraulic elevators are remarkably improved.

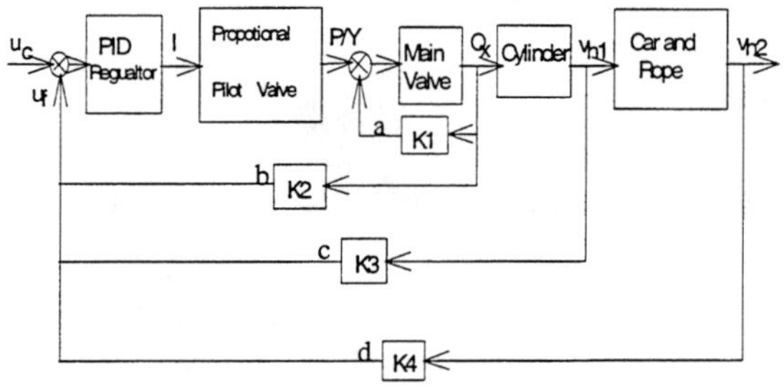

Fig. 1 Four types of feedback control system of hydraulic elevators

Four types of feedback control system of hydraulic elevators are shown in figure 1:

A. Electrohydraulic proportional flow control system with internal mechanical or hydraulic force feedback.

B. Electrohydraulic proportional flow control system with flow feedback.

C. Electrohydraulic proportional speed control system with hydraulic piston speed feedback.

D. Electrohydraulic proportional speed control system with elevator car speed feedback.

Both types 3 and 4 are speed feedback approaches, but their feedback points are different. Closed loop range of type 4 is wider than type 3, so the ride comfort and velocity control accuracy are significantly improved, but the difficulty for control system also increases. For elevator control system with load flow feedback, the temperature characteristics of flow meter have much influence on control systems. In order to keep well match between the flow meter and the flow control valve block, the flow meter usually is integrated in flow control valve block whose structure is complex and manufacture is difficult. With the car speed feedback, electrohydraulic proportional flow control system reduces the influence of dead zone and nonlinearities of flow valve, furthermore, simplifies the structure of the flow valve block and facilitates the manufacture.

Single-chip computer controlled units have characteristics of high integration and powerful functions, and in order to get the satisfied ride performance of the hydraulic elevator car, modern control algorithms are applied in systems to achieve parameters self-learning optimization and logical inference.

Also, with the car speed feedback, both the elevator car's speed and position are measured, so the elevator car direct landing with position criterion which is difficult to achieve in traditional flow feedback control systems may be realized.

4. Research on new type of flow control system of hydraulic elevators with car speed feedback

(1) Working principles of flow control system with elevator car speed feedback

Fig2 shows the working principles of the hydraulic elevator control system with car speed feedback. The system includes three parts: hydraulic system, single-chip computer controlled system and logical signal control system.

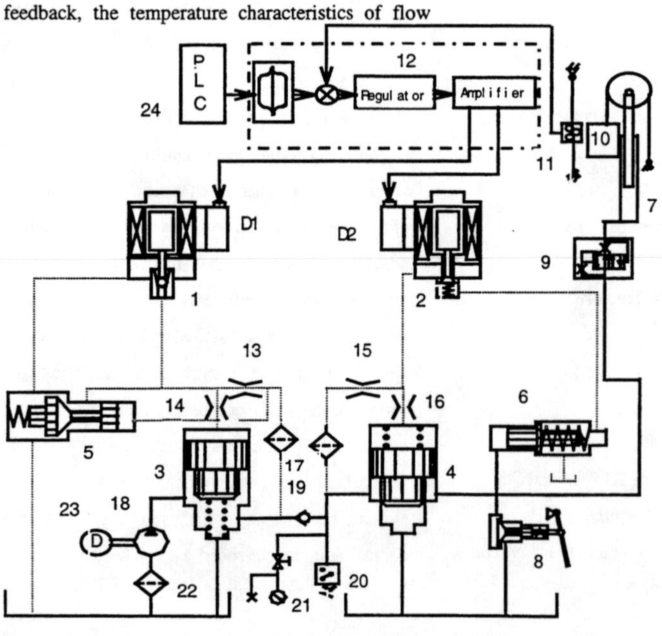

Fig. 2 Principles of hydraulic elevator control system with car speed feedback

Hydraulic system: By-pass orifice control circuit with three-way flow control valve is applied in up circuit, and return pipeline orifice speed control circuit with two-way flow control valve is applied in down circuit. As shown in figure 2, pump 18 driven by motor 23 outputs constant flow. Up main valve 3 and down main valve 4 control the up and down courses of the car separately. Check valve 19 isolates the up and down circuit. Rupture valve 9 can shut off automatically, quickly and smoothly when the hydraulic pipeline is broken or the car overspeed, thus prevents the loss of hydraulic oil in cylinder and improves the safety of hydraulic elevators.

Single-chip computer controlled system: Applying the Intel 80C196KC single-chip computer as the core, the system is comprised of a speed sensor (photoelectric encoder) which mounted on the car and a single-chip computer controlled unit. The speed sensor measures and processes the car speed data, then transmits them to the single-chip computer control unit where the error of the preset speed value and feedback value is calculated. Finally the error signal amplified in power amplify circuit is transmitted to proportional solenoid to control up and down main valves controlling the fluid feeding and discharging to or from the cylinder.

(2) Hardware and software design for Single-chip computer controlled system

Single-chip computer controlled system is shown in Fig 3, which includes: input circuits, output circuits, single-chip computer unit and power supply, etc.

Input circuits: (1) Speed sensor and interface circuit. (2) Parameters setting circuit. (3) Keyboard. (4) Command receiving circuit.

Output circuits: (1) PWM power amplifier circuit. (2) Hardware protection circuit. (3) Speed curves output and display circuit. (4) LED digital display unit.

Single-chip computer unit: (1) CPU and program memory. (2) 8155 programmable parallel I/O interface circuit. (3) Power supply.

The system software adopts module structure programs and includes following main parts: main program, input module subprogram, output subprogram, display subprogram, break module subprogram and computing module subprogram. The main program includes system initialization and subprogram transfer, etc.

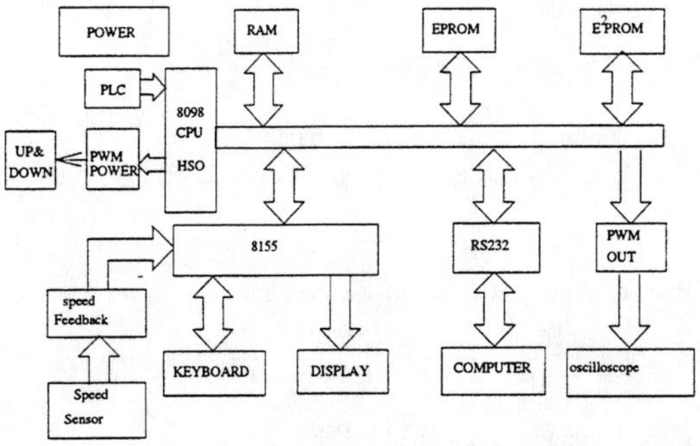

Fig. 3 Single-chip computer controlled system

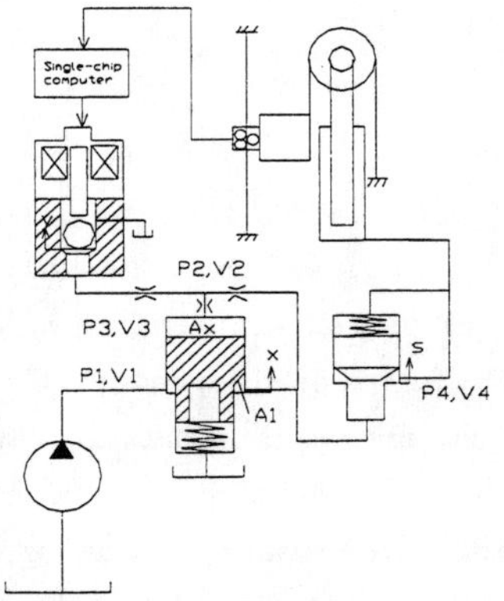

Fig. 4 Principles of up circuit for hydraulic elevators

5. Dynamic model of the innovation flow control system with car speed feedback

Working principles of control system of up circuit with car speed feedback in hydraulic elevators is shown in figure 4. According to Bernoulli's flow continuity equation and the 2nd Newton law, the dynamic mathematical model for the system is established.

Input voltage equation of proportional solenoid:

$$U(t) = K_P[e(t) + \frac{1}{T_i}\int_0^t e(\tau)d\tau + T_d\frac{de}{dt}] \cdots\cdots (1)$$

Force balance equation of proportional solenoid:

$$F_{em} = 80 * U(t) / R - 4 \cdots\cdots (2)$$

Force balance equation of spool of up pilot valve:

$$F_{em} - P_{2a} + F_{fy} = m_y\frac{d^2y}{dt^2} + d_y\frac{dy}{dt} \cdots (3)$$

Force balance equation of spool of main orifice valve:

$$P_1A_1 - P_3A_x - F_{fx} + K_{sx}(-x + 2x_0) = m_x\frac{d^2x}{dt^2} + d_x\frac{dx}{dt} \cdots\cdots (4)$$

Force balance equation of spool of check valve:

$$P_1A_s - P_4A_s - F_{fs} - K_{ss}(s_0 + s) = m_s\frac{d^2s}{dt^2} + d_s\frac{ds}{dt} \cdots (5)$$

Force balance equation of piston of hydraulic cylinder::

$$P_4A_h - m_hg - F_k = m_h\frac{d^2h}{dt^2} + d_h\frac{dh}{dt} \cdots\cdots\cdots\cdots (6)$$

Flow continuity equation of V1 chamber:

$$\frac{dP_1}{dt} = \frac{E_\beta(P_1)}{V_1}[Q_s - A_1\frac{dx}{dt} - G1\sqrt{P_1 - P_2} - B_{sx}\sqrt{P_1} - B_{ss}\sqrt{P_1 - P_4}\cdot sign(P_1, P_4)]$$

$$\cdots\cdots\cdots\cdots\cdots\cdots (7)$$

Flow continuity equation of V2 chamber:

$$\frac{dP_2}{dt} = \frac{E_\beta(P_2)}{V_2}(G1\sqrt{P_1 - P_2} - G2\sqrt{P_2 - P_3} - B_{yy}\sqrt{P_2}) \cdots (8)$$

Flow continuity equation of V3 chamber:

$$\frac{dP_3}{dt} = \frac{E_\beta(P_3)}{V_3}(-A_x\frac{dx}{dt} + G2\sqrt{P_2 - P_3}) \cdots\cdots (9)$$

Flow continuity equation of V4 chamber:

$$\frac{dP_4}{dt} = \frac{E_\beta(P_4)}{V_4}[B_{ss}\sqrt{P_1 - P_4}\cdot sign(P_1, P_4) - A_h\frac{dh}{dt}]$$

$$\cdots\cdots\cdots (10)$$

6 System simulation:

On the basis of the above-mentioned mathematical models, a set of friendly simulation software specialized in hydraulic elevator control systems is developed with the C computer language, it can plot, print, varies parameters, modify system mathematical model and add diversified restrain factors.

Key structure parameters variation simulation which have great influence on the control system is developed on the basis of mathematical models. The influence of main spring rigidity of down circuit is shown in figure 5. The harder the spring rigidity is, the shorter the step response time and the higher the amplitude. Because of limitation of static system design, the spring rigidity should get harder, i.e. if the spring rigidity is too soft, the system step response time will get longer obviously and system characteristics will become bad. The dynamic step response comparison of open loop system and closed loop system with PID regulator in up circuit is shown in figure 6. It indicates that open loop system in up circuit is a damping system which has severe vibration and is difficult to meet the requirements. With the compensation of PID control algorithm, the system delay and amplitude gets small, stability time also gets short, and excellent compositive dynamic characteristics are attained. The response of the up closed loop circuit which follows the preset speed curve is shown in figure 7, it indicates that the trailing characteristics of the system are excellent.

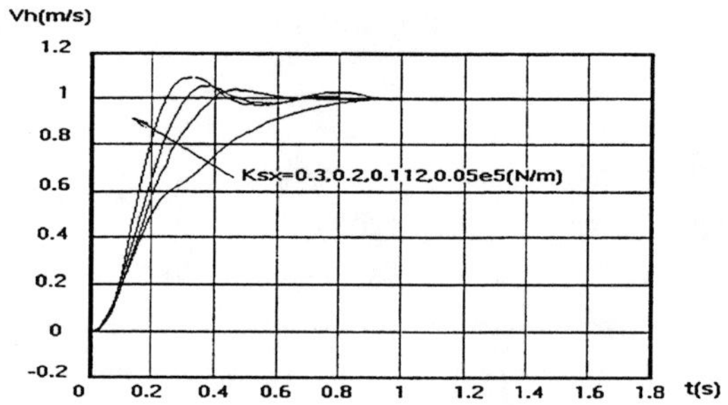

Fig. 5 The influence of spring rigidity

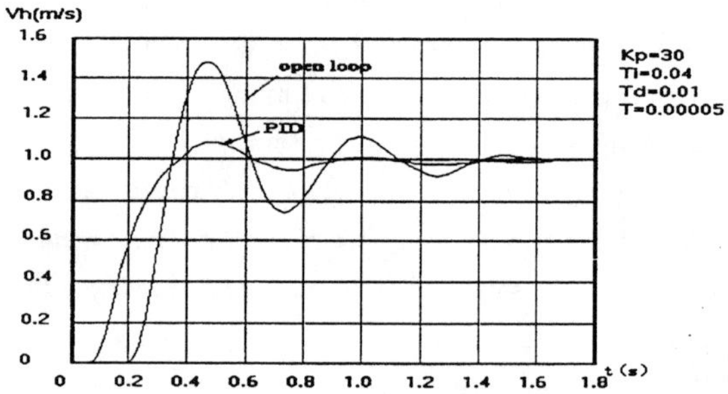

Fig. 6 Comparison of open loop and closed loop system

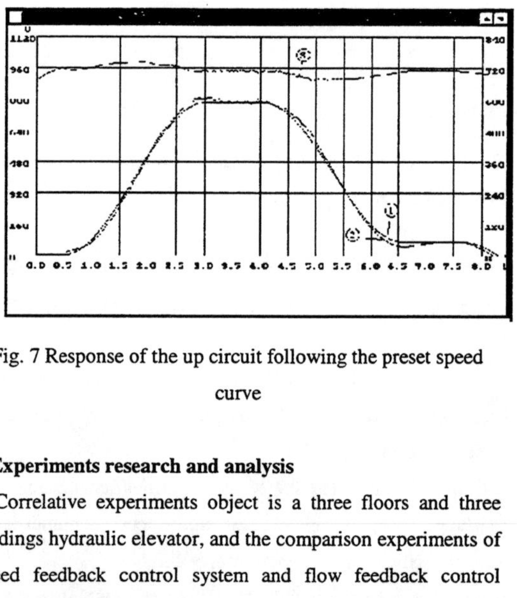

Fig. 7 Response of the up circuit following the preset speed curve

7 Experiments research and analysis

Correlative experiments object is a three floors and three landings hydraulic elevator, and the comparison experiments of speed feedback control system and flow feedback control system are made in detail.

The experiment results of the control system of hydraulic elevator with flow feedback are shown in figure 8. Curve 1 is a preset speed curve, and curve 2 is a practical car speed curve. At the start stage of the elevator car, the preset speed curve 1 outputs signal, but there is no practical speed output signal, when the preset speed value exceeds 140mm/s, there is practical speed output signal, thus the car vibration is high. At the stop stage of the elevator speed curve, the car speed reduces to zero suddenly rather than following the preset speed curve , then high impulsion will be felt by the passengers in the elevator car. Moreover, both the oil viscosity reduction due to the rise of oil temperature and difference between dynamic and static friction between cylinder and piston cause the car speed signal varies. So the regulation ability of control system of hydraulic elevators with flow feedback is little, thus causes the car impulsion during the start and leveling stage.

The experiment results of car speed feedback control system are shown in figure 9. Curve 1 is preset speed curve, curve 2 is practical car speed curve, and curve 3 is the output control curve of single-chip computer. At the start stage of elevator car, single-chip computer control unit has output signal, but there is no preset speed output signal. Elevator car doesn't move until the output signal value of single-chip computer control unit exceeds 870. After elevator car begin to move, single-chip computer control unit sends preset speed curve. As figure 9 shows, preset speed curve and practical speed curve are highly synchronous, and the acceleration value of the car is also same as the preset value, so the excellent riding comfort is obtained. During the acceleration and deceleration courses of the elevator car, there is a short delay. The elevator car levels slowly following the preset speed curve during the stop course, and at last the car speed reduces slowly from 60mm/s to zero.

As table 1 shows, in order to get more satisfied ride comfort or improve the control system performance, Fuzzy PID and DSMVS(Discrete Sliding Model Variable Structure) control strategies are proposed comparing with above-mentioned standard PID algorithm.

With the control system with car speed feedback, the soft start and soft stop of hydraulic elevators are attained, and the performance of hydraulic elevators is improved.

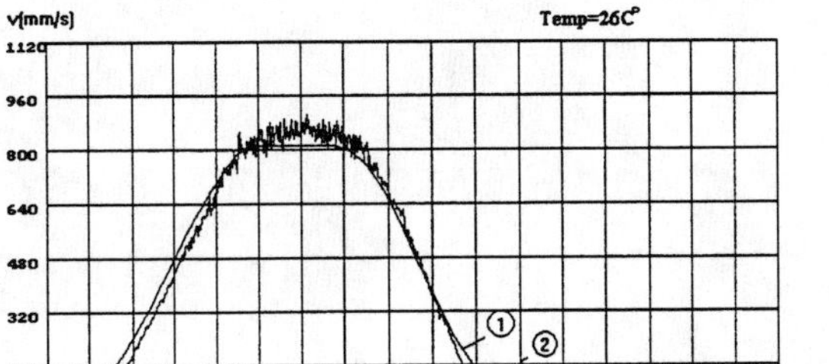

Fig. 8 Experiment results of hydraulic elevators with flow feedback

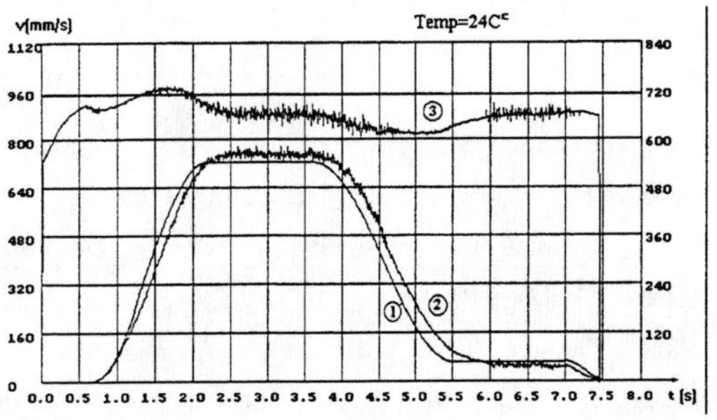

Fig. 9 Experiment results of hydraulic elevator with car speed feedback

8 Conclusion

Many comparison experiment investigations indicate that the performance of the soft start and soft level of the innovation control system with car speed feedback is better than the control system with flow feedback. The control system with car speed feedback avoids the adoption of the high accuracy flow meter so that reduces the cost of the electrohydraulic proportional integrated valve specialized in hydraulic elevators.

The single-chip computer controlled unit has excellent adaptability, and with it the intelligent control and direct car speed feedback can be easily achieved, so the difficult problems of the start and level of hydraulic elevator can be resolved effectively. The set of control system with car speed feedback will be widely applied in elevator industry.

Table 1 Comparison of performance index of different algorithms

	system equivalent mass m/kg	1500	2300	3100
Standard PID	**car position precision** $\Delta h / mm$	12	14	17
	speed tailing errors $T\sum_{i=1}^{800}\lvert v_d - v_p\rvert$	0.853	0.869	0.901
Fuzzy PID	**car position precision** $\Delta h / mm$	10	11	13
	speed tailing errors $T\sum_{i=1}^{800}\lvert v_d - v_p\rvert$	0.553	0.669	0.8
DSMVS	**car position precision** $\Delta h / mm$	11	15	19
	speed tailing errors $T\sum_{i=1}^{800}\lvert v_d - v_p\rvert$	0.785	0.925	1.0

9 References

Lu Yongxiang, Hu Dahong, 1988, *Electrohydraulic Proportional Technology,* Mechanical Industry Press.

Yang Huayong, Luo Jihao, 1996, *Hydraulic Elevators.* Mechanical Industry Press.

Zhao Guojun, 1995, *Design and Research on Microprocessor Controlled Hydraulic Eelvator system with Cabin Speed Feedback.* Master Thesis of Zhejiang Unviersity.

Xu Bing, 1995, *Research on Flow Feedback Electrohydraulic Valve Specialized in Hydraulic Elevators.* Master Thesis of Zhejiang Unviersity.

Yang Huayong, Xu Bing, 1998, "Introduction of Pump Station accessories Specialized in Hydraulic Elevators", *China Elevator,* pp34-37.

Yang Huayong, Zhang Jianmin, 1998, "Intelligent Control Theory and Hydraulic Elevators", *China Elevator,* pp5-8.

Ran Longlin, Yang Huayong, 1998, "An Elevator Hydraulic System Using Cabin Speed Feedback", *Elevator Technology 8, Proceedings of ELECON'98,* pp98-100.

SHA Daohang, YANG Huayong, 1998, "Research on Discrete Robust Optimum Control of Energy Saving Hydraulic Elevators", *2nd Tampere International Conference On Machine Automation,* ICMA'98, Tampere, Finland, pp15-18.

THERMAL CONSIDERATIONS IN FLUID POWER SYSTEMS MODELING

Raul Fernandez[1] and Robert L. Woods[2]

[1]Research Associate, Automation & Robotics Research Institute; Ph.D. Candidate
[2]Professor, Mechanical & Aerospace Engineering Department
The University of Texas at Arlington

ABSTRACT A systematic derivation of fluid capacitance for use in compressible and incompressible fluid power modeling is presented. Equations are stated explicitly for the control volume pressure and temperature states, which include a simple but realistic estimation of heat transfer effects based on a first-order approximation to the chamber wall thermal dynamics. Simulations representative of industrial hydraulic and pneumatic applications are presented, showing the expected fluid pressure and temperature responses for the control volume with various modeling assumptions, and ultimately justifying the use of simplified isothermal relations for hydraulics in general and for many low-pressure servopneumatic applications.

INTRODUCTION

Servopneumatics, defined as the control of compressible fluid media for power transmission, continues to penetrate new areas of application in manufacturing as well as in the medical, entertainment and human-assist fields. Research efforts have been underway at higher education institutions and industrial laboratories around the world for the past 40 years to help this technology establish its place alongside hydraulic and electric devices. Although steady advances have been made in control theory as well as in reducing pneumatic hardware cost and improving its reliability, the mathematical models that lie at the foundation of this effort, as first documented by Shearer[1] in the late 1950's, have remained very nearly unmodified to the present day. Much of the mainstream servopneumatics literature summarily neglects gas temperature variations while *simultaneously* declaring adiabatic conditions in gas capacitance modeling; the most commonly used approaches disregard both temperature variations associated with gas compression as well as heat transfer effects. A notable but seldom referenced cluster of publications[2-5] has correctly identified the equations which account for gas temperature as a state in analytical models and exposed several methods for approximating partial heat transfer conditions.

Though most derivations are strictly correct within their stated framework of assumptions, little if any insight on either side has been provided as to the magnitude of the errors involved in models that neglect temperature, or what operating regimes or heat transfer conditions warrant the more complete analyses. A void persists in understanding and justifying the use of isothermal and adiabatic assumptions.

The sections following present a full derivation from first principles of the pressure and temperature state equations for fluids. Starting from conservation of mass and conservation of energy, which apply regardless of compressibility, two separate sets of equations are eventually derived for the hydraulic and pneumatic case reflecting their different equations of state. Simulation examples are then shown to illustrate ideal and realistic fluid temperature profiles as controlled by several heat transfer modeling parameters, leading to the justification of simplified expressions for some engineering analyses.

MASS & ENERGY RELATIONS

The derivation of the fluid capacitance equation is based upon three equations or fundamentals in engineering: conservation of mass, conservation of energy, and an equation of state for the fluid. In this derivation both fluids will be considered — first a liquid, and then an ideal gas. The only difference between gases and liquids is the equation for flow in an orifice (compressible and incompressible), and the equation of state used to relate density, pressure and temperature.

Conservation of Mass

The conservation of mass in a control volume states that the net mass flow into the control volume must equal the time rate of change of the mass in the control volume.

$$\dot{m}_{in} - \dot{m}_{out} = \frac{d}{dt}\left[M_{cv}\right] \tag{1}$$

The mass in the control volume is the density times the volume. Considering that both the density and the volume could change, we can see the following.

$$\dot{m}_{in} - \dot{m}_{out} = \frac{d}{dt}\left[\rho_{cv}V_{cv}\right] = \dot{\rho}_{cv}V_{cv} + \rho_{cv}\dot{V}_{cv} \tag{2}$$

Normalizing with respect to the mass of the control volume, we get the following relation which is valid for liquids and gases.

$$\frac{\dot{\rho}_{cv}}{\rho_{cv}} = \frac{\dot{m}_{in} - \dot{m}_{out}}{\rho_{cv}V_{cv}} - \frac{\dot{V}_{cv}}{V_{cv}} \tag{3}$$

Conservation of Energy

The first law of thermodynamics for a control volume with fluid flowing in and out can be stated as follows.

$$Q_{heat}+\dot{W}+\left(h_{in}\dot{m}_{in}-h_{out}\dot{m}_{out}\right)=\frac{d}{dt}U_{cv} \tag{4}$$

Using the specific heat representations for enthalpy and internal energy, this equation can be stated as follows.

$$Q_{heat}+\dot{W}+C_p\left(T_{in}\dot{m}_{in}-T_{cv}\dot{m}_{out}\right)=\frac{d}{dt}\left[C_v M_{cv} T_{cv}\right] \tag{5}$$

The heat transfer is defined as positive into the control volume. Although heat exchange could occur with multiple sources, we will simply refer to it as Q_{heat} and treat incoming heat flow as positive.

The rate of work done on or by the control volume is the mechanical work rate, $P\dot{V}$. Positive power (work rate) would be work done on the control volume and would thus have a negative change in volume. Work done by (or power going out) the control volume would be expanding the control volume. We will consider only one source of mechanical work rate and by our definitions of power in and out, the mechanical power is the negative of the volume change rate, $\dot{W}=-P\dot{V}$.

$$\begin{aligned}&Q_{heat}-P_{cv}\dot{V}_{cv}+C_p\left(T_{in}\dot{m}_{in}-T_{cv}\dot{m}_{out}\right)=\\&C_v\left[M_{cv}\dot{T}_{cv}+\dot{M}_{cv}T_{cv}\right]\end{aligned} \tag{6}$$

Using the expression for the time rate of change of the mass of the control volume from the conservation of mass equation (1), we get the following simplification.

$$\begin{aligned}&Q_{heat}-P_{cv}\dot{V}_{cv}+C_p\left(T_{in}\dot{m}_{in}-T_{cv}\dot{m}_{out}\right)\\&=C_v M_{cv}\dot{T}_{cv}+C_v T_{cv}\left[\dot{m}_{in}-\dot{m}_{out}\right]\end{aligned} \tag{7}$$

$$\begin{aligned}\dot{T}_{cv}=&\left[\left(C_pT_{in}-C_vT_{cv}\right)\frac{\dot{m}_{in}}{C_vM_{cv}}-\left(C_pT_{cv}-C_vT_{cv}\right)\frac{\dot{m}_{out}}{C_vM_{cv}}\right]\\&-\frac{P_{cv}\dot{V}_{cv}}{C_vM_{cv}}+\frac{Q_{heat}}{C_vM_{cv}}\end{aligned} \tag{8}$$

Now, using the ratio of specific heats, $k=c_p/c_v$, which is true for liquids or gases, the energy equation can be stated as follows.

$$\begin{aligned}\dot{T}_{cv}=&\left[\left(kT_{in}-T_{cv}\right)\frac{\dot{m}_{in}}{M_{cv}}-T_{cv}\left(k-1\right)\frac{\dot{m}_{out}}{M_{cv}}\right]\\&-\frac{P_{cv}\dot{V}_{cv}}{C_vM_{cv}}+\frac{Q_{heat}}{C_vM_{cv}}\end{aligned} \tag{9}$$

If the heat transfer can be considered to be linear as would be true for either convection or conduction heat transfer, the relation between the heat transfer and the wall temperature — which will typically be close to the surrounding air temperature — can be expressed using a general thermal resistance, R_{th}.

$$Q_{heat}=\frac{T_w-T_{cv}}{R_{th}} \tag{10}$$

The thermal resistance results from convective heat transfer inside of the control volume.

$$R_{th}=\frac{1}{h_c A_{in}} \tag{11}$$

The following table shows the range of values for the convective coefficient h_c which may be used to represent different configurations.

Table 1. Approximate Convection Coefficients

Fluid and Condition	h_c, W/(m^2°C)
Air, free convection	6-30
Air, forced convection	30-600
Oil, free convection	30-150
Oil, forced convection	150-3000

Using this linear model for heat transfer simplifies the first law as follows. This equation is valid for liquids or gases.

$$\dot{T}_{cv}=\frac{1}{V_{cv}}\left\{\begin{aligned}&\left(kT_{in}-T_{cv}\right)\frac{\dot{m}_{in}}{\rho_{cv}}-\left(k-1\right)T_{cv}\frac{\dot{m}_{out}}{\rho_{cv}}\\&-\frac{P_{cv}}{C_v\rho_{cv}}\dot{V}_{cv}+\frac{\left(T_w-T_{cv}\right)}{R_{th}C_v\rho_{cv}}\end{aligned}\right\} \tag{12}$$

In summary, the volume of the control volume as well as the volume derivative will be determined by external forces or loads, and thus will be specified or will be a function of another state variable (e.g., piston position). Therefore, for the purposes of this modeling, the volume and volume derivative are inputs to the above equations. The temperature of the incoming mass flow rate, and the wall temperature for heat transfer will either be known (constant), or will be a state variable from another system. In particular, since wall capacitance is generally not negligible, an additional equation of state for T_w will be required to properly represent its effect. The mass flows in and out are determined by external pressures (which are known or are other state variables) and the control volume pressure. The mass flow will have to be calculated from other equations (e.g., orifice flow) depending upon the exact situation, but will be a function of known pressures and temperatures.

Therefore, the only unknowns in this control volume are the pressure, temperature, and the density. The conservation of mass equation gives us an equation for density derivative (3), and the first law gives us an equation for the temperature derivative (12). We must look to the equation of state to relate density and temperature to pressure. Whereas the conservation of mass and the first law are true for either a gas or a liquid, the equation of state is different for a liquid and a gas as presented in the following section.

DENSITY RELATIONS & DYNAMIC EQUATIONS

The density equation is an equation of state that relates density, temperature, and pressure. Unlike the conservation of mass and energy relations, this equation of state is different for a liquid and a gas. For a gas, we can use the ideal equation of state; however, for a liquid, there is no such theoretical equation of state and the density is related to pressure and temperature through a Taylor series expansion that gives the well-known coefficients for the bulk modulus and the cubical expansion coefficient.

Equation of State for Liquids

The equation of state for a liquid is stated in terms of a Taylor series expansion of the density as a function of the pressure and temperature around a reference state (ρ_O, P_O, and T_O). This expansion gives us two coefficients that are documented for liquids, the bulk modulus β and the cubical expansion coefficient α.

$$\frac{\rho_{cv}}{\rho_o}=\left[1+\frac{(P_{cv}-P_o)}{\beta}-\alpha(T_{cv}-T_o)\right] \tag{13}$$

The bulk modulus is the inverse of compressibility and has units of pressure.

$$\beta=\rho_o\left.\frac{\partial P_{cv}}{\partial \rho_{cv}}\right|_{P_o,T_o} \tag{14}$$

The cubical expansion coefficient is defined in terms of the negative of the slope of density versus temperature and has units of T^{-1}.

$$\alpha=-\left.\frac{1}{\rho_o}\frac{\partial \rho_{cv}}{\partial T_{cv}}\right|_{P_o,T_o} \tag{15}$$

Taking the derivative of the liquid equation of state and rearranging yields the following.

$$\frac{\dot{P}_{cv}}{\beta}=\frac{\rho_{cv}}{\rho_o}\frac{\dot{\rho}_{cv}}{\rho_{cv}}+\alpha\dot{T}_{cv} \tag{16}$$

Dynamic Equations for Liquids

Using the above equation from the density of a liquid and substituting the density derivative from the conservation of mass equation and the temperature derivative from the first law, gives the following expression for the pressure derivative.

$$\dot{P}_{cv}=\frac{\beta}{V_{cv}}\left\{\begin{array}{l}\left(\frac{\rho_{cv}}{\rho_o}+\alpha\left(kT_{in}-T_{cv}\right)\right)\frac{\dot{m}_{in}}{\rho_{cv}}\\ -\left(\frac{\rho_{cv}}{\rho_o}+\alpha(k-1)T_{cv}\right)\frac{\dot{m}_{out}}{\rho_{cv}}\\ -\left(\frac{\rho_{cv}}{\rho_o}+\frac{\alpha P_{cv}}{C_v\rho_{cv}}\right)\dot{V}_{cv}+\frac{\alpha\left(T_w-T_{cv}\right)}{R_{th}C_v\rho_{cv}}\end{array}\right\} \tag{17}$$

The above equation is the complete equation for pressure derivative for a liquid. However, for normal temperature and pressure variations, the density does not change more than about 3 to 4%. Therefore it is reasonable to make the following simplifications, noting further that for a liquid, the ratio of specific heats is close to unity.

$$\frac{\rho}{\rho_o}\approx 1 \tag{18}$$

$$\alpha\approx 0 \tag{19}$$

and the volume flow rate can be used.

$$\frac{\dot{m}}{\rho}\approx Q \tag{20}$$

Using these simplifications, the following pressure derivative for a liquid results, which is the well-known incompressible flow continuity equation.

$$\dot{P}_{cv}=\frac{\beta}{V_{cv}}\left\{Q_{in}-Q_{out}-\dot{V}_{cv}\right\} \tag{21}$$

Equation of State for an Ideal Gas

The equation of state for gases under pressures and temperatures typical to most pneumatic systems follow the ideal gas law.

$$\rho_{cv}=\frac{P_{cv}}{RT_{cv}} \tag{22}$$

Taking the derivative of this expression yields the following.

$$\dot{\rho}_{cv}=\frac{1}{RT_{cv}}\dot{P}_{cv}-\frac{P_{cv}}{RT_{cv}^2}\dot{T}_{cv} \tag{23}$$

or

$$\frac{\dot{P}_{cv}}{P_{cv}}=\frac{\dot{\rho}_{cv}}{\rho_{cv}}+\frac{\dot{T}_{cv}}{T_{cv}} \tag{24}$$

Dynamic Equations for Ideal Gases

Using the above equation for pressure derivative (24) and substituting the equations for density (22), density derivative (3) and temperature derivative (12), yields the following.

$$\frac{\dot{P}_{cv}}{P_{cv}}=\left\{\begin{array}{c}\dfrac{\dot{m}_{in}-\dot{m}_{out}}{\rho_{cv}V_{cv}} \\ -\dfrac{\dot{V}_{cv}}{V_{cv}}\end{array}\right\}+\left\{\frac{1}{V_{cv}}\left[\begin{array}{c}\left(k\dfrac{T_{in}}{T_{cv}}-1\right)\dfrac{\dot{m}_{in}}{\rho_{cv}} \\ -(k-1)\dfrac{\dot{m}_{out}}{\rho_{cv}} \\ -\dfrac{R}{C_v}\dot{V}_{cv}\end{array}\right]+\frac{\left(\dfrac{T_w}{T_{cv}}-1\right)}{R_{th}C_v M_{cv}}\right\} \tag{25}$$

The following relation is valid for an ideal gas.

$$\frac{R}{C_v}=k-1 \tag{26}$$

Even though the $R_{th}\ C_V\ M_{cv}$ term in equation (25) represents the thermal time constant, we can simplify the above equation even further by removing the pressure, density, and temperature terms and using the R/C_v expression.

Thus, the final simplification for the pressure derivative for an ideal gas is stated below.

$$\dot{P}_{cv}=\frac{kP_{cv}}{V_{cv}}\left\{\frac{T_{in}}{T_{cv}}\frac{\dot{m}_{in}}{\rho_{cv}}-\frac{\dot{m}_{out}}{\rho_{cv}}-\dot{V}_{cv}\right\}+\frac{(k-1)(T_w-T_{cv})}{R_{th}V_{cv}} \tag{27}$$

SUMMARY OF MODELING EQUATIONS

Concise restatements of the equations used for modeling hydraulic (liquid) and pneumatic (gas) systems are given below.

Hydraulic Systems

$$\dot{P}_{cv}=\frac{\beta}{V_{cv}}\left\{\begin{array}{c}\left(\dfrac{\rho_{cv}}{\rho_o}+\alpha\left(kT_{in}-T_{cv}\right)\right)\dfrac{\dot{m}_{in}}{\rho_{cv}} \\ -\left(\dfrac{\rho_{cv}}{\rho_o}+\alpha(k-1)T_{cv}\right)\dfrac{\dot{m}_{out}}{\rho_{cv}} \\ -\left(\dfrac{\rho_{cv}}{\rho_o}+\dfrac{\alpha P_{cv}}{C_v\rho_{cv}}\right)\dot{V}_{cv}+\dfrac{\alpha\left(T_w-T_{cv}\right)}{R_{th}C_v\rho_{cv}}\end{array}\right\} \tag{28}$$

$$\dot{T}_{cv}=\frac{1}{V_{cv}}\left[\begin{array}{c}\left(kT_{in}-T_{cv}\right)\dfrac{\dot{m}_{in}}{\rho_{cv}}-(k-1)T_{cv}\dfrac{\dot{m}_{out}}{\rho_{cv}} \\ -\dfrac{P_{cv}}{C_v\rho_{cv}}\dot{V}_{cv}+\dfrac{\left(T_w-T_{cv}\right)}{R_{th}C_v\rho_{cv}}\end{array}\right] \tag{29}$$

$$\rho_{cv}=\rho_o\left[1+\frac{\left(P_{cv}-P_o\right)}{\beta}-\alpha\left(T_{cv}-T_o\right)\right] \tag{30}$$

Using the simplifications given in (18-20), the above equations reduce to the following.

$$\dot{P}_{cv}=\frac{\beta}{V_{cv}}\left\{Q_{in}-Q_{out}-\dot{V}_{cv}\right\} \tag{31}$$

$$\dot{T}_{cv}=\frac{1}{V_{cv}}\left[\begin{array}{c}\left(kT_{in}-T_{cv}\right)Q_{in}-(k-1)Q_{out} \\ -\dfrac{P_{cv}}{C_v\rho_o}\dot{V}_{cv}+\dfrac{\left(T_w-T_{cv}\right)}{R_{th}C_v\rho_o}\end{array}\right] \tag{32}$$

Pneumatic Systems

$$\dot{P}_{cv}=\frac{kP_{cv}}{V_{cv}}\left\{\frac{T_{in}}{T_{cv}}\frac{\dot{m}_{in}}{\rho_{cv}}-\frac{\dot{m}_{out}}{\rho_{cv}}-\dot{V}_{cv}\right\}+\frac{(k-1)\left(T_w-T_{cv}\right)}{R_{th}V_{cv}} \tag{33}$$

$$\dot{T}_{cv}=\frac{1}{V_{cv}}\left\{\begin{array}{c}\left(kT_{in}-T_{cv}\right)\dfrac{\dot{m}_{in}}{\rho_{cv}}-(k-1)T_{cv}\dfrac{\dot{m}_{out}}{\rho_{cv}} \\ -\dfrac{P_{cv}}{C_v\rho_{cv}}\dot{V}_{cv}+\dfrac{\left(Tw-T_{cv}\right)}{R_{th}C_v\rho_{cv}}\end{array}\right\} \tag{34}$$

$$\rho_{cv}=\frac{P_{cv}}{RT_{cv}} \tag{35}$$

A simplified expression may easily be obtained for the pressure state under isothermal conditions, $\dot{T}_{cv}=0$, $T_w=T_{cv}$, by solving for Q_{heat} from (34) and substituting into (33). After some rearrangement, the following pressure equation valid for isothermal heat transfer results.

$$\dot{P}_{cv}=\frac{P_{cv}}{V_{cv}}\left\{\frac{\dot{m}_{in}}{\rho_{cv}}-\frac{\dot{m}_{out}}{\rho_{cv}}-\dot{V}_{cv}\right\} \tag{36}$$

OBSERVATIONS

The mathematical foundation for first and second order capacitance models is not new in the literature[4]; however, a concise compilation of explicit expressions for the pressure and temperature states for compressible and incompressible fluid modeling, including heat transfer effects, will be valuable to practitioners and researchers in the field.

The subject of heat transfer has been the source of creative effort as well as confusion. Usually in conjunction with an *isothermal* assumption, much of the industrial servopneumatics literature concurrently regards actuator motions to be fast enough for the use — in contradiction — of an *adiabatic* pressure formula (equation (36) with a *k* multiplying factor). As will be shown, however, even in "fast cycles" adiabatic conditions are unlikely to be sustained due to heat transfer to the *chamber walls*, regardless of external insulation.

A misconception commonly encountered is the use of the polytropic compression equation to predict temperature:

$$T_{cv} = T_{in}\left(\frac{P_{cv}}{P_{in}}\right)^{k-1/k} \tag{37}$$

This equation applies strictly under closed or steady-flow conditions, and its use is generally invalid for processes where arbitrary mass flows are added and subtracted from the control volume — which are typical of servopneumatic system design. Equation (34) instead applies.

Otis[2] has studied gas charged accumulators, replacing the selection of the polytropic exponent as either 1 (isothermal) or k (adiabatic) based on a purely arbitrary definition of the process as "fast" or "slow," with an empirical heat dissipation time constant in place of the Q_{heat} term found in equation (9):

$$\frac{T_{\infty} - T_{cv}}{\tau} = C_v(MT)_{cv} + P\dot{V}_{cv} \tag{38}$$

Otis obtains the time constant τ from constant-volume cooling data of a given accumulator size and precharge pressure, with the advantage that it will apply regardless of the work cycle, i.e., compression or expansion, isothermal or adiabatic. Opposite to this simple empircal method, Svodoba[3] et al take the approach of an in-depth analytical characterization to predict of thermal time constants for arbitrary accumulator geometries and precharge pressures.

Hullender and Woods[4,5] offer a much simplified *heuristic* method of accounting for heat transfer effects in the energy conservation equation. This consists simply of estimating the ratio r of the actual (Q_{heat}) to maximum (Q_{max}) heat transfer, the latter being that required to establish isothermal conditions. The drawback is that it is only applicable while fluid is charging or discharging, and does not work for dwell cycles.

A modeling approach that preserves simplicity and generality is the lumped thermal resistance and capacitance method, incorporated into equation (10). As illustrated in the examples, an additional equation of state is typically needed to represent the thermal dynamics of the chamber wall.

APPLICATION TO HYDRAULIC SYSTEM MODELING

In order to illustrate the models presented above, a hydraulic actuator driven by a fast-opening valve is analyzed. The actuator is single acting with a spring return (with no preload). The spring is stiff enough so that the actuator will not reach its extend limit at full pressure. The actuator is driving an inertial load and has no friction. In addition to the equations of state for pressure and temperature, we need an equation for the flow through the valve as well as the mechanical dynamics (from a force balance). The complete equations and coefficient values are stated below.

$$Q_{cv} = C_d A_{orifice}\sqrt{\frac{2\left|\delta P_s - \delta P_i\right|}{\rho_o}}\, sign(\delta P_s - \delta P_i) \tag{39}$$

$$V_{cv} = V_o + A_p z \tag{40}$$

$$\dot{P}_{cv} = \frac{\beta}{V_{cv}}\left[Q - A_p\dot{z}\right] \tag{41}$$

$$\dot{T}_{cv} = \frac{1}{V_{cv}}\left[(kT_{in} - T_{cv})Q_{in} - \frac{P_{cv}}{C_v\rho_{cv}}\dot{V}_{cv} + \frac{(T_w - T_{cv})}{R_{th}C_v\rho_o}\right] \tag{42}$$

$$\dot{z} = v \tag{43}$$

$$\dot{v} = \frac{A_p}{M}\delta P_i - \frac{K_{spring}}{M}z \tag{44}$$

Table 2. Hydraulic Modeling Coefficients

coefficient	value
ρ_o	868 kg/m^3
β	1862 MPa
α	0.000267 °C^{-1}
k	1.22
C_v	1543 J/(kg°C)
$D_{orifice}$	0.5 mm
C_d	0.6
K_{spring}	37,700 N/m
M	20 kg
D_{piston}	20 mm
ℓ	110 mm
h_{cin}	1000 W/(m^2°C)
R_{th}	0.133 °C/W
T_s	25 °C
δP_s	12 MPa

Figure 1 illustrates the dynamic response. The pressure rise looks exponential and is indicative of the actuator position. The temperature of the hydraulic fluid rises slightly due to the kT_s (which increases with temperature) and $P\dot{V}$ (which decreases with temperature) terms, and then decays due to the heat transfer. Performing this simulation resulted in exactly the same response from the simplified equations as from the full equations, validating assumptions (18)-(20). While the 3 °C rise exhibited in this example is unimportant, potentially significant temperature variations might occur if the supply temperature or the ambient temperatures were very much different from the internal fluid temperature, or if more work were done.

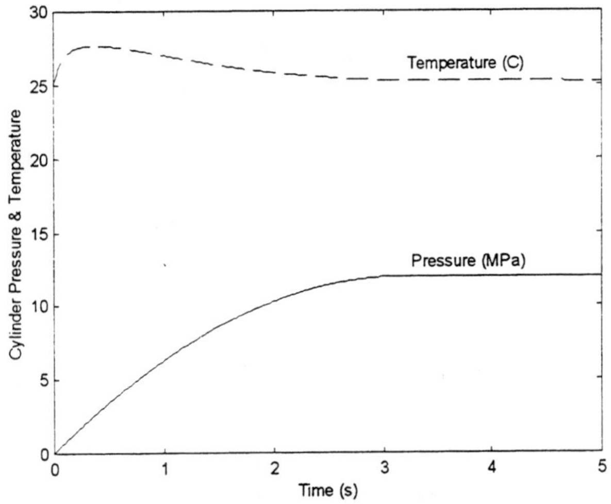

Figure 1. Dynamic Response of Spring-Loaded Hydraulic Actuator (Simulated)

APPLICATION TO PNEUMATIC SYSTEM MODELING

Accumulator Charge

The first of two compressible fluid examples involves the charging of an open, static chamber from atmospheric to line (0.7 MPa) pressure. This condition best illustrates thermal effects since no external work is involved. The chamber characteristics are as follows: 360 mm long PVC pipe section, 19 mm ID, 3 mm wall thickness. The controlling element is a stock 5/2 valve with a *pneumatic C_v rating* of 0.35, open to full flow at time 0.1 seconds. The fluid medium is air and all initial temperatures are at 20 °C. Orifice flow is modeled with the standard adiabatic compressible flow equation (ref. Woods[5]):

$$\dot{m}=C_d\,A_s\,P_u\sqrt{\frac{2}{R\,T_u}}\sqrt{\frac{k}{k-1}}\sqrt{\left(\frac{P_d}{P_u}\right)^{2/k}-\left(\frac{P_d}{P_u}\right)^{k+1/k}} \qquad (45)$$

Note that (45) is subject to choking (constant flow) at pressure ratios below critical — 0.528 in the case of air.

Under ideal, zero-heat transfer conditions — *including any heat transfer to the container walls* — the control volume will experience a dramatic temperature rise; indeed, the final temperature can be shown to be dependent solely upon the initial and supply pressure and temperature. For the example at hand, the theoretical final temperature is 116 °C. An instrumented setup will, however, record a peak of about 50 °C, which quickly dissipates to ambient. It is clear that a substantial amount of heat transfer occurs.

The model used to approximate the actual gas temperature has, as a starting point, the thermal resistance approach used in equations (10) and (11); this, however, must be supplemented by an additional state equation for wall temperature, since this element turns out to have a significant amount of capacitance. The additional state is described by the following equation, which is the thermal equivalent of a standard R-C-R filter.

$$\dot{T}_w=\frac{1}{C_w}\left[\frac{T_{cv}-T_w}{R_{thin}}-\frac{T_w-T_\infty}{R_{thout}}\right] \qquad (46)$$

Figure 2 below shows control volume temperature responses for the theoretical adiabatic process ($R_{thin}=\infty$), a linear resistance-only model (equation 34 including convection & conduction through the chamber walls), and the R-C-R model of (46). The latter most closely resembles experimental measurement. The predicted wall temperature rise for this example is less than 0.6 °C.

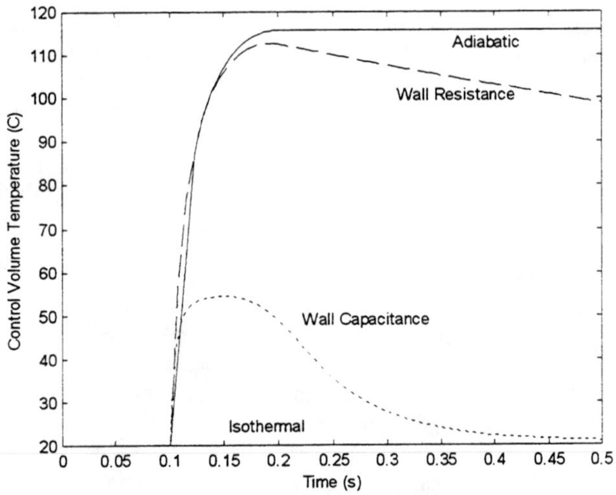

Figure 2. Temperature Response of Pneumatic Accumulator Charge (Simulated)

The dominant thermal variables and their effects on the control volume temperature may be summarized as follows:

- C_v, the specific heat of air, represents the gas internal capacitance and controls the initial rate of temperature rise. Because the mass of air is very small, the relative capacitance is also small — less than 1 J/°C — and correspondingly the temperature rise is fast. Only this phase, lasting a few tens of milliseconds, may be considered adiabatic.

- R_{thin}, the internal thermal resistance due to convection of air, determines how fast heat is transferred into the actuator walls. The entering flow pattern is highly turbulent and may be approximated with a high convective factor h_c around 500 to 600 W/m^2°C. A small internal resistance will have the effect of lowering the peak temperature reached by the control volume.
- C_w, the actuator wall capacitance, controls the amount of heat stored by the walls before it is given up to, or absorbed by, the surrounding atmosphere. This is a relatively large value for representative chamber geometries — 60 to 120 J/°C — and thus wall temperature varies only slightly. Unusual wall configurations with low capacitance and high resistance would warrant a second-order model with states for the inside and outside wall temperatures, lumping half the capacitance at each node.
- R_{thout}, the external convective resistance, controls the rate at which the actuator wall temperature is restored to ambient. Since the actuator wall capacitance is large and its temperature variation slight, R_{thout} will typically have little effect in the control volume temperature rate of change. Equivalently, any amount of insulation surrounding the chamber will not help preserve adiabatic conditions at the control volume due to the inevitable presence of wall capacitance.

Pressure plots for this example are shown in Figure 3. While temperature effects do not dominate the pressure response, high-pressure applications will see a non-trivial effect on settling time depending upon the heat transfer regime considered. Standard low-pressure force and motion control applications may, however, be represented with the isothermal approximation.

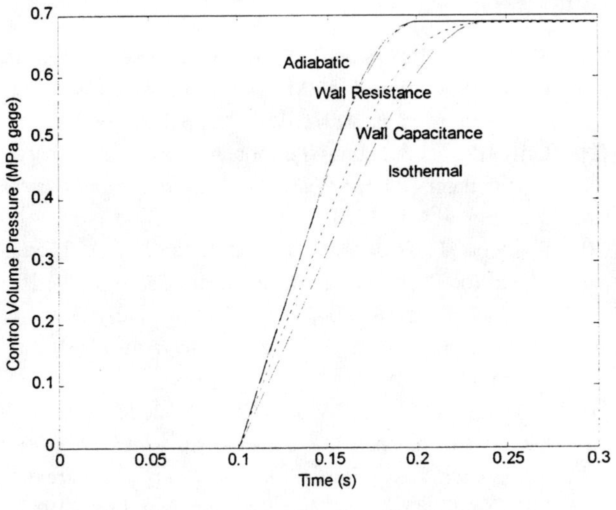

Figure 3. Pressure Response of Pneumatic Accumulator Charge (Simulated)

Actuator Extension

The second example involves the extension of a standard spring-loaded air cylinder under full line pressure, with the spring constant chosen to balance the pressure force at full extension and with zero preload. The piston characteristics are as follows: 100 mm stroke, 40 mm ID, 13 mm shaft OD, 0.5 mm steel wall thickness, 0.1 kg inertia, and a dead volume of about 15% of the swept volume. The same valve as in the previous example is used, with equal pneumatic C_v rating at the inlet and outlet orifices. External force balance equations are set up as in the hydraulic example shown above, except for a nominal amount of viscous friction (50 N-s/m) applied to damp out vibrations. As before, Figures 4 & 5 depict control volume temperature and pressure responses for the theoretical adiabatic, linear conductive/convective resistance through the wall, and wall capacitance cases.

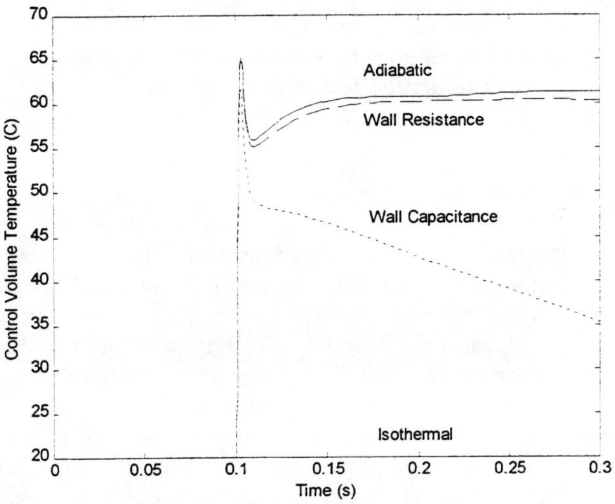

Figure 4. Temperature Response of Spring-Loaded Pneumatic Actuator (Simulated)

The initial temperature peak is caused by rapid warming of the small amount of gas in the dead volume when exposed to line pressure. Unlike the static charge case, a significant portion of the internal energy in the control volume is expended into the generation of work — expanding the boundary and compressing the spring. This has the effects of limiting the temperature rise in the actuator chamber, slowing down the pressure response, and diminishing the difference in pressure between the isothermal and adiabatic conditions.

An additional effect considered in the simulation is that, as the piston extends, an increasing amount of internal surface area becomes available for convective heat transfer from the fluid to the chamber walls; because this area is initially small, convective resistance is high and, at the beginning of the stroke, conditions resemble closely the adiabatic regime. As the extension progresses, thermal dissipation accelerates and the pressure response is restored to that predicted by the isothermal model, as can be seen in Figure 5.

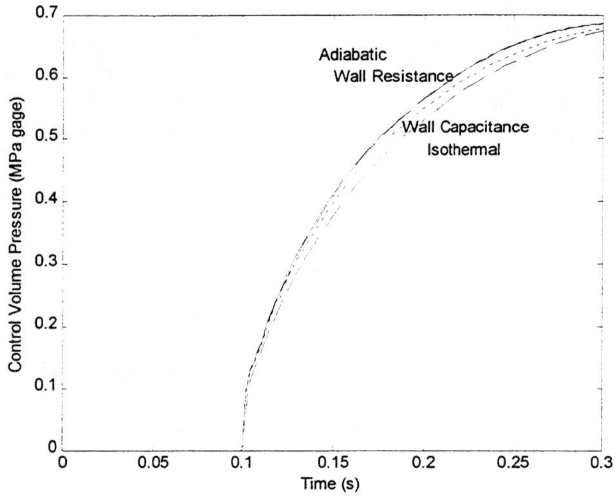

Figure 5. Pressure Response of Spring-Loaded Pneumatic Actuator (Simulated)

CONCLUSIONS

Explicit dynamic equations for control volume pressure and temperature states are derived, applicable to compressible and incompressible fluid modeling with heat transfer. A simple and powerful approach to heat transfer modeling using lumped thermal resistance and capacitance is used to estimate control volume temperatures, with a first-order wall capacitance model yielding the closest correlation with measured values. In general, the temperature and pressure responses will lie between the adiabatic and isothermal cases; however, the inevitable presence of wall capacitance and low convective resistance inside the actuator volume will result in non-negligible heat transfer and thus preclude the attainment of ideal adiabatic conditions to the point of favoring, in many cases, an isothermal simplification. The use of additional state equations for control volume and chamber wall temperature presented herein, absent in most published work, will be warranted in situations where large differences are known to occur between the incoming and control volume temperatures, for analyses demanding more precise estimations of control volume pressure, or if knowledge of the actual temperature of the fluid is important.

ACKNOWLEDGMENTS

This work was performed jointly between the Automation & Robotics Research Institute and the Mechanical and Aerospace Engineering Department of The University of Texas at Arlington. The authors gratefully acknowledge the use of ARRI's facilities and laboratory equipment.

NOMENCLATURE

α	Cubical expansion coefficient
β	Bulk modulus
ρ	Density
h	Specific enthalpy
h_c	Convection coefficient
k	Ratio of specific heats
$\dot{m}$	Fluid mass flow rate
r	Ratio of actual to maxmum heat transfer
v	Actuator speed
z	Actuator displacement
A_{in}	Actuator inside surface area
A_s	Servovalve geometrical orifice area
C_d	Coefficient of discharge
C_p	Constant pressure specific heat
C_v	Constant volume specific heat; pneumatic flow factor
C_w	Thermal lumped wall capacitance
M	Mass inside the control volume
P	Pressure
P_s	Supply pressure
P_u	Pressure upstream of servovalve orifice
P_d	Pressure downstream of servovalve orifice
Q	Volumetric flow rate
Q_{heat}	Rate of heat transfer
R	Gas constant
R_{th}	Thermal lumped resistance
T_{cv}	Control volume temperature
T_{in}	Supply temperature
T_u	Temperature upstream of servovalve orifice
T_w	Actuator wall temperature
T_∞	External (ambient) temperature
U	Internal energy
V	Volume
$\dot{W}$	Rate of mechanical or shaft work (power)

REFERENCES

[1] Shearer, J. L., "Study of Pneumatic Processes in the Continuous Control of Motion with Compressed Air," *Transactions of the ASME*, Vol. 78, pp 223-249, 1956.

[2] Otis, D., "Predicting Performance of Gas Charged Accumulators," *First Fluid Power Controls and Systems Conference*, May 1973.

[3] Svodoba, J., Bouchard, G., and Katz, S., "A Thermal Model for Gas-Charged Accumulators Based on the Heat Conduction Distribution," *Fluid Transients and Acoustics in the Power Industry,* ASME Winter Annual Meeting, San Francisco, 1978.

[4] Hullender, D. A., and Woods, R. L., "Modeling of Fluid Circuit Components," *Proceedings of the First Conference on Fluid Control and Measurement*, FLUCOME '85, Tokyo, London: Permagon Press, 1985.

[5] Woods, R. L., and Lawrence, K. L., *Modeling and Simulation of Dynamic Systems*, New York, New York: Prentice Hall, 1997.

DESIGN METHODOLOGY FOR FLUID POWER SYSTEMS USING HYPERMEDIA

Edmund J. Hughes, Thomas G. Richards and Derek G. Tilley

Engineering Design Centre in Fluid Power Systems
Department of Mechanical Engineering
University of Bath
Bath, BA2 7AY, UK
Phone: +44 1225 826131; E-mail: E.J.Hughes@bath.ac.uk

ABSTRACT

Fluid power system design is a skilful and complex task requiring significant practical experience and heuristic knowledge gained over many years. Industries specifying and/or designing fluid power systems in their products are at risk should they lose key personnel with this knowledge. This problem is compounded by a lack of formal training at the graduate level, creating a shortfall in qualified fluid power designers.

Design support tools that assist and guide designers in their work, as part of a Concurrent Engineering approach, should be capable of providing both good practice guidance and a framework into which product specific information can be stored for later reuse. This paper discusses the early development of a fluid power system design tool that seeks to offer this support using a methodology developed in hypermedia.

INTRODUCTION

Background

The basis of current thinking on the improvement of design methodologies, as part of a concurrent engineering approach, is the application of a systematic approach (Pahl & Beitz, 1996; Pugh, 1991). The approach advocates that the design process incorporate several prescribed stages in order to achieve a final design. These stages include specification, conceptualisation, detail design, manufacture, assembly, and test and development. Having made this assumption it is possible to make the relationships between each stage explicit, being defined upon key technical, functional, organizational and cost rules.

Carlson et al. (1996) describes the development of a concurrent engineering methodology for small companies based upon the systematic approach, and proposed separate methods depending upon whether the design undertaken was new, variant or adaptive (feature change). The adoption and successful application of theoretical methodologies is, however, dependent upon a receiving organization's design culture. Radcliffe and Harrison (1994) observed the working practices of designer's in a small hydraulics company and found:

- multiple tasks per employee;
- informal communication between personnel;
- little documentation of designs;
- concentration on short term goals;
- little formal project management and planning skills;
- design methods governed by rules-of-thumb.

These observations suggest that in order for the systematic approach to be accommodated the working practices of the company would need to be completely revised and new ones adopted. It is apparent that this would require the company's culture to change, creating upheaval in all functions. This is unlikely and could be one reason for the lack of adoption of concurrent engineering practices in small-medium enterprises (SME) where there are insufficient resources both in time and money for staff development and training.

The use of information technology is intrinsic to the adoption and development of concurrent engineering practices in industry. While personnel can be trained to adapt to new systems, it is suggested that computer systems also need to reflect and accommodate the current perception by key personnel of how the design process works, i.e., in a rather haphazard and discursive manner.

Fluid Power System Design

The design of a fluid power system involves the selection of a set of hydraulic components and the definition of the pipe-work connections between them, so as to produce a complete system that meets some given set of system requirements (Potter et al., 1997). The difficulty in designing fluid power systems is in understanding the interactions within the circuit and between components, a task that becomes increasingly difficult in more complex configurations.

It is therefore suggested that specific products such as fluid power systems require a structured methodology to encapsulate the idiosyncrasies and information formats commonly associated with the domain. This is particularly true for the fluid power industry where one of a kind design projects e.g., the design of a bridge operating system, are common. One of a kind design is also more likely to place greater emphasis on functionality and less perhaps on cost savings.

In general, for fluid power system design the requirements capture and design specification stages are the most important. It is a lack of emphasis placed on these stages by fluid power designers that results in the common problems and failures in the early stages of product commissioning. This maybe a result of a poor understanding of the consequences of design decisions on the overall circuit performance, however, with growing pressure to reduce lead times there is greater emphasis on a right-first-time approach being the norm rather than the exception. Greater qualification is therefore required before design decisions are made, and computer guidance would assist such an objective.

As SMEs make-up a significant proportion of the fluid power industry, both as users and suppliers, an approach that could be more easily adopted using current information technology systems was desirable. Use of design guidelines within a hypermedia framework was seen as such a method.

HYPERMEDIA FRAMEWORK

The current research work has made a comprehensive study of computer software to establish versatility, compatibility and universality with other software and systems. A full description of the investigation is given in Richards et al. (1999), however, for the purposes of this paper it can be stated that multimedia elements including animation, 2-D and 3-D simulation, video and 2-D circuit design have been developed using Java. These elements are incorporated into a hypermedia framework that features a search facility, menu system and hypertext links to access relevant resources within the framework and on the Internet. The framework will also permit the designer to add and recover specific information when required.

The framework also provides the designer with a list of subjects needing to be considered prior to, and for use in, producing a design. As the designer navigates through the framework issues will be highlighted enabling the designer to access and gather information to enable qualified decision making. This is a major advantage over other systems as the designer will follow links in a non-prescribed manner and so, it is argued, be more akin to actual design practice.

GENERIC DESIGN METHODOLOGY

Introduction

Tilley (1998) describes research at the University of Bath into the development of computer tools for several stages of fluid power system design. This highlighted how the use of computer tools would permit the development of detailed simulated solutions prior to commissioning. The advantage of such a system would be the evaluation of several iterations, and hence removal of major conflicts, prior to physical construction. Not all conflicts will be revealed in this process, however, the remaining minor conflicts should be such that they are easily resolvable as opposed to fundamental system performance problems. Development of a tool to aid in the design of solution circuits that could then be simulated was one objective of this work.

For this research various methods and approaches have been studied from the fluid power and other industries. The use of charts and/or hierarchical rules to develop a structured approach to the design process is not new. For example, Terry (1968) describes a design method developed for the United Kingdom's Atomic Energy Authority. There are, however, two major advantages of hypermedia over such "static" methods. Primarily, there is the capability to integrate the generic design methodology into a dynamic environment where links can be developed and defined by the designer. Secondly, it permits the designer to access external resources e.g., CD-Roms, web-sites, simulators, etc. As the designer searches for and defines the information within a hypermedia framework it is argued that it will encourage the collection of data, improve problem definition, and aid the search for, and definition of, sound solutions.

As stated, fluid power system design involves several key stages. These include:

- Idea/Request;
- Requirements Capture & Design Specification;
- Load Definition;
- Circuit Configuration;
- Component Selection & Sizing;
- Evaluation, Test & Development;
- System Commissioning.

Each stage of this design methodology is discussed with reference to the use of hypermedia in its execution and, if appropriate, benefits identified.

Idea/Request

Fluid power system design is a result of either market pull or technology push. Market pull is defined as market feedback from the marketing/sales departments, specific customer demands and expectations, and increased safety requirements. Technology push can be a result of either following a competitor who has a more technologically advanced system or an exploitable new technology being identified. New technology could enable a competitive advantage to be established through improved product performance or the opportunity to reduce costs by either minimizing the number and/or complexity of components.

Use of hypermedia will aid this process as it will permit identification of fluid power system design information from either industrial sources i.e., current products being marketed and on-line publications, or from research sources i.e., academic web-sites citing publications. The designer will also be able to undertake patent searches via the Internet.

Requirements Capture & Design Specification

Gause and Weinberg (1989) identified the relative cost of false assumptions made at each stage of the design process (Figure 1). This finding indicates the importance of the requirements capture phase in correctly defining the problem being solved. For fluid power systems correct definition primarily means both technically and functionally correct. However, whether this definition is commercially correct also depends upon issues such as production costs, competitor pricing and safety standards. Technical and functional correctness can only be obtained if system requirements are explicitly detailed, a task that is often incomplete, ambiguous or both. This explains the need to differentiate requirements capture and design specification. Without an effective requirements capture procedure the design specification produced will be based on incomplete information and so liable to be incorrect.

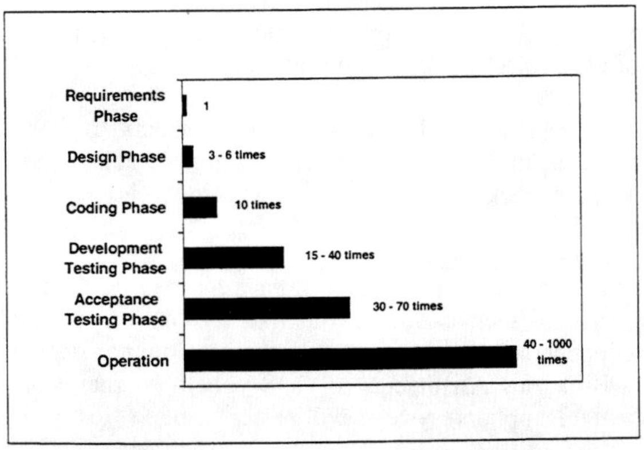

Fig. 1. Relative cost of false assumptions made at each stage of the design process. [Gause & Weinberg, 1989]

An industrial example of poor requirements capture resulting in poor design specification is used to exemplify this problem. A company contracted out the design and installation of a hydraulic ring main to power 50 machine tools. In the design specification the system pressure was specified by the company at 280±5bar. The contractor therefore designed a system that maintained the pressure between 275bar and 285bar. The client, however, for quality assurance purposes, required a constant pressure within the specified range. Subsequent random fluctuations in pressure during operating cycles resulted in expensive remedial investigation and re-modification work to be undertaken. This case study also re-emphasises the need for effective requirements capture in achieving a correct understanding of the problem.

Adoption and development of requirements capture tools such as Quality Function Deployment (QFD) within the fluid power industry has only been observed in large multinational companies. In many SMEs integration of formal design tools is slow to occur because it can be a potentially disruptive step into relatively unknown territory (Radcliffe and Harrison, 1994). Larger enterprises are also more capable of supporting the change process with resources for training, hardware, software and even new staff. An ability to offset costs over an extended period also makes the process a more viable option financially.

The method proposed consists of four main parts, and forms the basis of a design checklist:

- Design Assessment;
- System Design Specification;
- System Definition & Conceptualisation;
- Requirements of System Components.

Table 1 illustrates the topic areas covered. The meaning and intent of some topics is self-evident whilst others are deliberately left open to interpretation by the designer. It is argued that in working through the design checklist, establishing and recording appropriate answers, a thorough understanding of the problem is formed and appropriate solution(s) identified for further detailed investigation. In completing the checklist a design record will be created, useful for both quality assurance and design reuse purposes.

Hypermedia will permit the designer to access the checklist at the required point and provide links to relevant sites defined by the user. Hence, the system can evolve to match the specific needs of the designer and company enforcing the advantages described above.

Load Definition

Having made an initial design specification it is now possible to proceed to correct definition and analysis of the load. Consideration also needs to be given to pressure and temperature stability.

A significant advantage of hypermedia over current methods of load definition is that each load type i.e., resistive, over-running, or inertial can be examined and assessed using both searchable text and interactive graphics i.e., 2-D and 3-D simulation. Suitable techniques for analysis can also be demonstrated this way. Once analysis is complete the circuit to achieve the previously defined objectives can be configured.

Circuit Configuration

Several approaches to circuit configuration design have been developed using artificial intelligence (AI) including expert systems (Burton and Sargent, 1991, Westman, 1987) and neural networks (Potter et al., 1997). The objective of developing such tools is to examine whether circuit configuration can be automated. Currently this function is still very much a human one and it is the aim of this research to develop a computer tool to aid the human designer in producing solution circuits more efficiently and quickly.

Table 1. Fluid power system design checklist

Part I — Design Assessment

- ❑ Usage - *where and when, duration, pressure, temperature, sequence, maintenance, operation*
- ❑ Influences and Limitations - *environment, safety, future use & modification, manufacture, suppliers, test & installation, time scale, cost*
- ❑ Existing Resources - *previous designs, existing components, services available, design and production experience available*
- ❑ Risk - *liability, warranty costs, recall costs, loss of reputation/customer base*

Part II — System Design Specification

- ❑ Objective of Design
- ❑ System Performance
- ❑ Assumptions
- ❑ Limitations predicted/contingency measures
- ❑ Safety Features
- ❑ Environmental controls/limitations

Part III — System Definition & Conceptualisation

- ❑ Methods of Achieving Design Objective(s)
- ❑ Considerations & Conflicts - *theoretical, practical, components, production aspects, probable cost, development requirements*
- ❑ Alternative Systems

Part IV — Requirements of System Components

- ❑ Component
- ❑ Function & Method of Functioning - *controllers, sensors*
- ❑ Characteristics - *decided, undecided*
- ❑ Decision - *customer, project engineer, design engineer*
- ❑ Source of Supply - *stores, purchase, design, existing*

The current prototype incorporates hypermedia elements to allow the designer to configure circuits in both a 2-D (Figure 2) and 3-D (Figure 3) representation. 2-D will be the probable approach used by experienced designers to develop circuits using standard symbols. This format also gives scope for circuits to be downloaded into a simulation package e.g., Bath*fp* (Richards and Tilley, 1991) either on-site or remotely via the Internet should a company not have access to simulation software. Both inexperienced and experienced designers can use 3-D representation alike, with 3-D enabling inexperienced designers to learn about the different functionality of components. Importantly, it will also assist in checking if a design conforms to required spatial layout parameters e.g., on mobile equipment.

Component Selection & Sizing

Following circuit configuration the components will be selected and sized using information drawn either from company information stored within the framework or by using hypertext links to access product information for component suppliers. The benefit of an object-orientated hypermedia approach is that each component (object) within a configured circuit can be defined to incorporate hyperlinks to other resources.

Design is an iterative process and changes made here, for example, due to restrictions on the availability of components, could have an effect on the system design and performance. Use of hypermedia will give the designer scope to readily access the design specification to evaluate any design amendments made.

Evaluation, Test and Development

An objective of the current research is that the hypermedia framework be compatible with simulation software. Simulation is seen as the ideal method of reducing development time as fluid power system circuits can be rigorously analysed prior to prototype. Such capability should satisfy the reduced time-to-market objective of concurrent engineering as the number and severity of problems during the commissioning/signing-off phase of any project should be minimized.

It is also suggested that with improvements to Java's multimedia capabilities, designers will in future be able to have qualitative feedback for assessing system noise levels.

System Commissioning

The hypermedia framework will have the potential to link to other systems such as Bill-Of-Materials (BOM) software. By using client/server technology such as the Common Object Request Brokering Architecture (CORBA) both internal (sites) and external (suppliers) systems will be accessible.

CONCLUDING REMARKS

An industrially generic design methodology (IGDM) for fluid power systems has been described. The IGDM has been developed for integration into a computer based hypermedia framework containing multimedia elements and hypertext links. By providing a logical progression through the task structure, and through the use of guidelines, good design practice should prevail as all key factors are considered.

Use of a structured design methodology within a hypermedia framework will prove to be the basis of a practical,

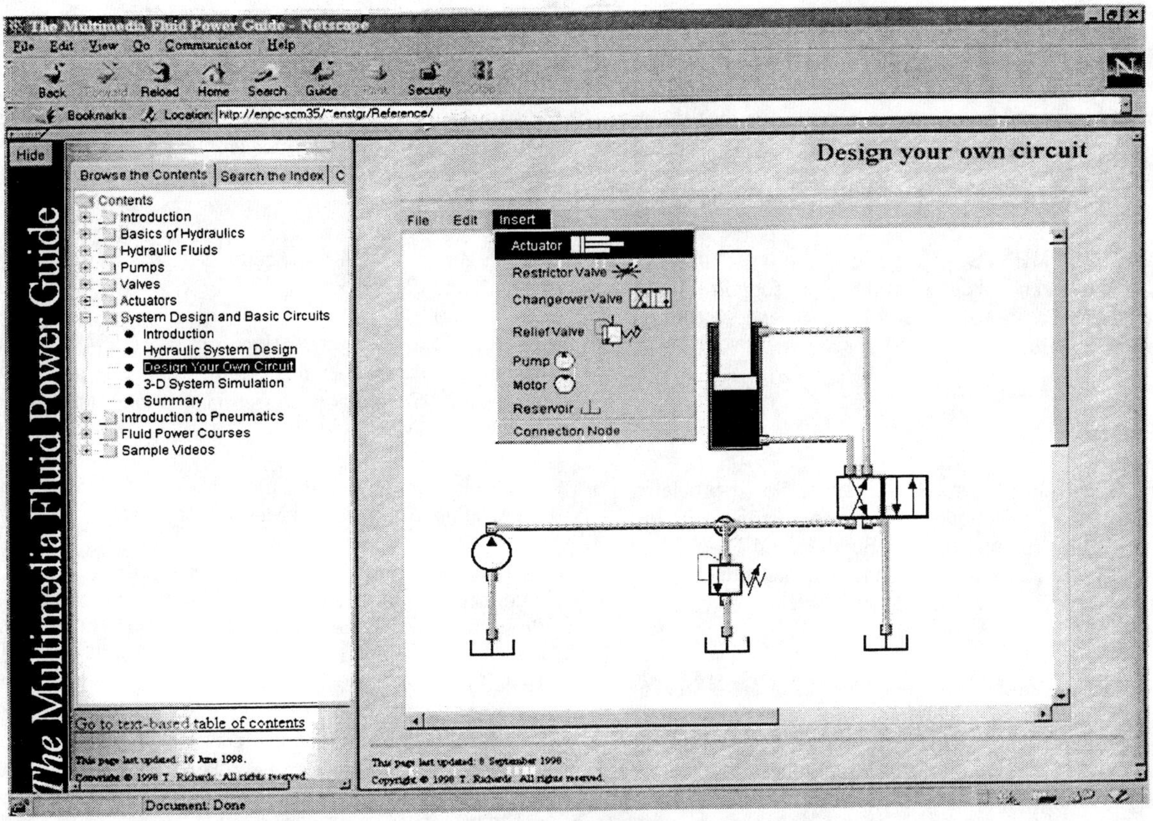

Fig. 2. Fluid power system design (2-D Java applet)

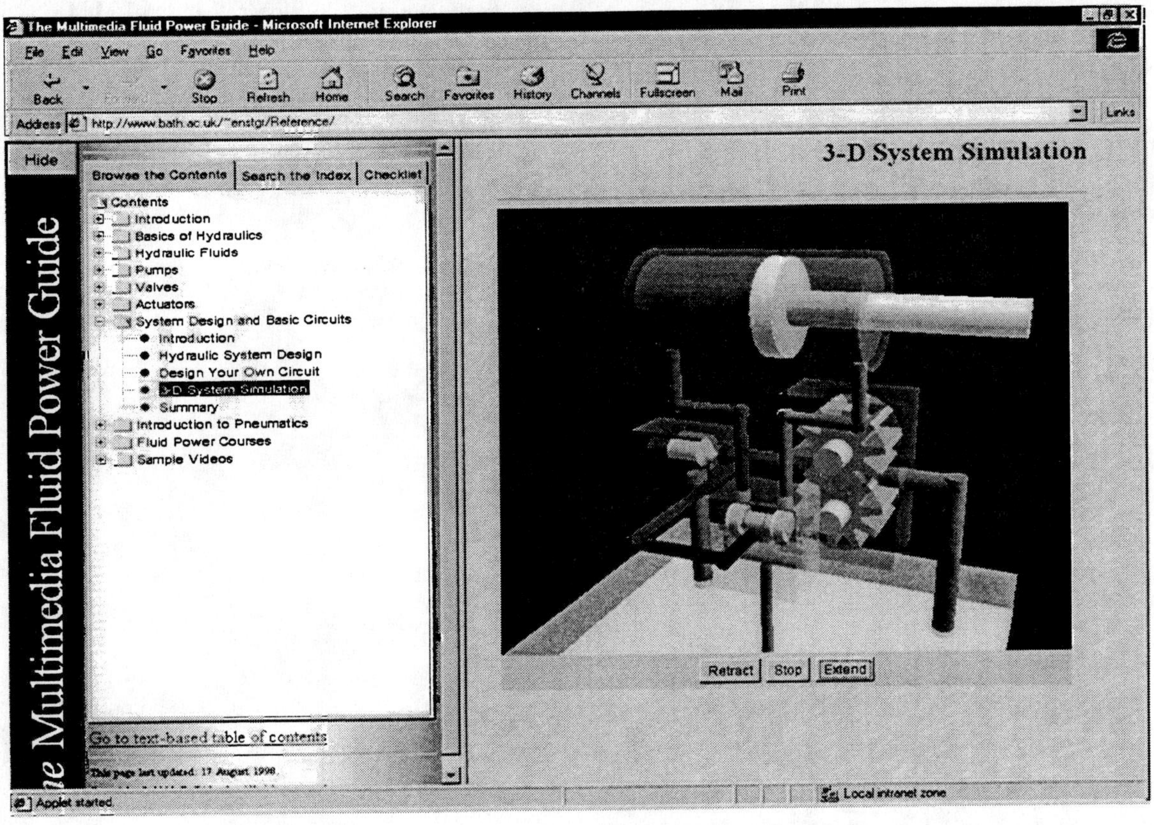

Fig. 3. Fluid power system design (3-D Java applet)

powerful and widely applicable fluid power design tool. As such its application could overcome several restrictions to the adoption of concurrent engineering methods by SMEs in the fluid power industry.

ACKNOWLEDGEMENTS

The authors acknowledge the financial support of the United Kingdom's Engineering and Physical Sciences Research Council (EPSRC) through grant GR/L26858.

REFERENCES

Burton, R.T. and Sargent, C.M. (1991) Some problems encountered in the development of an expert system for the design of hydraulic circuits. In C.R. Burrows and K.A. Edge, *Computers in Fluid Power.* 3rd Bath International Fluid Power Workshop, September 13-14, 1990 University of Bath, England, pp 137-151.

Carlson, S.E., Kemser, H-P. and Ter-Minassian, N., 1996, A concurrent engineering methodology for small companies. *Proceedings of the 1996 ASME Design Engineering Technical Conferences and Computing in Engineering Conference*, August 18-22, 1996, Irvine, Ca.

Gause, D.C. and Weinberg, G.M., 1989, *Exploring Requirements: Quality Before Design*, Dorsethouse, New York, NY. In A.B. Wooton, R. Cooper and M. Bruce, Requirements capture: where the front end begins? *International Conference Engineering Design ICED97*, August 19-21, 1997 Tampere, Finland, pp 75-80.

Pahl, G. & Beitz, W. (1996) Engineering Design – A Systematic Approach. 2nd edition. Springer-Verlag : London.

Potter, S., Chawdhury, P., Culley, S. and Burrows, C., 1997, Goal-based configuration design of fluid power systems using neural networks. *Proceedings of the 1997 ASME Design Engineering Technical Conferences*, September 14-17, 1997, Sacramento, Ca.

Pugh, S. (1991) Total Design. Addison-Wesley: Chichester, UK.

Radcliffe, D.F. and Harrison, P. (1994) Transforming design practice in a small manufacturing enterprise. In *Design Technology Conference: Design Theory and Methodology*, Vol. 68, 1994, ASME, pp 91-98.

Richards, C.W. and Tilley, D.G. (1991) System analysis with Bath*fp*. *Fluid Power*, March-April 1991, 3, 50-53.

Richards, T.G., Hughes, E.J. and Tilley, D.G. (1999) A multimedia approach to fluid power system design. *Proceedings of the IEEE International Conference on Multimedia Computing and Systems (ICMCS'99)*, 7-11 June 1999, Florence, Italy.

Terry, G.J. (1968) A chart system to help designers. *The Chartered Mechanical Engineer*, February, pp. 56-59.

Tilley, D.G., 1998, The Bath Engineering Design Centre: computer-based techniques to improve the design of fluid power systems. In S. Sivaloganathan and T.M.M. Shahin (Eds.) *Engineering Design Conference '98 – Design Reuse*, June 23-25, Brunel University, England, pp 113-122.

Westman, R.M. (1987) A knowledge-based system for hydraulic circuit design. PhD Thesis, University of Saskatchewan, August 1987.

FPST-Vol. 6, Fluid Power Systems and Technology — 1999
ASME 1999

PNEUMATICALLY OPERATED WHEELCHAIR FOR STEP DRIVE

Masatake SHIRAISHI
Hideyasu SUMIYA
Shinichi AOSHIMA

Faculty of Engineering, Ibaraki University
Hitachi, Ibaraki , Japan

ABSTRACT

A new wheelchair capable of 10 cm-step drive has been developed. It uses pneumatic pumps installed on the rear wheels of the wheelchair. The pneumatic pumps generate pressurized air, which is sent to the reservoir, in proportion to the rear wheel rotation: travel of about 12 m is needed to reach full pressure. The required air is introduced from the reservoir into two driving cylinders, fixed to the front support frames, that lift the front wheels. The lifting operation is performed by handling the braking levers. Our experimental design system shows the feasibility of using pneumatic pumps for this operation.

1. INTRODUCTION

Step traversing is one of the serious functional improvements and generally accomplished in the electric wheelchair. Eighty percent of user complaints on such powered wheelchairs were, however, in some way related to batteries [Sawka, et al., 1981; Brubaker, et al., 1984; Hess, et al., 1990; Cooper, R.A., 1993]. Most of the research associated with battery use in wheelchairs has included capacity, life, charging, chargers, discharge cycles, etc. Under these circumstances, the important factor for propulsion in powered wheelchairs is mechanical advantage. A representative lever-propelled wheelchair has been developed as a prototype [McLaurin, et al., 1986]. This uses a lever-actuated clutch, which provides control similar to that of hand rims. The advantage of the proposed system is its ability to use smaller-diameter drive wheels, permitting side transfers, wheelies, and the climbing of curbs of up to 15 cm. With these factors as background, this paper referes to a manually operated wheelchair that can drive a small step of up to 10 cm. A pair of pneumatic pumps generates and pressurized air to the reservoir proportional to the rear wheel rotation. When a step is encountered, air from the reservoir is introduced into the driving cylinders and lifts the front wheels to climb the step. This lifting operation is performed by using braking levers only. Experimental results show the feasibility of traversing steps of up to 10 cm.

2. SYSTEM CONSTRUCTION

The proposed wheelchair is composed of two pneumatic pumps (1) with contact rollers (2), driving cylinders (3) with rollers (4) at their rod ends, five-port valves (5) and a reservoir (6) as shown in Fig.1. Each pneumatic pump is connected to a rear wheel of the wheelchair through the contact roller and generates exhaust pressure by pressurizing the atmospheric air in accordance with a rear wheel rotation. Driving cylinders are fixed to the support frames of the front wheels for lifting operation. In normal use, as the wheelchair travels about 12 m, the two pneumatic pumps produce pressurized air of about 177 kPa in the reservoir.

When the front small wheels meet steps of about 10 cm, the user locks the wheelchair with the left-side braking lever and, at the same time, pulls the right-side braking lever which also activates the valves (5). The distance between the front wheels and the cylinder rollers is 15 cm and both cylinder rollers are above the step as shown in Fig. 1(a). Thereafter the five-port valve is automatically opened by the mechanical switch attached to the end of the right-side braking lever. Working pressure from the reservoir is then introduced into the driving cylinders, thus lifting the front wheels by their cylinder rods (Fig. 1(b)). On release of the left-side brake, the front wheels traverse the step, guided by the cylinder rollers on the step. Once the front wheels pass over the step, the right-side brake is released by the user and the driving cylinder rods move back to their initial positions accordingly (Fig. 1(c)). After this operation, the rear wheels can easily drive over the step under normal wheelchair operation.

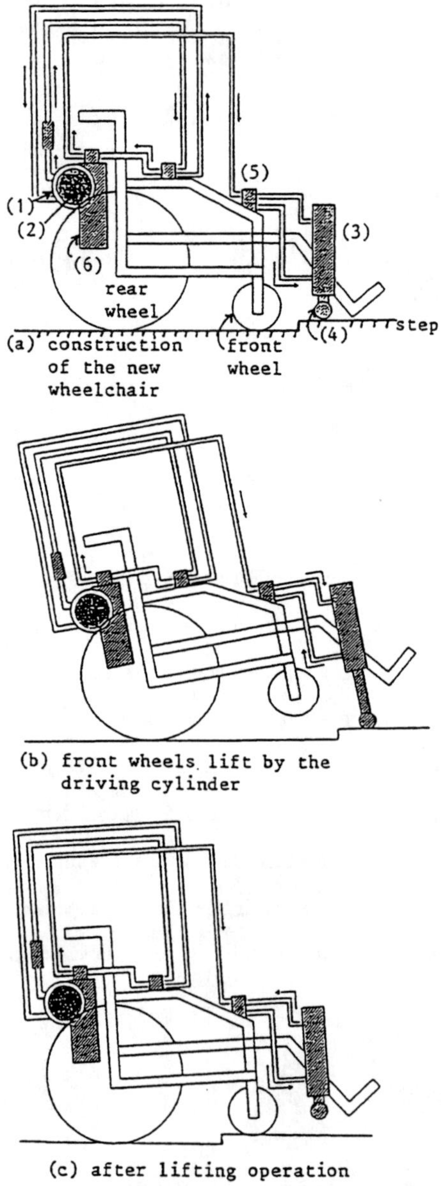

(1) pneumatic pump (4) roller
(2) contact roller (5) five-port valve
(3) driving cylinder (6) reservoir

Fig. 1 Construction of the Wheelchair

3. DESIGN PROCESS

The driving cylinders, reservoir, pneumatic pumps, and contact rollers between the pneumatic pump and the rear wheel must be determined for practical system construction.

3.1 Driving Cylinder and Reservoir. First, we must estimate the lifting force of the driving cylinder, using the model shown in Fig. 2. Assume that the total weight of the user and the wheelchair together is 735 N (user 588 N and wheelchair 147 N) and acts on the ground 200 cm away from the center of the rear wheel. The distance between the rear wheel center and the cylinder roller center is 670 cm. Let $F1$ be the driving force of the two cylinders for the lifting operation. Then, by considering the moment around the rear wheel center, we obtain

$$F1 \times 670 = 735 \times 200 \tag{1}$$

This yields $F1$ =220 N. The driving force for one cylinder is therefore 110 N.

Several types of driving cylinders were investigated, taking into account the requirement for this driving force and for a necessary 15-cm cylinder stroke. The reservoir has to provide sufficient power to keep the system in operation. It should be as compact and lightweight as possible. The cylinder with a 6.4 cm bore size and a 15 cm stroke, and the 3800 cm^3 reservoir were finally chosen for the practical system design.

Let P4 be the pressure in the driving cylinder. The required pressure for the lifting operation at one driving cylinder with a 6.4 cm bore size is then calculated by

$$(\pi \times 3.2^2) \times (P4 - 1) = \frac{F1}{2} \tag{2}$$

From Eqs.(1) and (2), $P4$ =134 kPa. To satisfy this pressure in the driving cylinder, the reservoir pressure Pr in Fig. 2 is governed by Eq. (3), assuming adiabatic change.

$$\text{Pr} \times V3^{\kappa} = P4(V3 + 2V4)^{\kappa} \tag{3}$$

where $V3$ and $V4$ denote the reservoir capacity and the driving cylinder volume, respectively, and κ is the adiabatic coefficient. Further, $V4$ is expressed by using the cylinder driving stroke $L2$ as

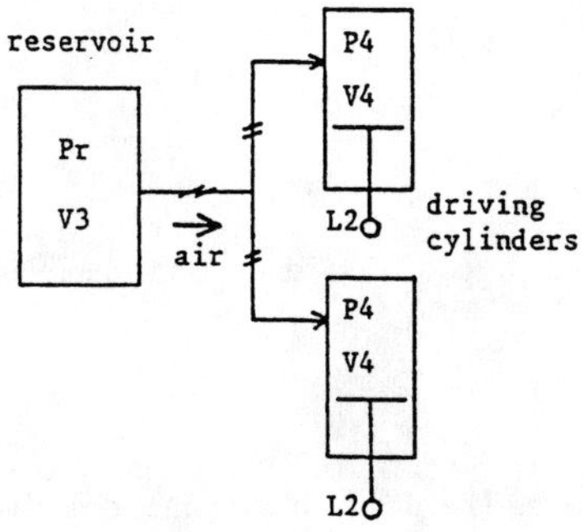

Fig. 2 Air flow from the reservoir to the driving cylinder

$$V4 = \pi \times 3.2^2 L2 \tag{4}$$

With the parameters of $V3 = 3800cm^3, \kappa = 1.4$, and $L2 = 15cm$, Eqs.(3) and (4) yield Pr = 184 kPa. This is the calculated pressure in the reservoir for the lifting operation.

3.2 Pneumatic Pump and Contact Roller. Before choosing a pneumatic pump, the diameter r of the contact roller must be determined. This roller, fixed to the pump shaft, is connecting to the rear wheel of the wheelchair (Fig. 1(a)). Assume that a wheel travel of 12 m produces a reservoir pressure of 184 kPa via the pneumatic pumps. Then the number of rotations of the rear wheel is calculated by using its diameter (560 cm) as 12/0.56 = 6.8. The number of rotations of the contact roller r is therefore $6.8 \times 560 / r$.

The pneumatic pump chosen has a 4.2 cm^2 area with a 1.31 cm stroke. One rotation of this pump can supply 16.5 cm^3 (= $4.2 \times 1.31 \times 3$ pistons) of air into the reservoir. Therefore a total air volume of 33.0 cm^3 (= 16.5×2 pumps) is fed to the reservoir at one rotation. By applying the equation of state for gas, we obtain Eq. (5) at n rotations of the two pumps with the atmospheric pressure of Po = 98 kPa:

$$98 \times (3800 + 33.0 \times n) = Pn \times 3800 \tag{5}$$

where Pn denotes the reservoir pressure at n rotations. This leads to

$$Pn = 98\left(1 + \frac{33.0}{3800}n\right) \tag{6}$$

By equating (5) and (6), n = 103.0. Also, by using the relashionship of

$$6.8 \times \frac{560}{r} = 103.0 \tag{7}$$

we obtain r =36.9 mm. The pump shaft diameter is about 15 mm. Therefore, by taking into account the rigidity between the pump shaft and the contact roller, the contact roller diameter was determined as r =30 mm.

4. SYSTEM ANALYSIS

The analysis of the new wheelchair is divided into two systems: the power supply system from the rear wheel to the reservoir through the pump, and the drive system from the reservoir to the driving cylinder. This pneumatic analysis is similar to that of the reference [Shiraishi, et al., 1996] and, therefore, only key equations and results are described in the following discussion.

The operational diagram of the power supply system is shown in Fig. 3. A dynamic equation is obtained by equating the inertia force to the sum of damper, pressure, coulomb-frictional, and crank forces acting on the pump piston, assuming that it has a displacement of $X1$ due to the rear wheel rotation.

$$\frac{M1}{g}\ddot{X}1 = -C1\dot{X}1 + (P1 - P2)A1 + F + Fcl\,\mathrm{sgn}(\dot{X}1)$$

$$F = \frac{F2}{Y(\theta)} \tag{8}$$

where $M1$: mass of pump piston, g : gravitational constant, $C1$: viscous damping coefficient of the pump piston, $P1, P2$: pressures denoted in Fig. 3, $A1$: cross-sectional area of the pump piston, F :

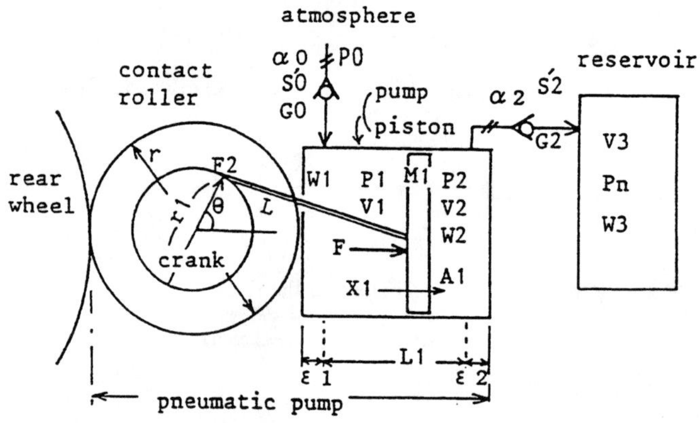

Fig.3 Power supply system

applied force to the pump piston due to the crank force $F2$, Fcl : coulomb-frictional force in the pump piston, $Y(\theta)$: pump characteristics. θ is the rotation angle of the pump crank. For a representative pneumatic pump consisting of three pistons with $\pm 120°$ phase difference, $Y(\theta)$ is written as

$$Y(\theta) = y(\theta) + y(\theta - 120^{\circ}) + y(\theta + 120^{\circ}) \tag{9}$$

$$y(\theta) = \frac{\sin\left\{\sqrt{1 - \left(\frac{r1}{2L}\sin\theta\right)^2} + \frac{r1}{2L}\cos\theta\right\}}{\sqrt{1 - \left(\frac{r1}{2L}\sin\theta\right)^2}} \tag{10}$$

where $r1$ and L are the crank diameter and the length of the connecting rod in the pump piston, respectively.

Pressure changes with respect to $P1, P2$ and Pn are expressed as

$$\dot{P}1 = \frac{RT}{(\varepsilon 1 + X1)A1}\left(Go - \frac{P1A1}{RT}\dot{X}1\right) \tag{11}$$

$$\dot{P}2 = \frac{RT}{(L1 + \varepsilon 2 - X1)A1}\left(\frac{P2A1}{RT}\dot{X}1 - G2\right) \tag{12}$$

$$\dot{P}n = \frac{RT}{V3}G2 \tag{13}$$

where R : gas constant, T : temperature, $\varepsilon 1, \varepsilon 2$: parameters shown in Fig. 3, $L1$: pump piston stroke.

Assuming adiabatic change, the flows Go and $G2$ are given by

$$Go = \frac{\alpha o S'oPo}{\sqrt{RT}}\sqrt{\frac{2g\kappa}{\kappa-1}\left\{\left(\frac{P1}{Po}\right)^{\frac{2}{\kappa}} - \left(\frac{P1}{Po}\right)^{\frac{\kappa+1}{\kappa}}\right\}},$$

$$G2 = \frac{\alpha 2S'2P2}{\sqrt{RT}}\sqrt{\frac{2g\kappa}{\kappa-1}\left\{\left(\frac{Pn}{P2}\right)^{\frac{2}{\kappa}} - \left(\frac{Pn}{P2}\right)^{\frac{\kappa+1}{\kappa}}\right\}} \tag{14}$$

where $\alpha o, \alpha 2$: flow rate coefficients, $S'o, S'2$: cross-sectional areas at flows Go and $G2$.

In the same way, from Fig. 4, dynamic and pressure equations are obtained for the drive system.

$$\frac{M2}{g}\ddot{X}2 = -C2\dot{X}2 + (P4 - P5)A2 - M2 - Fc2\,\mathrm{sgn}(\dot{X}2) \tag{15}$$

$$\dot{P}4 = \frac{RT}{(\varepsilon 3 + X2)A2}\left(G3 - \frac{P4A2}{RT}\dot{X}2\right),$$

$$\dot{P}5 = \frac{RT}{(L2 + \varepsilon 4 - X2)}\left(\frac{P5A2}{RT}\dot{X}2 - G5\right) \tag{16}$$

$$G3 = \frac{\alpha 3S'3Pn}{\sqrt{RT}}\sqrt{\frac{2g\kappa}{\kappa-1}}\left\{\left(\frac{P4}{Pn}\right)^{\frac{2}{\kappa}} - \left(\frac{P4}{Pn}\right)^{\frac{\kappa+1}{\kappa}}\right\},$$

$$G5 = \frac{\alpha 5S'5P5}{\sqrt{RT}}\sqrt{\frac{2g\kappa}{\kappa-1}}\left\{\left(\frac{Po}{P5}\right)^{\frac{2}{\kappa}} - \left(\frac{Po}{P5}\right)^{\frac{\kappa+1}{\kappa}}\right\} \tag{17}$$

where $M2$: total mass of the driving cylinder, $C2$: viscous damping coefficient of the driving cylinder, $P5$: pressure shown in Fig. 4, $A2$: cross-sectional area of the driving cylinder, $Fc2$: coulomb-frictional force in the driving cylinder, $\varepsilon 3, \varepsilon 4$: parameters shown in Fig. 4, $\alpha 3, \alpha 4$: flow rate coefficients, $S'3, S'5$: cross-sectional areas at flows $G3$ and $G5$.

As a result, the pressure Pn in the reservoir and the behavior of cylinder rod displacement $X2$ in the driving cylinder can be obtained by solving all the above equations.

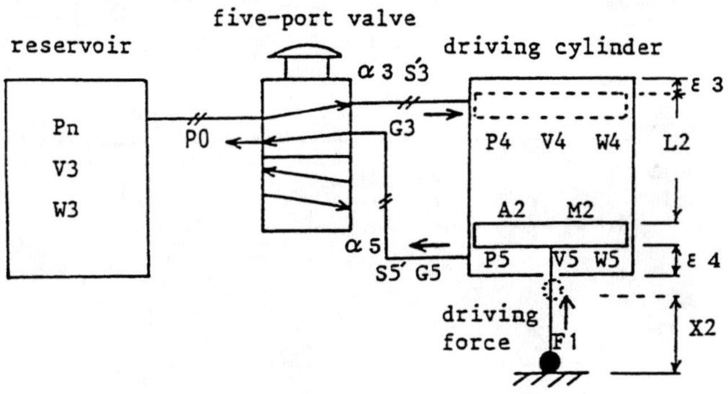

Fig.4 Drive system

5. SIMULATION AND EXPERIMENT

To evaluate the operating features of the new wheelchair, the computer simulation results based on the preceding equations were compared with experimental ones. Equations from (8) to (17) are calculated using the Runge-Kutta method. The parameters used for simulation are listed in Table 1. The experiments were performed by measuring the reservoir pressure Pn with a semiconductive-type pressure transducer and the driving cylinder stroke $X2$ with a linear potentiometer.

Figure 5 illustrates a variation of the reservoir pressure Pn obtained at different wheelchair speeds of (a) Vw = 0.8 m/s and (b) Vw = 1.2

Table 1 Parameters for simulation

parameters	value	parameters	value
A_1	4.2 cm^2	R	2929 $cm/{}^{\circ}K$
A_2	32.2 cm^2	S'_0, S'_2, S'_3, S'_5	0.1 cm^2
C_1, C_2	0.98 $N \cdot s/cm$	T	293 $^{\circ}K$
F	38.1 N	V_3	3800 cm^3
F_{c1}, F_{c2}	4.9 N	g	980.7 cm/s^2
L_1	1.28 cm	α_0, α_2, α_3, α_5	0.65
L_2	15.0 cm	ε_1, ε_2, ε_3, ε_4	0.5 cm
M_1	16.5 N	κ	1.4
M_2	2.2×10^2 N		
P_r	1.84×10^2 kPa		

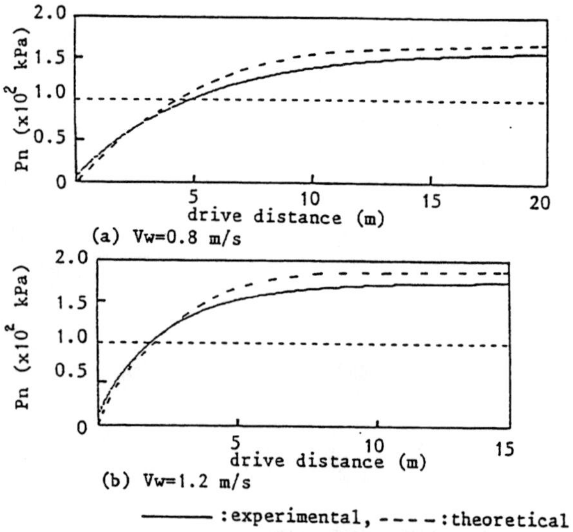

Fig.5 Variation of the reservoir pressur Pn at different wheelchair speeds Vw

m/s. According to our investigation, wheelchair speeds except for heavily disable users generally showed between 0.8 to 1.2 m/s. The measurements were undertaken without changing the initial speed Vw during one test. As suggested in Section 3.1, the calculated reservoir pressure Pr for lifting operation is 184 kPa within, ideally, a wheelchair travel of 12 m. In the case of Vw = 1.2 m/s, the experimental value indicates a final reservoir pressure of 177 kPa, which is nearly equal to the calculated pressure of 184 kPa. With the lower speed of Vw = 0.8 m/s, we obtain a 165 kPa reservoir pressure at a 15.4 m wheelchair travel. This may cause some problem in the lifting operation. As shown in Fig. 5, though the experimental results do not give the best fit to the theoretical curves due to several mechanical frictions, the overall trends agree relatively well.

As a wheelchair meets steps, its front wheels are quickly lifted 10 to 14 cm by the lifting operation. To reduce the user's anxiety due to this operation, it is preferable to control the lifting speed by the gate valve installed to the driving cylinder. Figure 6 shows a transient response of the driving cylinder stroke $X2$ with a user of 610 N at a stepwise pressure of 177 kPa from the reservoir, with the gate valve (a) 100% open and (b) 40% open. As shown in the figure, the lifting speed can be changed by reducing the open area in the gate valve. The experimental result with the gate valve at 100% open coincides approximately with the computer simulation.

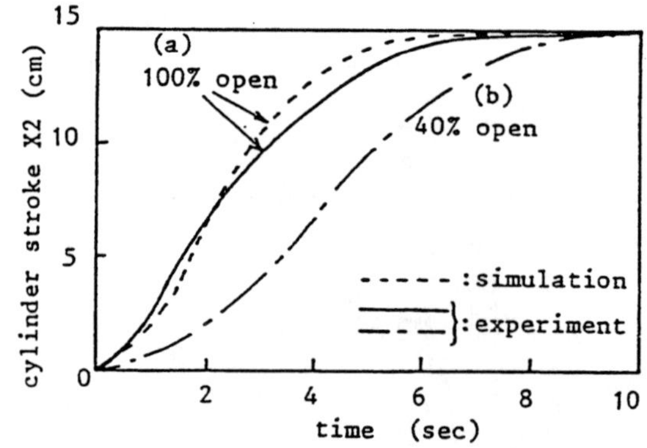

Fig.6 Transient response of the driving cylinder stroke

The field tests were performed for two steps with a 12 m distance between them (Table 2). The distance between the start and the first step is 15 m. The first step is 5 cm high and the second one is 8 cm high. The results with five users are shown in Table 2, where the performances are evaluated in terms of "satisfactory" and "unsatisfactory" with the reservoir pressure obtained. "Satisfactory" means the step drive was accomplished without difficulty.

Table 2 Test results

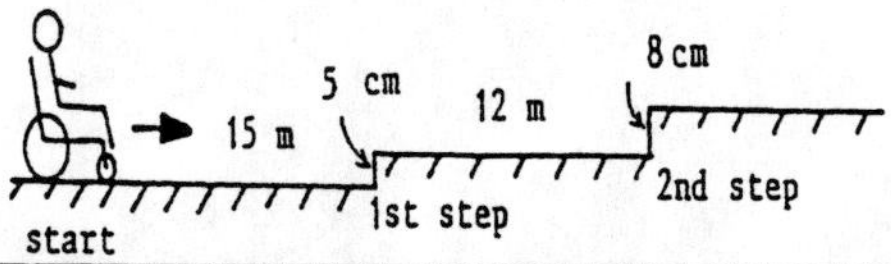

users	weight x10^2 N	wheelchair average speed m/s	reservoir pressure x10^2 kPa	1st step	reservoir pressure x10^2 kPa	2nd step
S1	6.37	1.2	1.78	sat.	1.76	sat.
S2	5.40	1.0	1.75	sat.	1.73	sat.
S3	5.0	0.6	1.67	unsat.	1.47	unsat.
S4	6.66	1.4	1.84	sat.	1.82	sat.
S5	6.08	0.8	1.74	sat.	1.61	unsat.
sat.= satisfactory, unsat.= unsatisfactory						

"Unsatisfactory" means that the operation could barely be done with the lifting stroke less than 5 cm or that the stroke was at most 4.5 to 5.0 cm because of insufficient reservoir pressure. Users S3 and S5 are under lower limb rehabilitation. The other users are all healthy men (S1 and S4) and woman (S2). Users S1, S2, and S4 could easily traverse the two steps with a reservoir pressure of more than 173 kPa as shown in Table 2. The lifting strokes obtained were 12.8 cm (S1), 11.2 cm (S2), and 14.4 cm (S4). User S4, especially, was able to produce high pressure with a high average wheelchair speed of 1.4 m/s. In this experiment, the highest step is 8 cm and, therefore, it is necessary to lift the front wheels at least 8.5 cm. At a lower wheelchair speed, however, achieving step drive within a 15 m travel distance was very difficult. For example, user S3 showed a 4.2 cm front wheels lift at the first step and user S5 showed a 7.1 cm front wheels lift at the second step. Judging from these results, the minimum reservoir pressure may be around 170 kPa for the experimental conditions under investigation. To overcome this problem, an additional reservoir or a more efficient pneumatic pump will be required in the system design.

6. CONCLUSION

A pneumatic wheelchair capable of up to 10 cm step drive has been developed. Pneumatic pumps installed to the rear wheel of the wheelchair generate pressurized air, which is supplied to the reservoir through contact rollers between the pumps and the rear wheels of the wheelchair. Pressurized air from the reservoir is fed to the driving cylinders to lift the front wheels. Analysis of the new system through step drive tests with five users showed the overall feasibility of traversing steps, although the issue of the relationship of necessary reservoir pressure to travel speed needs refinement. The key point in solving this problem is to develop a more efficient pneumatic pump that can provide high exhaust pressure within a short drive distance.

REFERENCES

Brubaker, C. E., 1988, "Advances in Wheelchair Technology," IEEE Trans. on Eng. in Medicine and Biology Magazine, Sept, pp. 21-24.

Cooper, R. A., 1993, "Stability of A Wheelchair Controlled by A Human Pilot," IEEE Trans. on Rehab. Engineering, Vol.1, No.4, pp. 193-206.

Hess, R. A., and Modjtakedzadeh, A., 1990, "A Control Theoretic Model of Driver Steering Behavior," IEEE Control Sys. Magazine, Vol. 10, No. 5, pp. 3-8.

McLaurin, C. A., and Brubaker, C. E., 1986, "A Lever Drive System for Wheelchairs," J. Rehab. Res. Dev., Vol. 23, pp. 52-56.

Sawka, M. N., Glaser, R. M., and Laubach, L. L., 1981, "Wheelchair Exercise Performance of the Young, Middle-aged, and Elderly," J. Appl. Physiol., Vol. 50, pp. 824-828.

Shiraishi, M., and Watanabe, H., 1996, "Pneumatic Assist Device for Gait Restoration," Trans ASME, J. of Dynamic Systems, Measurement, and Control, Vol. 118, No.1, pp.9-14.

FPST-Vol. 6, Fluid Power Systems and Technology — 1999
ASME 1999

A FUTURE MICRO HYDRO-PROPULSIVE SYSTEM

Javid Bayandor
Sir Lawrence Wackett Aerospace Centre

Sylvester Abanteriba
Department of Aerospace Engineering

Ian Bates
Faculty of Engineering

RMIT University
GPO Box 2476V, Melbourne, Victoria 3001, Australia
E-mail: bayandorj@asme.org

ABSTRACT

This paper discusses the development of production of a new type of hydro-propulsive system for the generation of the power. First, the principles which seek to quantitatively describe the behavior of this unconventional system are established. A brief description of the mechanical design is included. A parametric study of the major influences on the performance of the system has been carried out and discussed. Finally, the comparison of theoretical simulation and experimental is indicated and discussed in detail.

NOMENCLATURE

ds	elemental surface area per unit span
E	kinetic energy
F	force created by the fluid over the system
H	fixed control surface
I	momentum
M	moment
n	unit normal
P	time period
p	pressure
r	distance from the origin
S	system
t	time
V	volume
v	velocity field
Φ	potential function
ρ	density
Ω	fixed control volume

INTRODUCTION

The increasing demand for the development of an alternative non-Polluting and sustainable power generating system has led to a myriad of research projects worldwide to meet the goal. Although there is abundance of clean energy in various forms in nature, the biggest drawback in the utilization of this energy has been the very low efficiencies associated with its conversion.

The authors of this paper have established (Bayandor *et al.*, 1999a) from fundamental principles that a new conceptual hydropower system can be developed with an average conversion efficiency much higher than what the best of current hydropower devices can achieve by operating in a wide range of flow regimes. In this paper, a study underlying the effects of the parameters involved which enable the new type of hydropower system to ensure better performance will be reported.

As indicated by Abbasi and Abbasi (1997), mini-hydro and micro-hydro technologies can offer a harmless alternatives to the current dependence on pollution associated techniques of power generation. The new hydro-propulsive system provides a solution for the concerns initiated from utilization of depletable fossil fuels and nuclear energy.

The concept introduces the new hydro-propulsive system which comprises a combination of a multiplex propulsors and arms arrangement in a rotary device. The propulsors consist of a set of carefully designed aero-hydrodynamically formed blades connected to each other at their roots with different phase angles. The propulsors are mounted through flexible suspending systems to the extending radial arms. The arms themselves are attached to a central runner which carries the centerline of the system. Different arrangements of arms and their mounted propulsors together with different number of stages can be used in the system. This provides the system with the flexibility required for its wide range of applications in terms of power requirements and region.

Moreover, the flexible suspensions, connecting the propulsors to the arms, and the special configuration of the propulsors enable the system to be effectively used in a very wide range of fluid regimes including turbulent and random flows, for the purpose of efficient exploitation of the fluid energy. The significance of the propulsor design in capturing the energy of turbulent fluid particle movements and an analogy of the efficient system response to a random flow as proven by extensive experimental tests are discussed in detail by Bayandor *et al.* (1999b).

To validate the principles established for the new propulsion concept, a parametric study was carried out by simulation and experiment. This investigation also enabled a better clarification on the interacting effects between parameters involved as well as providing a guideline for conducting the process of optimization.

THEORY AND SIMULATION

Simulation of the concept required a solution consistent with the nature of the system operation which is based on the generation of external force fields around the rotating system. Thus, to construct a solution, hydrodynamics principles and relations which focus on external force fields in inviscid fluid flows are used.

As the occurring Reynolds numbers are reasonably high, the viscous forces are almost completely masked by the inertia forces and hence, the fluid is taken to be inviscid. Therefore, the resulting vorticities from the solution of the equation of motion are induced in the flow field with no prior reference to the fluid viscosity. However, as the vorticity is also linked to the viscosity of real fluids, where it causes generation of the aero-hydrodynamic forces at a propulsor or an arm, the viscosity is taken into consideration, and its effects on the flow behavior are acknowledged. The origin of singularities, including the trailing vorticities, flow separations and stall phenomena in the analysis are therefore well explained by applying a slight viscosity concept where critical. Incompressibility and linearization of induced velocity disturbances also introduce minor inaccuracy into final results.

In the analysis, the external forces are assumed to be the gradient of a potential function and as a result, they lose their ability to induce velocities in the flow. Therefor, their effect is only manifested in the variation of the pressure fields.

Hydrodynamic Forces and Moments Generated by the Flow on the System

A movement of incompressible and inviscid flow through a system consisting of an arrangement of lifting surfaces induces velocity and pressure fields in the fluid. The resulting pressure field will consequently generate external hydrodynamic forces and moments over the boundary of the submerged (semi-submerged) system.

Irrotationality nature of the fluid motion has to be presumed so that its velocity field v can be calculated based on the gradient of a potential function $\Phi = \Phi(x, y, z, t)$ at all points of the space outside the system (Anderson, 1991). At this stage, as a result of this assumption, any vorticity induction inside the fluid flow is neglected. The shedding of the vortices however, is reintroduced in a later stage of the analysis.

As noted:

$$v = grad\Phi \tag{1}$$

with components v_x, v_y and v_z in the x, y and z direction, respectively.

The velocity field experiences no divergence, therefore:

$$\text{div } v = \Delta\Phi = \left(\frac{\partial^2}{\partial x^2} + \frac{\partial^2}{\partial y^2} + \frac{\partial^2}{\partial z^2}\right)\Phi = 0 \tag{2}$$

A control surface $\tilde{H}$, which is coupled to the fluid particles, is considered around the system S. It is assumed that the control surface floats with the fluid. The momentum theorem can be applied within a control volume $\tilde{\Omega}$, on the fixed fluid amount surrounded by $\tilde{H}$, between the inner surface of $\tilde{H}$ and the boundary ∂s of the system S. By replacing $\tilde{H}$ by a fixed control surface H and defining a fixed control volume Ω, which is bounded by H and ∂S, the momentum equation can be simplified (Fig. 1.). As a result, to balance the equation, it will then be required to add the rate of change of fluid momentum within Ω and the flux of momentum leaving Ω through the outer surface of H. It is also necessary to subtract the incoming flux from the other momentum terms of the equation to justify the replacement of $\tilde{\Omega}$ by Ω.

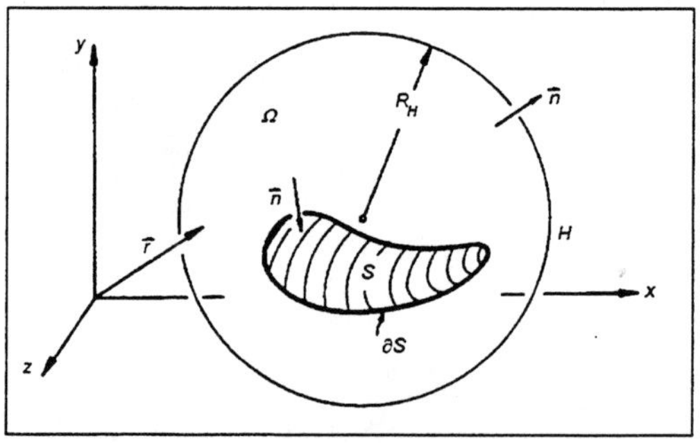

Figure 1. System and control volume Ω (adapted from Sparenberg, 1984)

H is assumed to have a sphere shape with R_H being its large given radius. Its center located on a close neighborhood of S. The momentum of the fluid inside Ω can therefore be shown as:

$$I(t) = \rho \int_{\Omega} \text{grad}\Phi \text{dV} = \rho \int_{\partial S+H} \Phi_n \text{d}s \tag{3}$$

where the unit normal $\boldsymbol{n} = (n_x, n_y, n_z)$ points out of the region Ω.

In order to calculate the force $F = F(t)$ exerted by the fluid on the system S during its motion, it is imperative to introduce the force $F_H = F_H(t)$ exerted by the fluid outside H at the fluid inside H. Thus, the balance of the momentum can be written as (Sparenberg, 1984, and Prandtl and Tietjens, 1957):

$$F_H - F = \rho \frac{\text{d}}{\text{d}t} \int_{\partial S+H} \Phi_n \text{d}s + \rho \int_H v(v \cdot \text{n})\text{d}s \tag{4}$$

where the last term is the mentioned momentum flux through H.

The force F_H also can be written as:

$$F_H = -\int_H p\text{n}\text{d}s \tag{5}$$

From Bernoulli's equation for unsteady motion:

$$p + \tfrac{1}{2}\rho|v|^2 + \rho\frac{\partial\Phi}{\partial t} = p_\infty + \tfrac{1}{2}\rho|v_\infty|^2 \tag{6}$$

where p_∞ and v_∞ are the pressure and the velocity at infinity, respectively. Equation (5). can then be re-written as:

$$F_H = \rho \int_H \left[\frac{\partial \Phi}{\partial t} + \tfrac{1}{2}|v|^2 \right] \mathrm{n} \mathrm{d}s \tag{7}$$

where for any sufficiently smooth closed surface H:

$$\int_H \mathrm{n} \mathrm{d}s = 0 \tag{8}$$

As an extension to the scalar equality the following vector relation can be derived:

$$\int_H \left[f_1(y,z) n_x + f_2(x,z) n_y + f_3(x,y) n_z \right] \mathrm{d}s = 0 \tag{9}$$

where f_i $(i = 1, 2, 3)$ is an arbitrary function.

With respect to the coordinate system H is at rest, hence:

$$\int_P \frac{\partial \Phi}{\partial r} \mathrm{n} \mathrm{d}s = \frac{\mathrm{d}}{\mathrm{d}t} \int_H \Phi \mathrm{n} \mathrm{d}s \tag{10}$$

By substitution of Eqs. (7). and (10). into Eq. (4):

$$F = -\rho \frac{\mathrm{d}}{\mathrm{d}t} \int_{\partial H} \Phi \mathrm{n} \mathrm{d}s + \rho \int_H \left[\tfrac{1}{2}|v|^2 \mathrm{n} - v(v \cdot \mathrm{n}) \right] \mathrm{d}s \tag{11}$$

In a similar way the moment which originates from the hydrodynamic pressures at ∂S can be derived with respect to the origin O.

The moment of momentum of the fluid about O in Ω is:

$$\rho \int_\Omega (r * v) \mathrm{d}V = \rho \int_{\partial S + H} \Phi (r * n) \mathrm{d}s \tag{12}$$

where $*$ denotes the vector product. $M_H = M_H\ (t)$ are the moments resulted from the hydrodynamic forces exerted by the fluid outside H about O at H.

The sum of the resulting moments from external forces about O acting over an amount of fluid equals to the change of the moment of momentum of that amount of fluid about O per unit of time, therefore:

$$M_H - M = \rho \frac{\mathrm{d}}{\mathrm{d}t} \int_{\partial S + H} \Phi \cdot (r * n) \mathrm{d}s + \rho \int_H (r * v)(v * \mathrm{n}) \mathrm{d}s \tag{13}$$

where the last term denotes the moment of momentum flux through the fixed surface H.

Using Eqs. (6). and (9). the moment M_H can be written as:

$$M_H = \rho \int_H \left\{ \frac{\partial \Phi}{\partial t} + \tfrac{1}{2}|v|^2 \right\} (r * \mathrm{n}) \mathrm{d}s \tag{14}$$

Using an equivalent formula from Eq. (10)., substituting Eq. (14). into Eq. (13). and taking the limit $R_H \to \infty$ Eq. (15). is obtained:

$$M = -\rho \frac{\mathrm{d}}{\mathrm{d}t} \int_{\partial S} \Phi \cdot (r * \mathrm{n}) \mathrm{d}s \tag{15}$$

Work Done by an External Force Field on the System

With respect to a Cartesian coordinate system (x, y, z) it is assumed that the fluid is at rest at infinity. The presence of the moving system $S(t)$, which changes its shape periodically, inside the fluid flow generates an external aero-hydrodynamic force field $F(x, y, z, t)$. The rate of change of the fluid kinetic energy E can be determined:

The nonlinear equations of motion can be written in the form of Eq. (16) (Batchelor, 1974).

$$\frac{\mathrm{d}v}{\mathrm{d}t} = \frac{\partial v}{\partial t} + (v \cdot \mathrm{grad}) v = -\frac{1}{\rho} \mathrm{grad} p + \frac{1}{\rho} \mathrm{F} \tag{16}$$

Where the conservation of mass principle will lead to the following equation:

$$\mathrm{div}\, v = 0 \tag{17}$$

Assume for the finite time dependent region $T(t)$ surrounded by the large spherical control surface H, consisting of the moving system $S(t)$, and fixed in space, $F \neq 0$. It is assumed that the region $T(t)$ and the system do not overlap. Again, Ω is H with the exception of S:

$$\rho \int_\Omega v \cdot \left(\frac{\partial v}{\partial t} + v_x \frac{\partial v}{\partial x} + v_y \frac{\partial v}{\partial y} + v_z \frac{\partial v}{\partial z} \right) \mathrm{d}V = -\int_\Omega v \cdot \mathrm{grad} p \mathrm{d}V + \int_T v \cdot \mathrm{F} \mathrm{d}V \tag{18}$$

According to Eq. (17):

$$\frac{\rho}{2} \int_\Omega \left[\frac{\partial v^2}{\partial t} + \mathrm{div}(v^2 v) \right] \mathrm{d}V = -\int_\Omega v \cdot \mathrm{grad} p \mathrm{d}V + \int_T v \cdot \mathrm{F} \mathrm{d}V \tag{19}$$

The left hand side integral can be written in the form of its partial integral components:

$$\frac{\rho}{2} \int_\Omega \frac{\partial v^2}{\partial t} \mathrm{d}V = \frac{\rho}{2} \frac{\mathrm{d}}{\mathrm{d}t} \int_\Omega v^2 \mathrm{d}V - \frac{\rho}{2} \int_{\partial S + H} v^2 (v \cdot \mathrm{n}) \mathrm{d}s = \frac{\mathrm{d}E}{\mathrm{d}t} - \frac{\rho}{2} \int_{\partial S + H} v^2 (v \cdot \mathrm{n}) \mathrm{d}s \tag{20}$$

where E is the kinetic energy of the fluid inside H.

The second part of the integral can be written as:

$$\frac{\rho}{2} \int_\Omega \mathrm{div}(v^2 v) \mathrm{d}V = \frac{\rho}{2} \int_{\partial S + H} v^2 (v \cdot \mathrm{n}) \mathrm{d}s \tag{21}$$

Therefore, for Eq. (14):

$$\frac{\rho}{2} \int_\Omega \left[\frac{\partial v^2}{\partial t} + \mathrm{div}(v^2 v) \right] \mathrm{d}V = \frac{\mathrm{d}E}{\mathrm{d}t} \tag{22}$$

Using partial integral equations the first term at the right hand side of Eq. (18). can be re-written as:

$$-\int_{\Omega} \operatorname{div} p v \mathrm{dV} = -\int_{H} p(v \cdot \mathrm{n}) \mathrm{d}s - \int_{\partial S} p(v \cdot \mathrm{n}) \mathrm{d}s \tag{23}$$

For $R_H \to \infty$ the integral over H tends to zero. Therefore, Eq. (18). can be simplified to:

$$\frac{\mathrm{d}E}{\mathrm{d}t} = \int_{T} v \cdot \mathrm{F} \mathrm{dV} - \int_{\partial S} p(v \cdot \mathrm{n}) \mathrm{d}s \tag{24}$$

At the right hand side of Eq. (24)., the first term refers to the work done by the external force field. The integral determines the scalar product of the velocity v with the infinitesimal forces $F\,dV$. The work done by the system S appears in the second term.

For extending the results to the case of a rotational movement, a mean velocity V_∞ is assumed for the external force field $F(x, y, z, t)$ in the positive x direction with the motion period of P. Assuming the validity of the linearized equation of motion, a system of periodic vortices extended to infinity can be introduced behind the external force field.

If it is assumed that the force field has been activated at infinity, this means that the neighboring velocity field of the force region is fully periodic. Therefore, according to Sparenberg (1984), the work done by F, which is also the increase of kinetic energy per period of the motion, equals the kinetic energy between two planes s_1 and s_2 (Fig. 2.) parallel to the (y, z) plane, at a period P apart and situated infinitely far behind F.

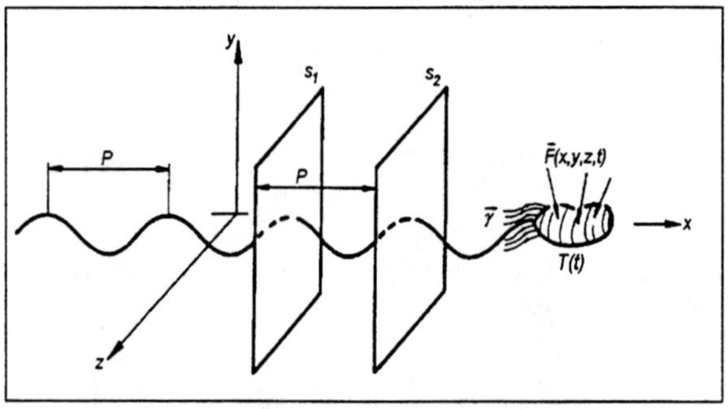

Figure 2. Periodically moving force field $\bar{F}(x,y,z,t)$ (from Sparenberg, 1984)

NUMERICAL RESULTS AND DISCUSSION

Having determined the external force field and its resultant work over the system, the system was simulated and its performance analyzed.

The analysis has shown that a simplified model of the new system consisting of one arm and one flexible suspension together with its mounted propulsor, during the period of one full rotation, is capable of generating positive torque (Fig. 3.). This indicates that the addition of arms and propulsors, as expected, will increase the production of torque and power output.

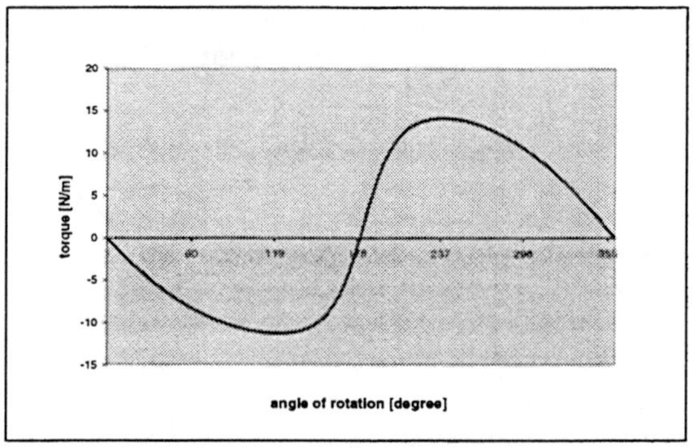

Figure 3. Torque generated vs angle of rotation (u=3 m/s)

The arm length also contributes to the power output of the system considerably. The simulation results show that by increasing the length of the arm from *0.3 m* to *0.6 m* the power output doubles. Further increase of the arm length however, results in reduction of the power output as the propulsor would not be able to produce a large enough force to handle a higher inertia resulting from a greater weight of the system (Fig. 4.).

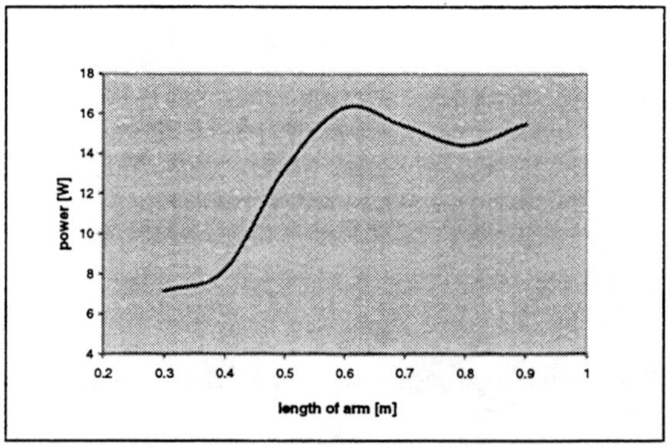

Figure 4. Power output vs arm length (u=3 m/s)

It is clear that any increase of the upstream flow velocity can result in the increase of the net torque and power output (Fig. 5.).

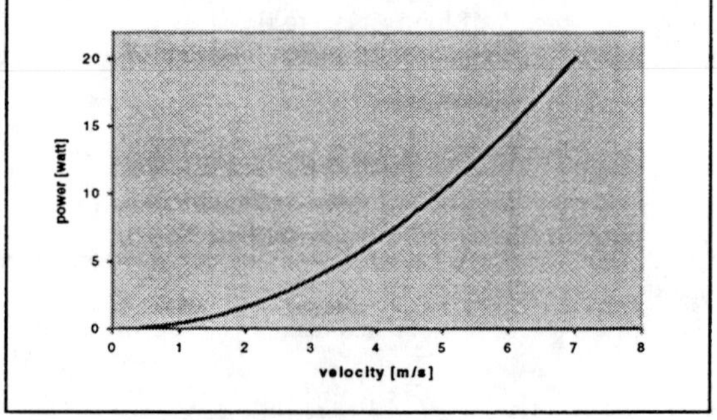

Figure 5. Power output vs upstream flow velocity

However, the structural limitations of the system, specifically the yielding points of the flexible suspending components, are the factors defining the maximum power output and the failure point of the system, beyond which more rigid materials need to be used.

The simulation was re-run using several types of flexible suspensions with different overall modules of elasticity and moments of inertia. The simulation allowed a better understanding of the behaviors of the suspending components used, while assisting the process of selecting the specifications for the final suspension system at the optimization stage. Figure 6. compares the effects of six different categories of flexible suspensions on the power output when assembled on the new system.

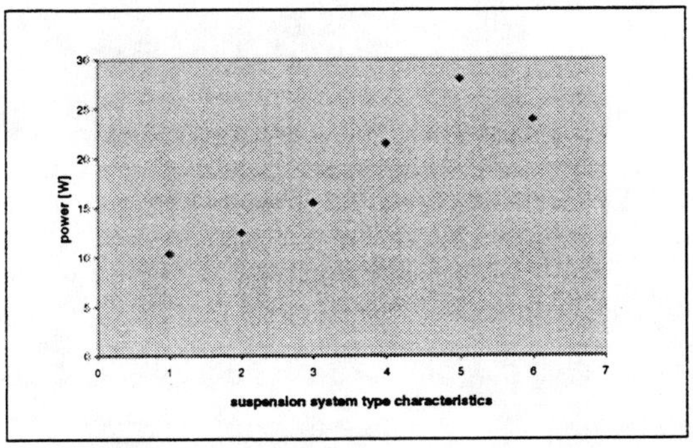

Figure 6. Power generated vs suspension system type characteristics

Arrangement of the blading system of the propulsors is critical in capturing the flow particle movements and the efficient power generation of the system. The simulation results illustrate the fact that increasing the number of blades in the propulsors do not essentially produce a higher power output. On the contrary, as an example, having a three bladed propulsors with *120* degrees phase angle between its consecutive blades, creates a relatively lower power output in a straight upstream flow condition (Fig. 7.). This is due to the fact that only the images of the blade areas in the horizontal plane are capable of producing the favorable lift force required for rotation of the system which is less than a two bladed case. In the mean time, the third blade also increases the drag force applied on the system which results in the retardation of its rotational motion.

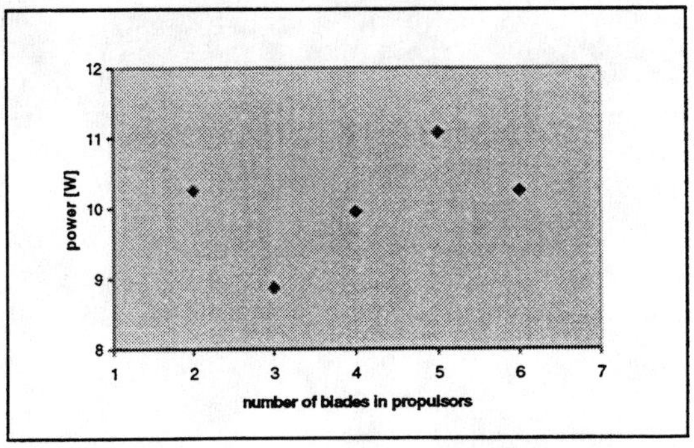

Figure 7. Power output vs number of blades in propulsors

Another example indicates that a four bladed propulsor –in which the blades correspond to the horizontal and vertical directions of the Cartesian coordinates- produces less power than a two bladed propulsor and a six bladed one. This is caused by the generation of a greater drag force in a four bladed propulsor, while the creation of the lift force is similar to the case of two bladed one. In the case of six bladed propulsor, the increase of the total wet area of the blade combination results in a higher lift force generation. In addition, this configuration imposes the bi-plane effect which generates and maintains an extra lifting force on the lower blades. However, the six bladed propulsor causes a greater disturbance in the flow as well as producing a greater drag force that prevents the system to reach a considerably higher output.

OPTIMUM DESIGN OF A PROTOTYPE

The simulation results were used in developing an optimized design for the system prototype. The prototype was constructed using rapid prototyping and stereolithography. The scaled down prototype has the following dimensions and specifications:

- diameter: *0.26 m*
- flexible suspension: number *5*
- number of propulsor blades: *5*
- possible number of arms per stage: *5, 7, 9*
- max number of stages: *3*

Other dimensions and required blade and arm profiles were also analyzed and included in the design.

COMPARISON OF TEST RESULTS WITH SIMULATION

The prototype was tested under different upstream flow conditions corresponding with those of the simulation to provide a basis for the evaluation of the numerical model. Different arrangements of the prototype represented the various modes of the simulation. These arrangements were constructed by the alteration of the number of arms and stages as well as the usage of a range of suspension systems.

The comparisons made show a good level of agreement between the experiments and simulation. The deviation factor between these two sets of data varies from just over *11%* at the critically low upstream flow velocity of *0.3 m/s* to less than *7%* at the flow velocities greater than *0.46 m/s*. For the flow velocities equal or greater than *0.3 m/s*, the instrumentation used was able to effectively detect the power output of the system.

The slight difference between the experimental and numerical data to a limited extent originates from the linearization of the velocity disturbances involved in the equations of motion and assumptions such as inviscid fluid and irrotationality nature of the flow. However, the assumption of non-buoyant system as well as the effects of scaling down and boundary layer, in particular, contribute to the discrepancy at low upstream flow velocities.

SUMMARY

An investigation in the form of a parametric study of the new hydro-propulsive system has been carried out to illustrate the mechanisms which enable the system to achieve a higher conversion rate. The simulation output was closely analyzed and based on the optimum calculated data a prototype was designed and constructed.

Using the prototype, different configurations and arrangements of the system in free-stream flow conditions were examined. The experimental results validate the simulation data and hence by extension the principles propounded for the operation of the new system.

REFERENCES

Abbasi, S. A., and Abbasi, N., 1997, "Environmental impacts of renewable energy sources," *International Journal of Energy-Environment-Economics*, vol. 5, no. 1, pp. 1-49.

Anderson, J. D., Jr., 1991, "*Fundamentals of Aerodynamics*," 2nd ed., McGraw-Hill, New York.

Batchelor, G. K., 1974, "*An Introduction to Fluid Dynamics*," Cambridge University Press.

Bayandor, J., Abanteriba, S., and Bates, I., 1999a, "Principles of a new zero head hydro-propulsive system," *International Joint Power Generation Conference and Exposition 1999: Proceedings of the IJPGC'99*, San Francisco, California, July 25-28, 1999, American Society of Mechanical Engineering, 1999.

Bayandor, J., Abanteriba, S., and Bates, I., 1999b, "A conceptual hydro-propulsive system for more efficient exploitation of fluid power," *World Renewable Energy Congress 1999: Proceedings of the WREC'99*, Kuala Lumpur, Malaysia, June 8-11, 1999, Ministry of Energy, Telecommunications and Posts, Malaysia, World Renewable Energy Network and UNESCO.

Prandtl, L., and Tietjens, O. G., 1957, "*Applied Hydro- and Aeromechanics*," Dover, New York.

Sparenberg J. A., 1984, *Elements of Hydrodynamic Propulsion*, Mechanics of Fluids and Transport Processes, Moreau R. J. (Editor), Martinus Nijhoff Publisher, The Hague, The Netherlands.

FPST-Vol. 6, Fluid Power Systems and Technology — 1999
ASME 1999

Two-level Hierarchical Control of Hydraulic Elevator Speed

Ran Longlin Yang Huayong
Institute of Fluid Power Transmission & Control
Zhejiang University

Abstract

The speed control of hydraulic elevators is usually carried out by PID controllers. But as the system properties slowly changes with the working situation, the control result is not satisfactory after long time operation. In this paper, a two-level hierarchical controller consist of an intelligent PI controller and a simple upper level expert controller is introduced. The expert controller is based on the expert knowledge accumulated from research experiences on the speed control of hydraulic elevators. The controller is realized on a small chip computer system, which is cheap and flexible to use.
Keywords: Hierarchical control, Expert controller, Intelligent control

1 Introduction

A typical hydraulic elevator system controlled by a proportional valve is shown in Fig. 1. The purpose of the control is to ensure the cabin to move according to the ideal speed curve as shown in Fig. 2, so as to give riding comfort and accurate leveling to the stop. The system's flowchart of control can be seen in Fig. 3. In the system, the cabin speed changes with the cylinder speed. The cylinder speed is decided by the flow rate to and from the cylinder. The flow rate is controlled by a proportional valve which adjusts the flow rate according to the electrical current i, which is proportional to the control value of the single chip computer controller. There is an optical encoder fastened on the cabin, the controller acquires cabin speed feedback by counting the impulses of the encoder.

The controller of the system is usually a PID controller. Because the effects of static friction, compressibility of the oil and relatively big mass of the cabin, the parameters of the PID controller have to be carefully selected to ensure the stability of the system and these parameters usually have very small adjustable scope.

Another problem of using a normal PID controller is that the controller is not adaptive to the change of the system properties. In a hydraulic elevator system, one of the most important effective factor is the variation of oil temperature. As in hydraulic elevator, most energy that input into the system converts into heat to increase the oil temperature. When oil temperature increases with the operating time, it causes the system properties changes. The changing oil temperature can make the preset PID parameters not suitable for the changing situation and causes control quality deteriorate.

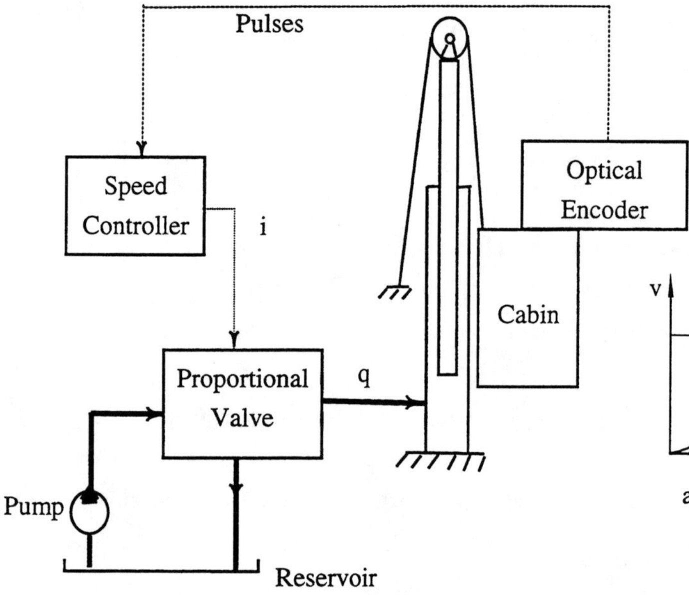

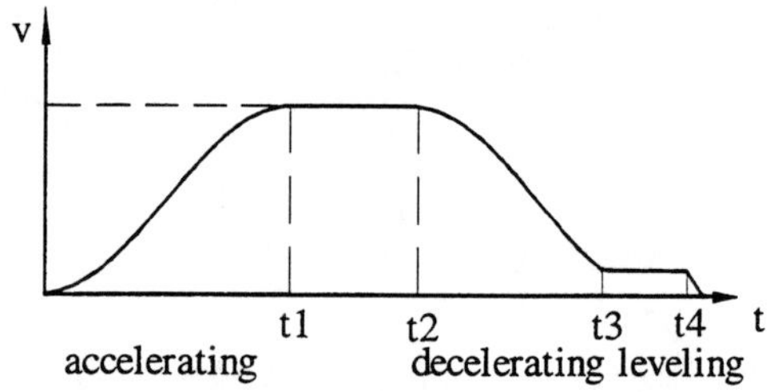

Fig. 2 Ideal Speed Curve

Fig. 1 A Hydraulic Elevator System

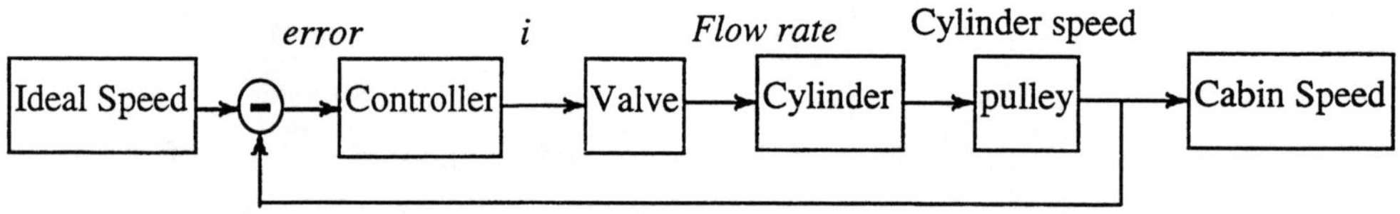

Fig.3 Control Flowchart of Hydraulic Elevator

2 Intelligent PI controller

The intelligent PI controller is developed from a normal PI controller. Its advantage compared with a PID controller, is the stability. The deference between an intelligent PI controller and a normal PI controller can be seen in Fig. 4. There is an switch 'K' in the intelligent PI controller, which is switched on when $e \cdot \dot{e} \geq 0$. The realization of the intelligent PI controller is:

$$u(k) = u(k-1) + \Delta u(k) \qquad (1)$$

$$\Delta u(k) = \begin{cases} K_p \cdot [e(k) - e(k-1)] + K_i \cdot e(k) & \text{if } e \cdot \dot{e} \geq 0 \\ K_p \cdot [e(k) - e(k-1)] & \quad e \cdot \dot{e} < 0 \end{cases} \qquad (2)$$

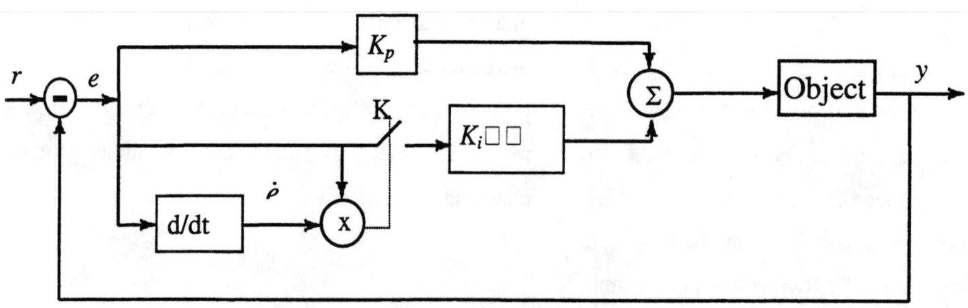

Fig. 4 An intelligent PI Controller

3 Performance of the Intelligent PI controller

When the intelligent PI controller is applied to control speed of hydraulic elevator. The experiment result shows that if the coefficients Kp and K_i is properly selected in accordance with the system condition, the controller does offer good output speed curve. But the problem caused by oil temperature sustains.

Usually, when system output quality deteriorates, the ideal speed curve input into the system or Kp, K_i of the controller have to be adjusted manually. Experiments shows that the best way of adjust the system is adjusting Ki. If in some extreme situation, adjusting K_i can not make the output satisfactory, the accelerating time and decelerating time of ideal speed have to be prolonged or shortened.

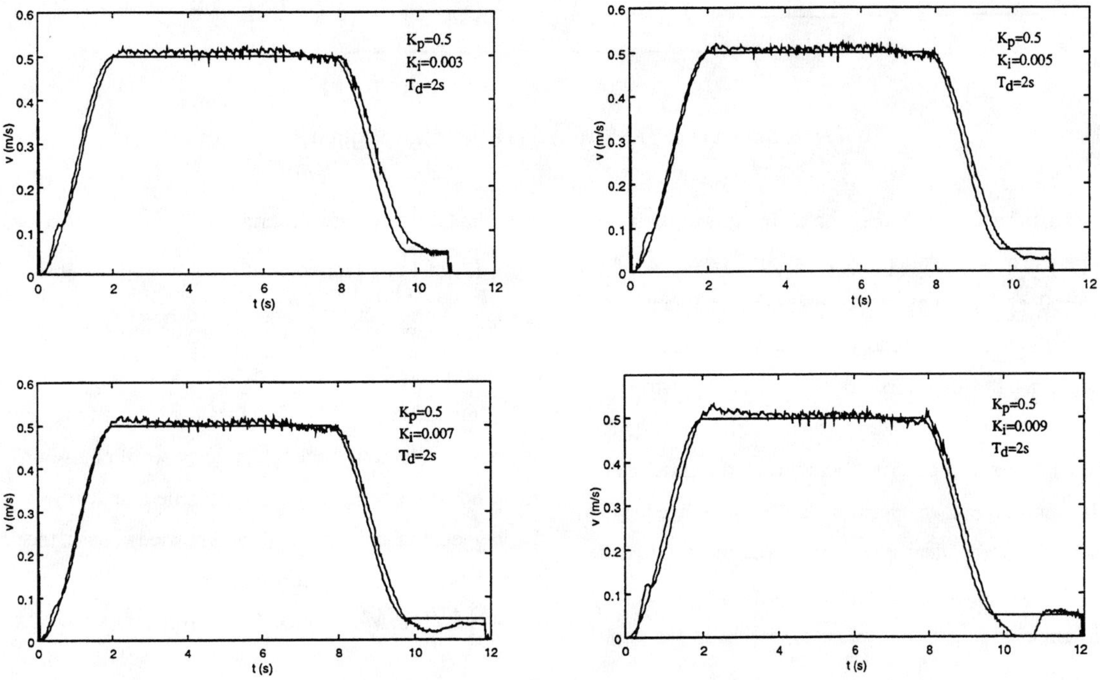

Fig. 5 Control Result of Intelligent PI Controller with Different Parameters

Fig. 5 shows the effect of K_i on the response of the system. It can be seen that increasing K_i causes the system response fast, but also causes chance of overshot and even creep of the cabin. But if the value of K_i is too small, the speed can't be reduced to the expected leveling speed when the cabin have to stop, so that the leveling accuracy can't be reach. The value of K_i, therefore, has to be carefully selected.

Two Level Hierarchical Controller

In order to make the control system adaptive to the change of the operating conditions, a supervisory expert controller is introduced into the system to form two-level hierarchical control. The purpose of using the expert controller is to adjust the intelligent PI controller and the ideal speed curve of the system so that to ensure the system output satisfactory speed curve in any operating conditions.

The structure of the two-level hierarchical controller can be seen in fig. 6. The supervisory expert controller works like an expert who adjusts the parameters of the intelligent PI controller or the ideal speed curve of the system according to a certain criteria of output quality. The criteria will be introduced on the left part of the paper.

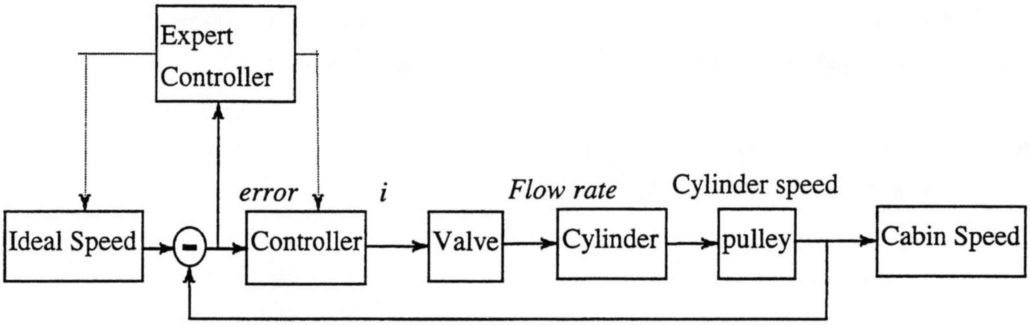

Fig. 6 Structure of The Two-level Hierarchical Controller

It must be noticed that here in this two-level hierarchical controller, the upper level expert controller does not work on the intelligent PI controller's control process of each operation, i.e. each time the elevators operation from on stop to another. It works after each operation. It evaluates the control result of each operation of the intelligent PI controller, then decides if the intelligent PI controller or the ideal speed curve need to be adjusted for the next operation. This is effective because the oil temperature and the system properties is changing very slowly.

4 The Expert Controller

The expert controller is shown in Fig. 7. It consist of information collector, knowledge base, deducing device, and control rule set.

Times of experiment show that the error during the leveling period of the elevator operation is a typical criteria to evaluate the system's response quality. Integration of error during the period can reflect the response quality of the operation. So the expert controller adjust Ki and/or accelerating, decelerating time of the ideal speed curve for the next operation according the following criteria:

$$U = \int_{t3}^{t4} e(t)dt \quad \text{......(3)}$$

The knowledge base including definition of fuzzy set of system's response characteristic U; fuzzy set of adjusting value of K_i, □K_i; fuzzy set of adjusting value of accelerating time, □T_a; fuzzy set of adjusting value of decelerating time, □T_d; smallest limit of K_i, SL; biggest limit of $K_{i,\ BL}$. In the knowledge base, the fuzzy sets have the save form such as:

$$U = \{NB, NM, NS, 0, PS, PM, PB\} \quad \text{......(4)}$$

But each element of each fuzzy set reflects a certain scope of the value.

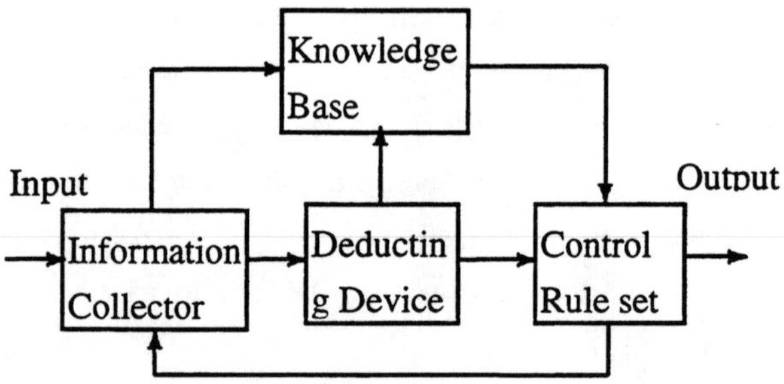

Fig. 7 Structure of Expert Controller

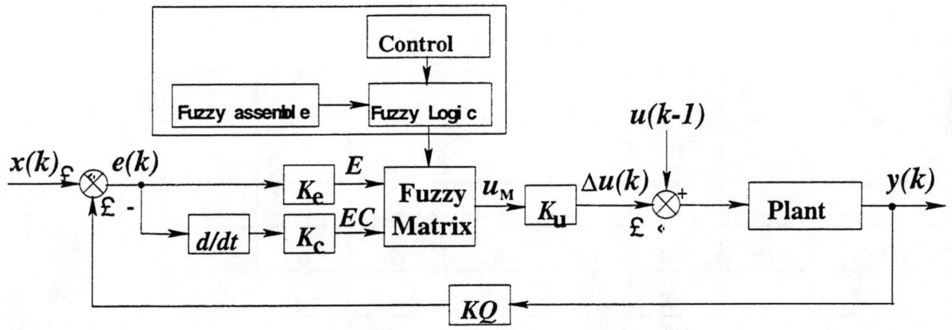

Fig. 8 Structure of a Fuzzy controller

(1) Defining Domain

Input variable error E and error differential EC are the fuzzy variable, and Fuzzy assemble domain of E□EC are defined:

□□6□□5□□4□□3□□2□□1□0□1□2□3□4□5□6 □

Fuzzy assemble domain of u_M is defined:

□□7□□6□□5□□4□□3□□2□□1□0□1□2□3□4□5□6□7 □

After E□C *and* u_M are defined□subject function should be proposed, i.e. fuzzy variable u_M shown in table 2.

Table 1 The value of Fuzzy variable E

E	-6	-5	-4	-3	-2	-1	0	1	2	3	4	5	6
PB	0	0	0	0	0	0	0	0	0	0.1	0.4	0.8	1.0
PM	0	0	0	0	0	0	0	0	0.2	0.7	1.0	0.7	0.2
PS	0	0	0	0	0	0	0.3	0.8	1.0	0.5	0.1	0	0
O	0	0	0	0	0.1	0.6	1.0	0.6	0.1	0	0	0	0
NS	0	0	0.1	0.5	1.0	0.8	0.3	0	0	0	0	0	0
NM	0.2	0.7	1.0	0.7	0.2	0	0	0	0	0	0	0	0
NB	1.0	0.8	0.4	0.1	0	0	0	0	0	0	0	0	0

Table 2 The value of Fuzzy variable u_M

u_M	-7	-6	-5	-4	-3	-2	-1	0	1	2	3	4	5	6	7
PB	0	0	0	0	0	0	0	0	0	0	0	0.1	0.4	0.8	1.0
PM	0	0	0	0	0	0	0	0	0	0.2	0.7	1.0	0.7	0.2	0
PS	0	0	0	0	0	0	0	0.4	1.0	0.8	0.4	0.1	0	0	0
O	0	0	0	0	0	0	0.5	1.0	0.5	0	0	0	0	0	0
NS	0	0	0	0.1	0.4	0.8	1.0	0.4	0	0	0	0	0	0	0
NM	0	0.2	0.7	1.0	0.7	0.2	0	0	0	0	0	0	0	0	0
NB	1.0	0.8	0.4	0.1	0	0	0	0	0	0	0	0	0	0	0

□2□Precision variable fuzzied

$$E=\text{int}[e(k)/K_e+0.5] \quad (5)$$

$$EC=\text{int}[(e(k)-e(k-1))/K_c+0.5] \quad (6)$$

Where□$e(k)$□$e(k$-1) are the precision value of domain E□EC separately, and complying with the following rules:

if $E>6$□ then $E=6$□

if $E<-6$□ then $E=-6$□

if $EC>6$□ then $EC=6$□

if $EC<-6$□ then $EC=-6$□

□3□Established fuzzy control table. As shown in table 3 and 4, output values are gave:

Table 3 Fuzzy control rule table

CE \ E (U_M)	NB	NM	NS	O	PS	PM	PB
NB	NB	NB	NM	NM	NS	O	O
NM	NB	NM	NS	NS	O	O	O
NS	NM	NS	NS	O	O	O	PS
O	NM	NS	O	O	O	PS	PM
PS	NS	O	O	O	PS	PS	PM
PM	O	O	O	PS	PS	PM	PB
PB	O	O	PS	PM	PM	PB	PB

Table 4 Fuzzy control table

CE \ E (u_M)	-6	-5	-4	-3	-2	-1	0	1	2	3	4	5	6
-6	-7	-7	-6	-5	-4	-4	-4	-3	-3	-1	-1	0	0
-5	-7	-6	-6	-5	-5	-4	-4	-3	-2	-1	0	0	0
-4	-7	-6	-5	-5	-4	-4	-3	-3	-2	-1	0	0	0
-3	-7	-6	-5	-4	-4	-3	-3	-2	-1	0	0	0	1
-2	-6	-5	-4	-3	-2	-2	-1	-1	0	0	1	1	1
-1	-5	-4	-3	-3	-2	-1	-1	0	0	1	1	2	2
0	-4	-3	-3	-2	-1	-1	0	0	1	1	2	3	4
1	-2	-2	-1	-1	0	0	1	2	3	3	4	4	5
2	-1	-1	-1	0	0	1	1	2	3	3	4	5	6
3	-1	0	0	1	2	3	3	4	4	5	5	6	7
4	0	0	0	1	2	3	4	4	5	5	6	6	7
5	0	0	0	1	2	3	4	4	5	5	6	6	7
6	0	0	1	1	3	3	4	4	4	6	6	7	7

□4□Fuzzy variable defuzzying

After a proportional constant multiplying fuzzy control variable u_M □the precision of domain u_M can be calculated□

$$u(k) = K_u u_M \quad \text{(7□}$$

The control rule set including the following rules:

1. If *U=NB* and $K_i>SL$ then □$K_i=NB$
2. If *U=PB* and $K_i<BL$ then □$K_i=PB$
3. If *U=NB* and K_i□*SL* then □$K_i=0$ and □*Ta=NB*, □*Td=NB*
4. If *U=PB* and K_i□*BL* then □$K_i=0$ and □*Ta=PB*, □*Td=PB*

...

The control rules reflect expert decision that if the response is too fast, then reduce the value of Ki. If Ki is adjusted to the limit, then the ideal speed curve is reset.

The deducing device, i.e. the chip computer adjust K_i and / or the accelerating time, decelerating time according to these control rules. The adjustment also depends on the definition of the values and fuzzy sets in the knowledge base.

5 Performance of The Two-level Hierarchical Controller

The two-level hierarchical control is realized on an Intel 80C196 chip computer system. As in the real time control process of the elevators moving process, only the intelligent PI controller works, and the expert control only works after the real time control of the intelligent PI controller, the calculation amount is small and the program is simple.

Because in a hydraulic elevator system, the drift of system properties is very slow. The application of the expert controller on adjusting the value of K_i, accelerating time and decelerating time of ideal speed in limited scopes is successful. Experiments shows when this two-level hierarchical controller is used, the criteria U can be controlled in a small section near zero and the control result of the speed curve is indirectly ensured. Even in extremely bad operating conditions, the cabin will not creep.

6 Conclusion

In systems using simple controllers such as intelligent PI or normal PID controller, if relationship between controller parameters and system response can be achieved, expert controller based on rules can be used to improve the control adaptability by properly defined criteria of system response quality.

References:

[1] Li Shiyong□1996, *Fuzzy, Neural and Intelligent Control*□ Harbin Institute of Technology publication.

[2]. Ran Longlin; Yang Huayong, Guo Shuiming, 1997, "An Elevator Hydraulic System Using Cabin Speed Feedback", *Elevator Technology 8, The Proceedings of the ELEVCON 97'*, Shanghai.

[3]. Moudgal, Vivek G.; Passino, Kevin M.; Yurkovich, Stephen, 1994, "Expert Supervisory Conrol for A Two-link Flexible Robot". *Proceedings of The 1994 IEEE International Conference on Rotobics and Automation*, San Diego, USA.

FPST-Vol. 6, Fluid Power Systems and Technology — 1999
ASME 1999

Design and Control of a Spherical Air Bearing System for multi-DOF Ball-joint-like Actuators

Kok-Meng Lee*, Dan E. Ezenekwe and Tian He***

*Georgia Institute of Technology, The George W. Woodruff School of Mechanical Engineering, Atlanta GA 30332-0405
**Corning Incorporated, Telecommunications Products Division, 14556 Highway 601 South, Midland, NC 28107

Abstract –This paper presents an analytical investigation on the design method and modeling of an air bearing system for ball-joint-like actuators. It addresses the method of regulating the three DOF translations of the rotor introduced by the air bearing system and thereby improving the dynamic performance of the orientation motion manipulation. The kinematics, which relates the rotor displacement and the air gaps, are essential for design optimization, dynamic simulation and motion control, are derived in closed-form. With a detailed modeling of the pressure-flow relationship as a function of the rotor position, the forces and dynamics of the system are formulated, and design methods of regulating the rotor displacement have been explored analytically. Simulation results suggest that the fluid forces could be generated to passively stabilize the otherwise open loop unstable electromagnetic system. It is expected that this research will be a basis for designing and evaluating an improved VR spherical motor with enhanced torque capability by eliminating mechanical friction.

1. INTRODUCTION

Many applications in the industry require the use of fluid bearings to overcome friction, provide more precise location and alignment of components, and enable smooth movement of large components from one point to another point. In other applications, precise alignment of parts is very critical to the quality and specifications of the product. As advances in technology continue to demand more accurate and precision devices, there are opportunities to utilize the unique attractive features of air bearing systems to meet these challenges.

We explore the use of a non-contact, spherical air bearing system for multi-DOF spherical actuators [1] which operates on the principles of variable reluctance (VR). The motor combines the roll, pitch and yaw motion in one joint making it attractive for many applications. Besides the capability of three degree-of-freedom (DOF) motions in one joint, the motor has large range of motion, isotropic properties and no singularities within the workspace. These flexible design features make the VR spherical motor suitable for a wide range of applications such as water-jet cutting, laser cutting, painting, welding, material handling, coordinate measurement etc., where smooth uniform manipulation of the end-effector is required.

Three basic techniques used in non-contact bearing designs; namely, fluid bearing, magnetic levitation, and the Meissner effect. Non-contact air bearing is considered here since it has the potential to enhance the motor performance for more advanced applications, has cooling effect on interacting components and does not interfere with actuator electromagnetic system. Many researchers [2] [3] have investigated bearing designs and characteristics. Other authors [4] [5] [6] have contributed to improving bearing stability of fixed orifices. Recently, there has been some work on control of air-bearing devices [8]. These works have concentrated on single axis bearing devices. In this paper we are interested in the control of multi-DOF spherical devices.

The reminder of this paper is organized as follows: The next section discusses the bearing design configuration for a ball-joint-like spherical device. This is followed by the characterization of airflow through the bearing in Section 3, which generates the forces on the rotor. Then, the system dynamics is presented in Section 4 and followed by design methodology and simulations in Section 5. Finally, a conclusion is presented.

2. BEARING FOR BALL-JOINT-LIKE DEVICES

Consider an arbitrary displacement of the two spherical surfaces as shown in Figure 1, where $\vec{r}_o$ is the rotor displacement with respect to the center of the stator.

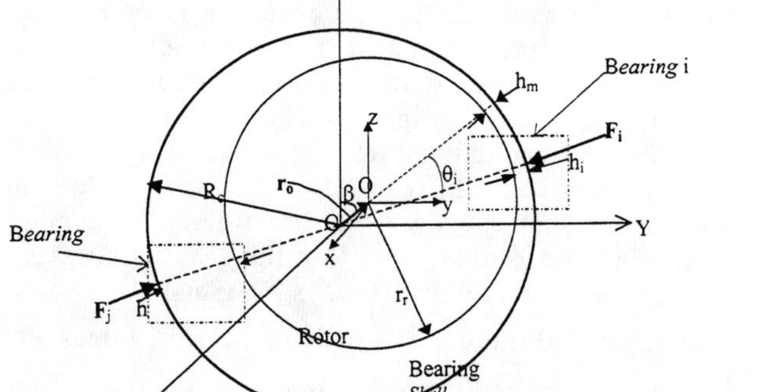

Figure 1 Conceptual schematics illustrating spherical bearing

In Figure 1, the reference frame *XYZ* is defined at the center of the stator, *O*, with the *Z*-axis pointing towards the opening for the stator and *xyz* is fixed at the center of the rotor, *O*', with its z-axis along the rotor output shaft. At equilibrium, the coordinate frames *XYZ* and *xyz* have a common origin at their centers. Thus, any position on the rotor can be expressed with respect to the *XYZ* coordinate system using a 4X4 homogeneous transformation matrix, [*T*]:

$$[P]_{XYZ} = [T][P]_{xyz} \tag{1}$$

where $[P]_{XYZ}$ and $[P]_{xyz}$ are the position of a point on the rotor measured with respect to XYZ and *xyz* respectively.

2.1 Bearing Configurations

The design concept is to use pressurized air to regulate the rotor such as the rotor displacement $\bar{r}_o$ is null regardless of any "disturbances" caused by the electromagnetic actuation or an external force. Multiple independent spherical bearings are strategically designed to support the rotor. In general, the more the number of bearings, the larger the load the system is capable of supporting. The challenge, however, is to design a compact yet efficient air bearing system to fit into the limited surface area of the rotor. In addition, the bearings should be designed so that they do not interfere with the electromagnetic poles of the spherical motor, which are located following the pattern of a regular polygon.

Theoretically, the minimum number of non-coplanar forces required to achieve bi-directional position control of the spherical rotor in a three-dimensional space is four. One possible arrangement of a minimum number of bearings is to place the bearings at locations corresponding to the vertices of a regular tetrahedron. The arrangement of minimum number of bearings is not necessarily attractive since four independent actuators are required, and that the four point bearings can not be grouped in pairs such that the couples exert equal and opposite forces along their lines of action.

An attractive alternative is to design so that pressurized air passes through its center enabling the unit to serve as a bearing. The advantages of joint magnetic-pole/bearing units are twofold: (1) The air jet will provide cooling effect to the coil windings. (2) It will optimize the stator surface by maximizing the size of a bearing, thereby enhancing load-bearing capacity. In the following discussion, we shall consider the case where six or more bearing points are evenly spaced on the spherical bearing such that bearing forces can always be grouped in pairs. As illustrated in Figure 1, the pair of bearing forces $\bar{F}_i$ and $\bar{F}_j$ exert equal but opposite forces through the center of the stator.

2.2 Forward Kinematics

The gap between the stator and rotor along a pair of forces, $\bar{F}_i$ and $\bar{F}_j$, can be determined with the aid of Figure 2. The net force, $\bar{F}_{ij} = \bar{F}_i + \bar{F}_j$, can be described by:

$$\bar{F}_{ij} = F_{ij}\bar{e}_{ij} \tag{2}$$

where F_{ij} and $\bar{e}_{ij}$ are the magnitude and the unit vector (known) of the resultant force. As shown graphically in Figure 2b, the minimum air gap between the rotor and the stator is in the direction of $\bar{r}_o$. Thus, the included angle between $\bar{F}_{ij}$ and $\bar{r}_o$ is

$$\theta_{ij} = \cos^{-1}(\bar{e}_{ij} \cdot \bar{e}_{ro}) = \cos^{-1}\left(\frac{\bar{F}_{ij} \cdot \bar{r}_o}{F_{ij}\, r_o}\right) \tag{3}$$

where r_o is the magnitude and $\bar{e}_{ro}$ is the unit vector of $\bar{r}_o$. For a given rotor displacement, the air gaps, h_i and h_j are given by Equations (4a) and (4b):

$$h_i = r_s - \left\{\sqrt{r_r^2 - r_o^2 \sin^2\theta_{ji}} + r_o \cos\theta_{ji}\right\} \tag{4a}$$

and

$$h_j = r_s - \left\{\sqrt{r_r^2 - r_o^2 \sin^2\theta_{ji}} - r_o \cos\theta_{ji}\right\} \tag{4b}$$

where r_s and r_r are the radii of the stator and rotor at the interface respectively.

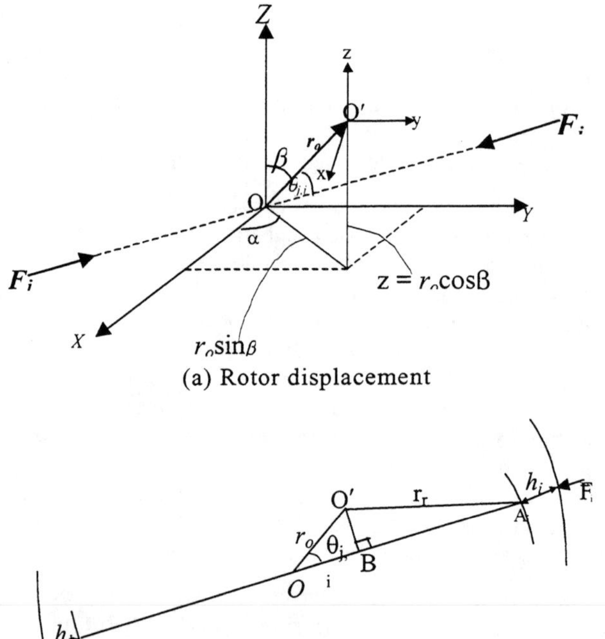

(a) Rotor displacement

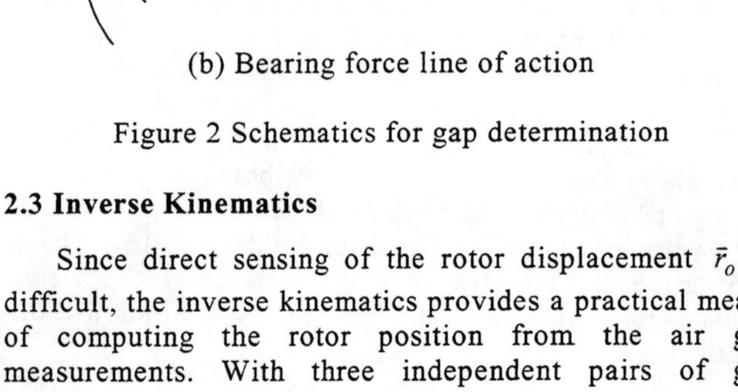

(b) Bearing force line of action

Figure 2 Schematics for gap determination

2.3 Inverse Kinematics

Since direct sensing of the rotor displacement $\bar{r}_o$ is difficult, the inverse kinematics provides a practical means of computing the rotor position from the air gap measurements. With three independent pairs of gap measurements, the three orthogonal components of the rotor displacement can be determined as follows.

Subtracting Equation (4a) from Equation (4b), we have

$$r_o \cos\theta_{ji} = \bar{r}_o \cdot \bar{e}_{j,i} = \frac{h_j - h_i}{2} \tag{5a}$$

Two other similar equations can be obtained from two other pairs of bearing forces.

$$\bar{r}_o \cdot \bar{e}_{j-1,i-1} = \frac{h_{j-1} - h_{i-1}}{2} \tag{5b}$$

$$\bar{r}_o \cdot \bar{e}_{j+1,i+1} = \frac{h_{j+1} - h_{i+1}}{2} \tag{5c}$$

Equations (5a), (5b), and (5c) can be written in matrix form:

$$\begin{bmatrix} \bar{e}_{j-1,i-1} & \bar{e}_{j,i} & \bar{e}_{j+1,i+1} \end{bmatrix} \begin{bmatrix} r_x \\ r_y \\ r_z \end{bmatrix} = \frac{1}{2} \begin{bmatrix} h_{j-1} - h_{i-1} \\ h_j - h_i \\ h_{j+1} - h_{i+1} \end{bmatrix} \tag{6}$$

where r_x, r_y *and* r_z are the components of $\bar{r}_o$ in the direction of X, Y and Z respectively.

3. AIR BEARING DYNAMIC MODEL

Figure 3 shows a schematic of a simple, pocketed, orifice compensated bearing. Air enters from a pressure source, passes through an orifice restriction of diameter d_o, then expands isentropically into the pocket of diameter R_p and recess d_p, and finally exhausts to the atmosphere through the annulus which consists of two parallel surfaces of spacing h.

3.1 Flow characteristics

The following assumptions are made in deriving the dynamic model: (1) The pressure in the pocket is uniform. (2) The air is isothermal. (3) Changes in air density are attributed mainly to variations in pressure and the ideal gas law, $p = \rho RT$ where p, R and T are the pressure, gas constant and temperature of the air respectively, is assumed to hold throughout. Thus, the force acting on the rotor is given by integrating the pressure over the bearing surface as follows:

$$f = 2\pi \left[\int_0^{R_p} p_p r dr + \int_{R_b}^{R_b} p r dr \right] \tag{7}$$

where p_p and p are the pressures in the pocket and the annulus respectively. The mass m contained in the bearing is a function of the rotor displacement along the direction of the actuating force as well as the air density and the state of the air. The time rate of change of the air is the difference between the inflow and the outflow, or

$$\frac{dm}{dt} = q_R - q_o \tag{8}$$

where m is the mass of the air contained between the bearing surfaces; and q_R and q_o are the mass flow rate through the orifice restriction and the exhaust respectively. The mass m is given by

$$m = 2\pi \left[\int_0^{R_p} (d_p + h)\rho r dr + \int_{R_p}^{R_b} h\rho \, r dr \right]$$

or

$$m = \frac{p_p}{RT} \pi R_p^2 (d_p + h) + \frac{2\pi}{RT} \int_{R_p}^{R_b} hp \, r dr \tag{9}$$

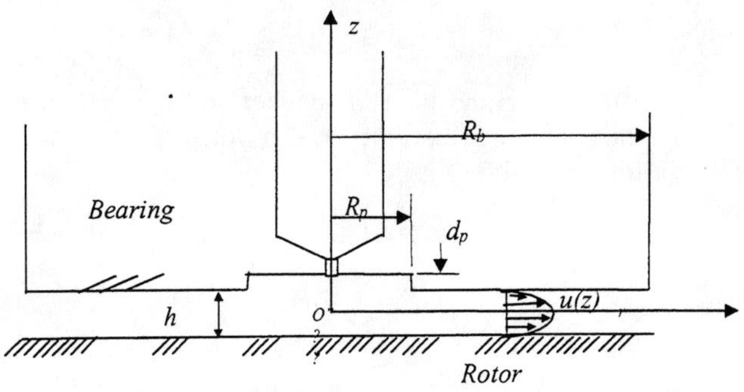

Figure 3 Orifice compensated air bearing

Since the gap in the annulus is very small and the pressure variation in the z-direction is negligible, the flow between the surfaces is laminar. Thus, the flow-pressure relationship through the annulus, which is essentially laminar flow between two parallel surfaces, can be shown to be

$$p^2 = p_a^2 + \frac{12\mu q_o RT}{\pi h^3} \ln\left(\frac{R_b}{r}\right)$$

which can be re-arranged for the flow rate as a function of pressure and the air gap follows:

$$q_o = \frac{(p_p^2 - p_a^2)\pi h^3}{12\mu RT \ln\left(\dfrac{R_b}{R_p}\right)} \tag{10}$$

The flow through an orifice has been modeled by several authors [8]. A particular form known as Fliegner's approximation [9] has been chosen for this analysis due to its convenience for analytical and computational purposes.

$$q_R = 2\left(\frac{\pi d_o^2}{4}\right) \left[\frac{\gamma}{RT} \left(\frac{2}{\gamma+1}\right)^{\frac{\gamma+1}{\gamma-1}} \left(p_s p_p - p_p^2\right) \right]^{\frac{1}{2}} \tag{11}$$

3.2 Perturbation model

Note that the flow-pressure relationships given by Equations (10) and (11) are highly non-linear. For the study of the dynamics due to a small deviation of the rotor displacement about the equilibrium as shown in Figure 2b, a perturbation model about an equilibrium operating condition (where the rotor is concentric with the stator), is derived. For this purpose, the pressure in the annulus is approximated by a linear relationship as follows:

$$p = p_p - (p_p - p_a)\frac{r - R_p}{R_b - R_p} \tag{12}$$

where the subscript 'e' denotes the variables are evaluated at the steady state. The total force acting on the rotor is obtained by integrating the pressure over the bearing surfaces.

$$\tilde{f} = 2\pi\left[\int_0^{R_p}(\tilde{p}_{pi} - \tilde{p}_{pj})rdr - \int_{R_p}^{R_b}(\tilde{p}_{pi} - \tilde{p}_{pj})\frac{r - R_p}{R_b - R_p}rdr\right] \tag{13}$$

where the first integrand is at the pocket and the second is from the bearing annulus; the ~ sign denotes variation from equilibrium values.

$$\tilde{f} = \tilde{p}_p A_{eq} \tag{14}$$

where $\tilde{p}_p = \tilde{p}_{pi} - \tilde{p}_{pj}$

and $A_{eq} = \pi R_{eq}^2 = \frac{\pi}{3}(R_b^2 + R_b R_p + R_p^2)$.

In Equation (14), the deviation of the pocket pressure p_p depends on the flow-pressure characteristics of the bearing. Since there is no contact between the surfaces and frictional effect of air is negligible, the equation of motion for the rotor along the direction of the actuating force is

$$m_r \ddot{\tilde{h}}_i = A_{eq}\tilde{p}_p . \tag{15}$$

At equilibrium, since the gap h is equal to h_e and there is no change of air stored in the bearing, the flow through the orifice is equal to that through the annulus, q_e. To the first degree of Taylor series approximation, the small deviation of the flow rate through the restrictor about the equilibrium can be written as

$$\tilde{q}_R = -a_1\tilde{p}_p + a_2\tilde{p}_s \tag{16}$$

where $\tilde{p}_s = \tilde{p}_{si} - \tilde{p}_{sj}$; $\tilde{q}_R = \tilde{q}_{ri} - \tilde{q}_{Rj}$; and

$$a_1 = \left(\frac{\partial(q_R)}{\partial p_p}\right)_{\substack{p_p = p_{pe} \\ q_R = q_e}} = \frac{q_e}{2p_s}\left(\frac{2 - \frac{p_s}{p_{pe}}}{\frac{p_s}{p_{pe}} - 1}\right)$$

$$a_2 = \left(\frac{\partial q_R}{\partial p_s}\right)_{\substack{q_R = q_e \\ p_p = p_{pe}}} = \frac{q_e}{p_s - p_{pe}}$$

The corresponding linear approximation of the flow rate through the annulus about the equilibrium condition can be derived from Equation (10b), which yield

$$\tilde{q}_o = a_3\tilde{p}_p + a_4\tilde{h}_i \tag{17}$$

$$a_3 = \left.\frac{\partial(q_o)}{\partial p_p}\right|_{\substack{q_o = q_e \\ p_p = p_{pe}}} = \frac{2q_e p_{pe}}{(p_p^2 - p_a^2)}$$

$$a_4 = \left(\frac{\partial(q_o)}{\partial h_i}\right)_{h_i = h_e} = \frac{3q_e}{h_e}$$

where $\tilde{q}_o = \tilde{q}_{oi} - \tilde{q}_{oj}$ and note that $\tilde{h}_j = -\tilde{h}_i$.

As shown in Equation (8), the difference between the flow through the restriction and the annulus is stored in the bearing. Substituting p from Equation (12) into Equation (9), we have

$$m = \frac{[h_i p_p A_{eq} + d_p\, p_p \pi R_p^2 + h_i p_a(\pi R_b^2 - A_{eq})]}{RT} \tag{18}$$

the linear approximation of which is given as follows:

$$\frac{d\tilde{m}}{dt} = a_5\dot{\tilde{p}}_p + a_6\dot{\tilde{h}} \tag{19}$$

where $a_5 = \left(\frac{\partial m}{\partial p_p}\right)_e = \frac{A_{eq}h_e + d_p \pi R_p^2}{RT}$

$$a_6 = \left(\frac{\partial m}{\partial h_i}\right)_e = \frac{A_{eq}(p_{pe} - p_a) + \pi R_b^2 p_a}{RT}.$$

Hence,

$$a_5\dot{\tilde{p}}_p + a_6\dot{\tilde{h}}_i = (-a_1\tilde{p}_p + a_2\tilde{p}_s) - (a_3\tilde{p}_p + a_4\tilde{h}_i) \tag{20}$$

To obtain a dynamic equation in term of h_i explicitly, we eliminate the pressure p_p by substituting it and its time derivative from Equation (15) into Equation (20). The resulting equation of the rotor motion is given by

$$\dddot{\tilde{h}}_i + \left(\frac{a_1 + a_3}{a_5}\right)\ddot{\tilde{h}}_i + \left(\frac{A_{eq}a_6}{a_5 m_r}\right)\dot{\tilde{h}}_i + \left(\frac{A_{eq}a_4}{a_5 m_r}\right)\tilde{h}_i = \left(\frac{A_{eq}a_2}{a_5 m_r}\right)\tilde{p}_s \tag{21}$$

Using Routh Hurwitz stability criteria, the condition for an asymptotically stable system implies $\frac{(a_1 + a_3)a_6}{a_4 a_5} > 1$,

which yields $\frac{1}{3}\left[\frac{\frac{3}{2}p_s - p_{pe}}{p_s - p_{pe}}\right] > 1$ or $p_{pe} > \frac{3}{4}p_s$. In addition, all the coefficients must be positive for stability implies that $p_{pe} < p_s$. Hence,

$$\tfrac{3}{4}p_s < p_{pe} < p_s . \tag{22}$$

4. DESIGN METHODOLOGY

Consider the case where the supply pressure is constant ($\tilde{p}_s = 0$). In other words, the air bearing system is essentially a passive regulator.

4.1 Design Tradeoff's

Note that the system is third-order and thus at least one of the characteristic roots is real which implies that Equation (21) can be written in the following form:

$$(s+\sigma)(s^2+2\xi\omega_n s+\omega_n^2)=0 \tag{23}$$

By expanding Equation (23) and equating its coefficients that the corresponding terms in Equation (21), the design parameters (σ, ξ and ω_n) can be related to the system parameters as follows:

$$\sigma+2\xi\omega_n=\frac{C_2}{\left(1+C_a\dfrac{d_p}{h_e}\right)}\frac{q_e}{h_e} \tag{23a}$$

$$\omega_n^2+2\xi\omega_n\sigma=\frac{C_1}{h_e\left(1+C_a\dfrac{d_p}{h_e}\right)} \tag{23b}$$

$$\sigma\omega_n^2=\frac{C_o}{\left(1+C_a\dfrac{d_p}{h_e}\right)}\frac{q_e}{h_e^2} \tag{23c}$$

where

$$C_2=\frac{RT}{A_{eq}}\left[\frac{2p_{pe}-p_s}{2p_{pe}(p_s-p_{pe})}+\frac{2p_{pe}}{p_{pe}^2-p_a^2}\right] \tag{24a}$$

$$C_1=\frac{A_{eq}(p_{pe}-p_a)+\pi R_b^2 p_a}{m_r} \tag{24b}$$

$$C_o=\frac{3RT}{m_r} \tag{24c}$$

$$C_a=\frac{\pi R_p^2}{A_{eq}} \tag{24d}$$

Note that A_{eq} is generally constrained by the spherical actuator geometry. Thus, Equations (23a), (23b) and (23c) represent the design trade-off's among the three design variables, q_e, h_e, and d_p for a specified dynamic response, which can be expressed as

$$q_e=\frac{1+n}{1+4n\xi^2}\frac{C_1}{C_2\omega_n} \tag{26}$$

$$h_e=\frac{1+n}{n}\frac{C_o}{C_2\omega_n^2} \tag{27}$$

$$d_p=\frac{C_o}{C_aC_2\omega_n^2}\left(\frac{C_1C_2}{C_o}\frac{1}{1+4n\xi^2}-\frac{n+1}{n}\right) \tag{28}$$

where $n=\dfrac{2\xi\omega_n}{\sigma}$. The effect of the third pole could be reduced if σ is large compared to $\xi\omega_n$ (or n>>1). However, for a practical air bearing system, the pocket depth must be real, finite and positive and thus imposes a constraint on the choice of ω_n, σ, and ξ.

4.2 Design Example

Table 1 lists the parameters determined for the spherical motor [1] where the stator poles are located following the pattern of an Icosahedron.

Table 1: Parameters

Parameters	Value
Mass m_r, (kg)	0.2
Outer radius R_b, (mm)	12.7
R_b/R_p	16
Supply pressure, p_s (KN/m^2)	420
P_{pe}/p_s	0.92

Figures 4(a) and (b) show the pocket depth plotted as a function of natural frequency and damping ratio respectively. As h_e, and d_p are inversely proportional to ω_n^2, we chose ω_n=*240 Hz* (or 1,508rad/sec), $\sigma=2\xi\omega_n$ (or *n=1*) and ξ=*0.5* to provide a reasonable physical size for the pocket depth and the air gap and acceptable dynamic response. The corresponding calculated parameters are q_e = 1.281e-5 Kg/s, h_e = 0.069mm and d_p = 0.019mm.

The air bearing and electromagnetic system are simulated using MATLAB for two cases. The first case is one degree-of-freedom with only a pair of bearings. Figures 5(a) and (b) show the rotor displacement in the direction of actuating force for an initial displacement of 5.8μm respectively. Although the electromagnetic system is open-loop unstable, the air bearing system is shown to dominate the magnetic system and stabilizes the combined system.

The second case is a three degree-of-freedom with five pairs of bearings located on ten of the stator poles without the top and bottom vertices of the Icosahedron. The 3-D simulation discussed below computes the displacement, velocity and acceleration vectors for multiple bearings (5 pairs) and decomposes them into the orthogonal directions. Since each air bearing system is linear, we compute the resultant forces on the rotor dynamics in 3-D by using vectors to sum the individual contributions.

5. CONCLUSIONS

We have presented the method for design of a practical air bearing system for a VR spherical motor. Specifically, this paper addresses the following fundamental issues of the bearing system design:

(1) The method of generating the necessary rotor support forces with externally pressurized air and the strategic arrangement of point bearings is discussed.

(2) The forward and inverse kinematics between the rotor displacement and the individual air gaps at positions round the stator are developed in closed-forms, which are essential for design, dynamic simulation and control purposes.

(3) Along with the pressure-flow relationship as a function of the rotor position, the paper presents a detailed dynamic model of the air bearing. Trade-off's between the design parameters and the dynamic performance of the air bearing regulator system have been discussed with using a design example.

The studies have led to the design of a potentially useful air-bearing system capable of eliminating frictions in ball-joint-like actuators. The dynamic performance of the air bearing system has been evaluated analytically by simulation.

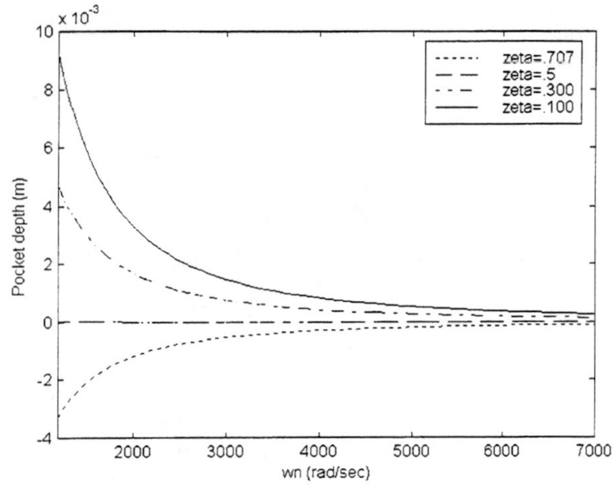

(a) Pocket depth Vs natural frequency (n = 1)

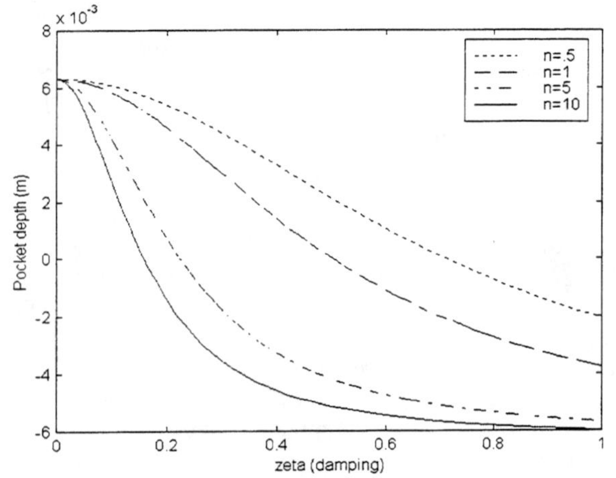

(b) Pocket depth Vs damping ratio (ω_n = 240 Hz)

Figure 4 Effect of ω_n and ξ on the choice of pocket depth

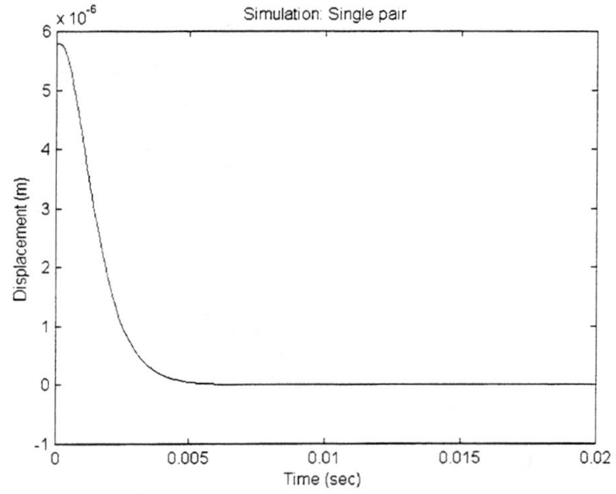

(a) Rotor displcement (1-DOF)

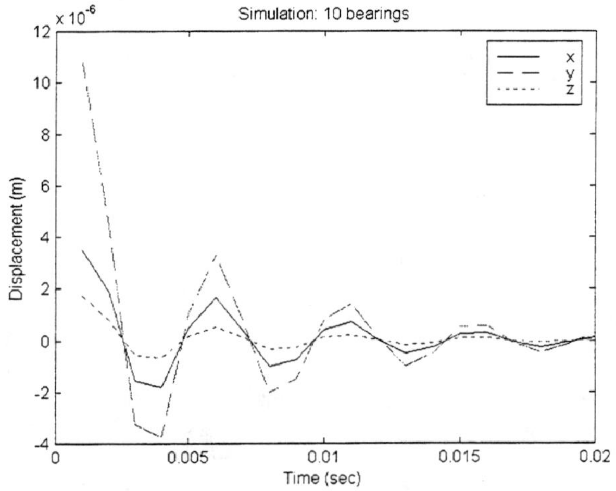

(b) Rotor displcement (3-DOF)

Figure 5 Simultion results of rotor displacement

REFERENCES

[1] Lee, K.-M., Roth, R., and Zhou, Z., "Dynamic Modeling And Control Of A Ball-Joint-Like Variable Reluctance Spherical Motor," ASME Journal Of Dynamics Systems, Measurements, And Control, Vol. 118, No. 1, March 1996, pp.29-40.

[2] Fearing, R. S., "Micro Structures and Micro Actuators for Implementing Sub-millimeter Robots," excerpts from Precision, Sensors, Actuators and System, Kluwer Academic Publishers, Netherlands, 1990, pp. 39-72.

[3] Hamcock, B. J., "Fundamentals of Fluid Film Lubrication," McGraw-Hill Inc., NY, 1994.

[4] Licht, L., Elrod, H., "A Study of the Stability of Externally Pressurized Gas Bearings", ASME Journal of Applied Mechanics, Vol. 82, 1960, pp. 250-258.

[5] Mori, H., Mori, A. Kaneko, R., Yoshida, K., "Stability Element for External Pressurized Gas Bearing (2nd Report, Stabilizing Element Inserted into the Gas Supply Line)", Transactions of JSME, Vol. 32, No. 244, 1966, pp. 1883-2064.

[6] Ohsumi, T., Ikeuchi, K., Mori, H., Haruyama, H. J., Matsumoto, Y., "Characteristics of a Hydrostatic Bearing with a Controlled Compensating Element", Wear, Vol. 105, April, 1991, pp. 177-194.

[7] Sato, Y., Murata, K., Harada, M., "Dynamic Characteristics of Hydrostatic Thrust Air Bearing with Actively Controlled Restrictors", Transactions of ASME (tribology Conf.), Vol. 110, No. 1, Jan., 1988, pp. 156-161.

[8] Licht, L., Fuller, D. D., Sternlicht, B., "Self-Excited Vibrations of an Air-Lubricated Thrust Bearing", Transactions of ASME, Vol. 80, 1958, pp.411-414.

[9] Slocum A. H., "Precision Machine Design", Prentice-Hall Englewood Cliffs, NJ, 1992.

NONLINEAR FEEDFORWARD-PLUS-PID CONTROL FOR ELECTROHYDRAULIC STEERING SYSTEMS

Hongchu Qiu, Graduate Assistant
Qin Zhang, Assistant Professor
John F. Reid, Professor
Duqiang Wu, Visiting Scholar

Department of Agricultural Engineering
University of Illinois at Urbana-Champaign
Urbana, IL 61801
qiu@age.uiuc.edu
Tel.: 217-333-8595 Fax.: 217-244-0323

ABSTRACT

This paper presents the development of a nonlinear feedforward-plus-Proportional-Integral-Derivative (FPID) controller for electrohydraulic (E/H) steering on wheel-type tractors. An E/H steering system is a typical nonlinear system with deadband, saturation, asymmetric flow gain, time delay, and other nonlinear behaviors. Conventional PID controllers are incapable of achieving accurate steering control effectively on such nonlinear systems. In this research, an FPID controller was developed for effective and accurate steering control. The feedforward loop in this controller was designed to compensate for the deadband of the E/H system. The PID loop was designed to compensate the tracking error in steering control. A coordinated nonlinear gain function was designed to change the PID loop gain based on the level of the tracking error. This FPID controller has significantly improved the steering accuracy comparing with that from a PID controller. Test results showed that the maximum tracking error in steering angle was less than 0.5° corresponding to a sinusoid steering command of ±5° at the command frequency of 0.1 Hz. The maximum overshoot was less than 12% and the rise time was less than 0.25 s corresponding to a steering command of 5° step input. This FPID controller achieved effective and accurate steering control on agricultural tractor E/H steering systems.

KEYWORDS: Feedforward-plus-PID control, nonlinear system, electrohydraulic steering, steering control.

INTRODUCTION

A steering controller for a vehicle adjusts the steering angle based on a steering input signal. Agricultural tractors often operate on the unprepared, changing and unpredictable terrain. Therefore, the steering controller for agricultural tractors should be capable of implementing appropriate steering in response to the variation in tire cornering stiffness, travel speed, and vehicle-terrain interaction.

Modern agricultural tractors commonly use hydraulic steering system. The high power-to-weight ratio and the design flexibility of a hydraulic steering system are capable of providing effective steering to heavy-duty agricultural tractors. Recent development in electrohydraulic (E/H) technology also provides the base for adopting advanced control technologies in tractor steering control. Van Der Lely (1985) developed an automatic guidance system using automatic E/H actuated steering. The performance of E/H steering controllers can improve the steering accuracy.

The design of a steering controller depends on many factors including E/H steering components and vehicle dynamics. With regarding to the affects of hydraulic components on an E/H system controller, Erbach *et al.* (1991) stated that neither negligible nor constant friction could produce significant and unpredictable sideslip. Taking the vehicle dynamics into the consideration of steering controller design,

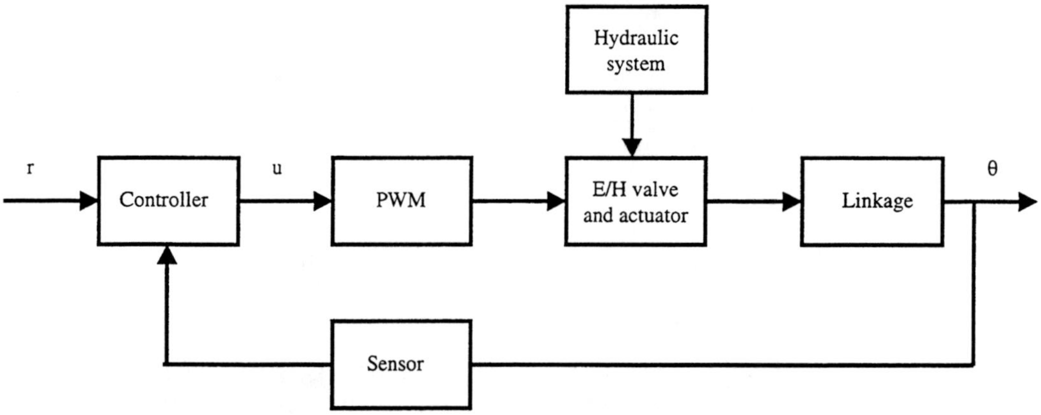

Figure 1. Block diagram of an electrohydraulic steering system (closed-loop).

Julian (1971) developed a transfer function for the turning (yaw) rate of a tractor. O'Connor *et al.* (1996) developed a hybrid controller for agricultural vehicle. This hybrid controller consisted of a linear controller for small angle steering, and an "on-off" controller for large angle steering. Two look-up tables, converting either the steering angle into a heading rate or the heading rate to steering angle, were used in this controller.

Benson (1998) developed a PID controller for small angle steering of an agricultural tractor. However, a PID controller was shown to be incapable of achieving high-performance control on nonlinear systems (Wu *et al*, 1998). The fixed-gain PID controller has a conflict of low overshoots with slow responses or quick responses with high overshoots. E/H steering systems are typical nonlinear with large deadband, saturation, hysterics, and other time-varying characteristics. These factors result in difficulties for controlling an E/H steering system satisfactorily (Stombaugh, 1997). Among these factors, deadband has the largest effect on controller's behavior, and should be compensated efficiently for better steering performance (Recker, 1991; Wu *et al*, 1998). Seraji (1997) developed a method of using a nonlinear gain function to change the gains in a PID controller and achieved better control performance.

The objective of this research was to develop a feedforward-plus-PID (FPID) controller for implementing accurate steering control on a nonlinear E/H steering system. This paper presents the characteristics of this E/H steering system, the design of the FPID controller, and the implementation of the developed steering controller on a wheel-type agricultural tractor.

SYSTEM DESCRIPTION

A Case-IH Magnum[1] 8920 MFD agricultural tractor was modified as the test platform for automated steering control by adding an electrohydraulic steering system in parallel with the original tractor steering system. The modified E/H steering system consisted of an electrohydraulic proportional direction control valve, a pulse width modulated (PWM) driver, two hydraulic steering actuators, and a steering linkage (Fig. 1). The steering command signal, r, was the desired steering wheel angle in degrees. The control input signal, u, was the voltage signal to the PWM driver in volts. The system output was the steering rate of tractor front wheel in degrees per second.

The E/H steering control valve was a four-way proportional direction control valve. This valve controlled both the flow direction and flow rate so that the steering actuators could steer the front wheels to make desired turns. Since the E/H steering system was added in parallel to the existing handpump steering system, this modified platform could be steered either manually or automatically.

A High Country Tek's (Nevada City, CA) dual-coil PWM driver card was used to drive the solenoids to implement valve control. This PWM driver was driven by 12V DC power. A computer-based controller sent steering control signal to the PWM driver, which converted the signal into duty cycles and then drove a solenoid actuator to push the valve spool so that both the flow direction and the flow rate could be controlled accurately.

To provide steering angle feedback to the PID controller, a linear potentiometer was used as the wheel angle sensor. This potentiometer was mounted on the right-side hydraulic steering actuator to measure its displacement. The relationship between the wheel angle and the actuator displacement was determined based on a kinematic model of the tractor steering linkage. This kinematic model could provide the gain of the linkage.

$$\varphi_a = 0.0005 \cdot \Delta L^2 - 0.3335 \cdot \Delta L \quad (1)$$

$$\Delta L = -43.365 \cdot V + 100.605 \quad (2)$$

[1] Case-IH and Magnum are trademarks of Case Corporation, used with permission. Mention of a trade name, proprietary product, or specific equipment does not constitute a guarantee or warranty by the University of Illinois, and does not imply the approval of the named product to the exclusion of other products that may be suitable.

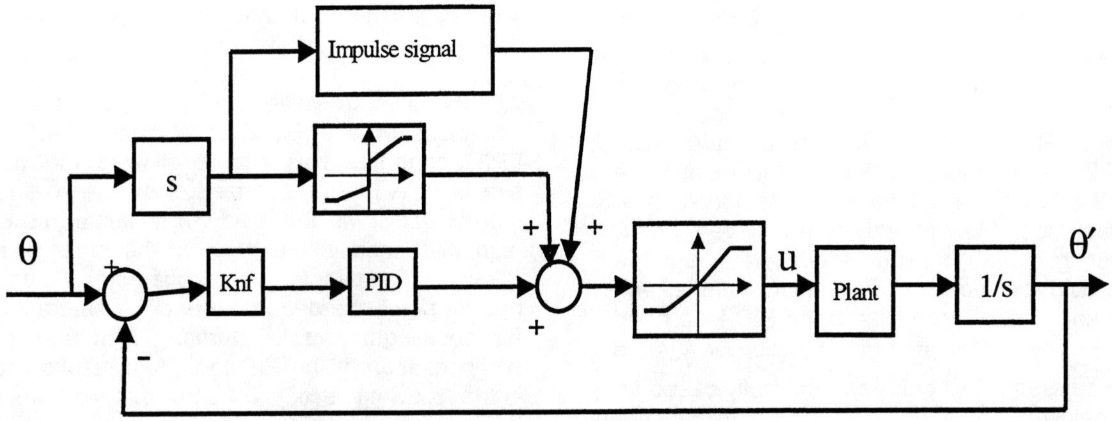

Figure 2. Block diagram of FPID control algorithm.

where φ_a is Ackerman steering angle of the vehicle, ΔL is the displacement of the hydraulic actuator, and V is the output voltage of the potentiometer. Equation (1) describes the relationship between Ackerman steering angle and the hydraulic actuator displacement (Qiu, *et al.* 1999). Equation (2) presents the calibration equation for the potentiometer with regarding to the sensor output in volts.

The modeling and identification of this E/H steering system has been reported in a separate paper (Qiu, *et al.* 1999a). The open loop E/H steering system could be presented and linearized as a forth order system for position control with assumptions of incompressible fluid, no leakage, and a critical-lapped valve.

$$\frac{\Theta(s)}{U(s)} = \frac{K}{s(Ts+1)(s^2+2\varsigma\omega_n s+\omega_n^2)}e^{-\tau s} \quad (3)$$

where K, T, ω_n and ς are, respectively, system constants and τ the time constant.

CONTROLLER DESIGN

A FPID controller was designed to compensate the large deadband using an inverse valve transform (IVT) based feedforward loop, and to compensate tracking error using a PID loop. Cobo (1996) successfully developed a feedforward-plus-PI controller for achieving smoother E/H implementing control on a wheel type loader. In that system, the feedforward IVT provided a large initial gain to overcome the deadband. In this research, a Feedforward-plus-PID (FPID) controller was developed to improve control performance of a nonlinear E/H steering system on an agricultural tractor. The goals for this FPID controller were to eliminate the delay in steering response caused by the large deadband of the E/H system and to improve steering accuracy. Figure 2 shows the block diagram of this FPID controller.

This FPID controller has a feedforward loop and a PID loop. The feedforward loop consists of an impulse signal to overcome the static friction and an IVT to determine the baseline steering signal based on the steering rate. A differential operator was used to convert the reference signal of steering angle into steering rate due to the fact that an E/H system is naturally a velocity control system. The valve transform is the relationship between the control signal to the E/H steering valve and the steering rate. It is commonly obtained from open-loop steering control tests. The IVT used in the feedforward loop was derived from the relationship between the control input signal u and the steering rate $\dot{\theta}$ using Eqs. (4) and (5).

$$\dot{\theta} = f(u) \quad (4)$$

$$u = f^{-1}(\dot{\theta}) \quad (5)$$

An impulse signal was inserted to add a large, short duration, initial input value to the PWM driver to overcome the static friction of the valve spool. The impulse signal was tuned and tested to determine the appropriate level and duration. A 0.05 s impulse signal was sufficient for this particular test platform.

The PID loop consisted of a coordinated gain schedule block, a PID controller, and a steering angle sensor. The feedback steering angle signal was original the linear displacement of the steering actuator, and converted into steering angle based on the linkage gain.

The PID gains, K_p, K_i, and K_d in the FPID controller were tuned when the feedforward loop was disabled. The overall PID gain was determined by the equation below.

$$K(s) = k_p + \frac{k_i}{s} + k_d s \quad (6)$$

The coordinated gain factor, K_{nf}, was a gain adjusting function for the PID loop. Under a constant steering signal, the steering rate was zero, and the feedforward loop was inactive. To ensure the effectiveness of FPID controller, a relatively large gain was required for the PID loop. Since the PID gains were tuned while the feedforward loop was disabled, a gain factor of 1 was used. When the system received a non-constant steering signal, both the feedforward and the PID loops would be active. To ensure stable and accurate steering control, the gains of the PID loop need to be reduced. The coordinated gain function, K_{nf}, was introduced to adjust the PID gains automatically based on the equation below.

$$K_{nf} = \begin{cases} 1 & Err \le E_0 \\ K & \text{otherwise} \end{cases} \tag{7}$$

Err is the error between the desired and the actual steering angles, E_0 is a threshold of the steering error, and *K* is a reduced gain factor, $0 \le K \le 1$. Noticed that K_{nf} could have more piecewise segments to provide a finer tuning relation, which could result in better tracking performance.

IMPLEMENTATION AND RESULTS

TUNING OF THE FEEDFORWARD LOOP

The tuning of the feedforward loop determined the inverse valve transform (IVT) with respect to the steering rate. The IVT was derived from the valve transform, which was obtained experimentally. The valve transform presents the relationship between the input steering signal in volts and the actual steering rate in degrees per second. The valve transform represents the characteristics of the deadband, saturation, and nonlinear flow gains of the E/H steering system. Figure 3 shows the valve transform obtained on the tractor test platform. The system deadband was from –0.9 V to +1.0 V, and the saturation ranges were either lower than –4.0 V or higher than 6.0V.

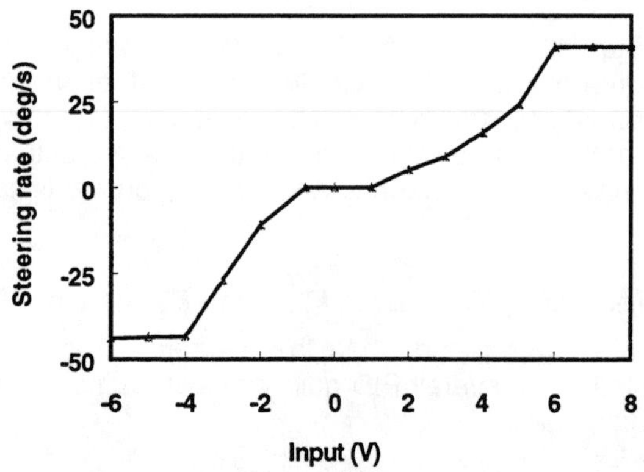

Figure 3. Valve transform of the E/H steering valve obtained from tractor platform experimentally.

The inverse valve transform (IVT) was derived using Eqs. (4) and (5). One valuable advantage of the FPID controller was that it allowed the use of a simplified IVT with linearized or piecewise linear flow control gains. In this study, a linearized flow control gain was used in the IVT. In the other words, the derived IVT presented only 5 characteristic points: the neutral point, the deadband and the saturation points for both right and left turns. With tracking error compensation of the PID loop, test results presented in the following section will show that such a simplified IVT was capable of proving satisfactory deadband and saturation compensation.

TUNING OF THE COORDINATE GAIN FUNCTION

The tuning of the coordinate gain function, K_{nf}, determined an appropriate schedule of PID gains for a smooth steering. This function coordinated the FPID controller gains in such a way that the PID gains should be decreased to a lower level when the feedforward loop was functional, and adjusted to normal values when the feedforward loop was disabled.

The K_{nf} was tuned on the tractor using sinusoidal and step input steering signals. The tuning criterion was to achieve a smooth steering operation under all operating conditions. Since the feedforward loop was capable of providing an appropriate steering control input that would result in a limited steering angle error, a steering error threshold was used to determine whether the feedforward loop was in functional or not.

With an allowable steering angle error of 0.3 degrees, the optimal K_{nf} for this tractor platform was 0.3 when the feedforward loop was functional (Fig. 4). This coordinate gain function would thus allow the PID loop dominate the controller if the steering error was greater than the threshold error value E (0.3 degree), and the feedforward loop would dominate if the tracking error was small.

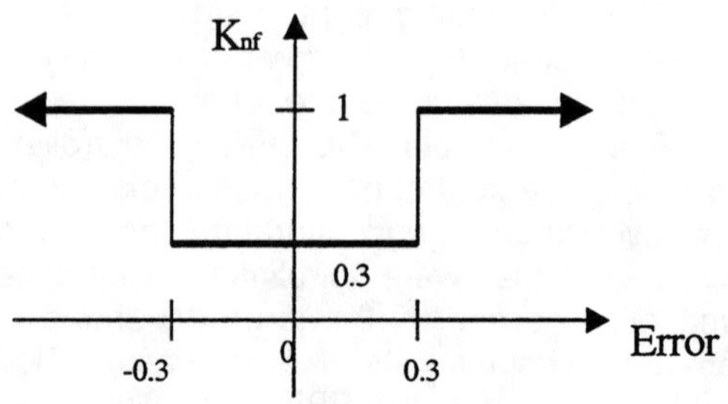

Figure 4. Combined coordinate gain K_{nf}.

TESTS RESULTS

The FPID controller was evaluated on the tractor platform with different patterns of steering commands, including sinusoidal, step input, combined, and path tracking tests.

Sinusoidal tests were performed to evaluate the capability of the FPID for compensating the nonlinearity, especially deadband under changing steering commands. Test results indicated that the FPID controller could compensate the nonlinearity of the E/H steering system effectively. The FPID controller provided accurate steering angle tracking based on a series of sinusoidal steering tests on prepared road surface. Figure 5 shows the steering angle tracking results with a sinusoidal steering command of ±5° under a sinusoidal frequency of 0.1 Hz. The maximum error in steering angle tracking was less than 0.4 degrees. Further tests on sinusoidal steering commands with different frequencies showed that the tracking error would increase with the increase of the command frequency. This E/H steering system would not able to response to a sinusoidal steering command of 4 Hz or higher. Theoretically, an E/H steering system will be not able to response to a control signal with a higher frequency than the system's natural frequency.

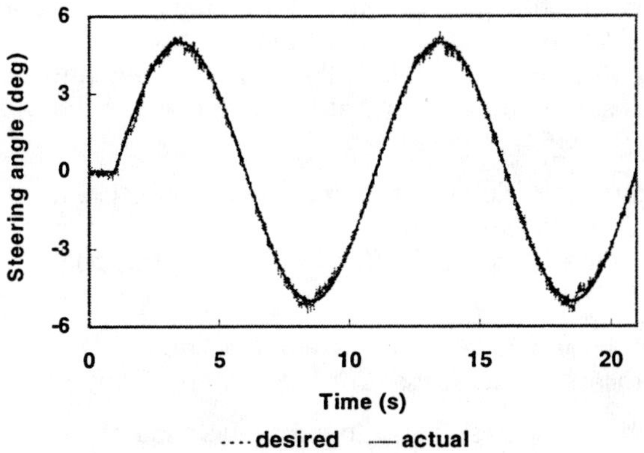

Figure 5. Steering angle tracking accuracy to a sinusoid steering command of ±5°and 0.1 Hz from the FPID controller.

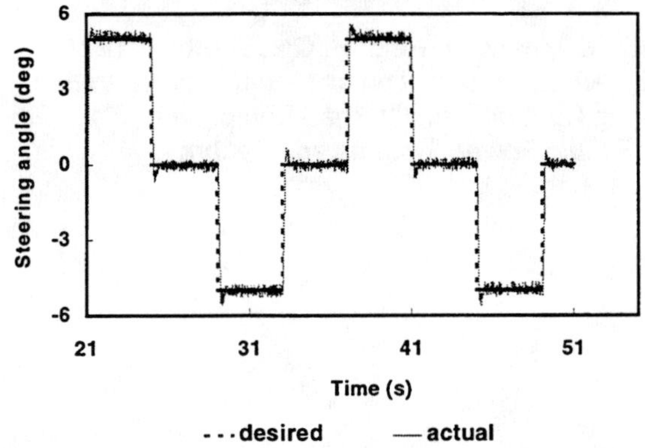

Figure 6. Steering angle tracking accuracy to a step steering command of ±5° from the FPID controller.

Similarly, step input tests were performed to evaluate the dynamic response of the FPID controller. Figure 6 shows that the actual steering angle matched with the command angle very well under a step input 5° steering command from this controller. The maximum overshoot observed from a series of tests with different step command angles was less than 12% on prepared road surface. Both the rise time and the settling time were less than 0.25 s. The FPID controller was capable of providing smooth and accurate steering control under step steering commands.

Figure 7 shows steering angle tracking error recorded from an evaluation test with a combined steering command of sinusoidal and step input of ±5°. Since the feedforward loop was mostly disabled in the period corresponding to the step steering command, the results showed in Fig. 7 indicated the transition performance of the steering controller while switching from FPID mode to PID mode. This controller could be switched from one mode to another smoothly.

Path tracking tests were conducted to evaluate the steering accuracy in actual path tracking operation. This test was implemented on the Case tractor platform guided by a machine vision based automated guidance system on a prepared road surface. Figure 8 shows the tracking error obtained from this automatically guided path-tracking test at a traveling speed of 1.8 m/s. The command steering angle (dashed line) was given by the machine vision automated guidance system. The solid line was the actual steering angle (Ackerman angle). The tractor was placed some distance away from the desired path at the start point of the path. There were also two curved segments on the path. One was at the beginning and the other was near the end of the path. The results indicated that the FPID controller was able to steer the tractor quickly into the desired path, and then follow the desired path accurately.

CONCLUSIONS

A FPID controller provided satisfactory steering control for agricultural tractors with an E/H steering system. The feedforward loop of the controller compensated for the deadband and saturation, inherent of E/H steering systems. The PID loop improved the steering accuracy. Both the feedforward loop and the PID loop were developed and tuned separately. The coordinated gain function was important role for adjusting the contribution of the feedforward loop and the PID loop to ensure the smooth operation of the system.

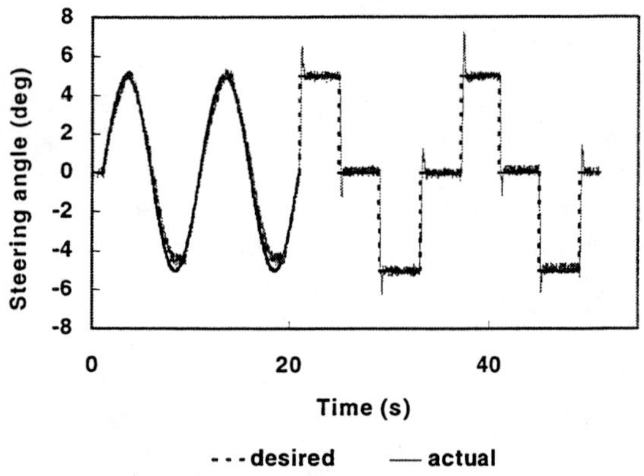

Figure 7. Tracking a sinusoid and step input reference signal with a tuned PID controller.

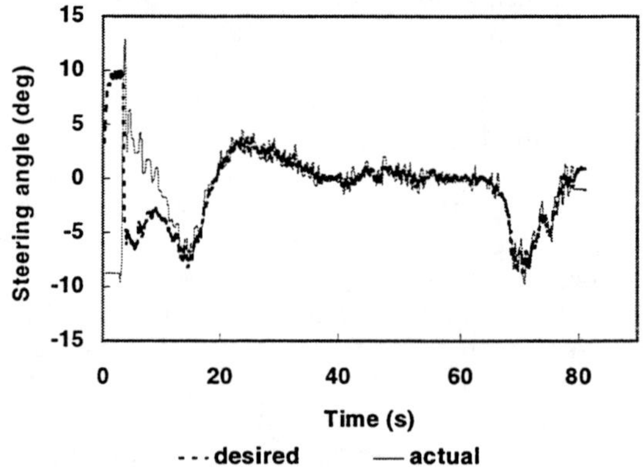

Figure 8. Path tracking with the FPID controller guided by the machine vision system on an agricultural tractor.

ACKNOWLEDGEMENTS

This research was supported by the University of Illinois at Urbana-Champaign Research Board, Illinois Council on Food and Agricultural Research (98-069 AE), and Case Corporation. All of the mentioned support are gratefully acknowledged.

REFERENCES

Benson, E., 1998, "An Evaluation of the Suitability of A Geomagnetic Direction Sensor (GDS) for Vehicle Guidance in Precision Agriculture Applications," *MS Thesis*, University of Illinois at Urbana-Champaign Library, Urbana, IL.

Cobo, M.A., 1996, "Velocity Servo Control System Design for Wheel Type Loader Earthmoving Equipment," *MS thesis*, University of Illinois at Chicago Library, Chicago, IL.

Erbach T.C., C.H. Choi, and K. Noh. 1991. "Automated Guidance for Agricultural Tractors. In: *Proceeding automated agriculture for the 21*st *century*," ASAE Publication No. 11-91: 182-191.

Homaifar, A. M. Bikdash, and V. Gopalan, 1997, "Design Using Genetic Algorithms of Hierarchical Hybrid Fuzzy-PID controllers of two-link Robotic arms," *Journal of Robotic systems* 14(b), pp449-463.

Julian, A. P. 1971, "Design and Performance of a Steering Control System for Agricultural Tractors," *Journal of Agricultural Engineering Research* 16(3):324-336.

O'Connor, M., T. Bell, G. Elkaim, and B. Parkinson, 1995, "Automatic Steering of Farm Vehicles Using GPS," Papers presented at *the 3*rd *International Conference on Precision agriculture.* Minneapolis, MN June 23-26.

Qiu, H. Q. Zhang, J.F. Reid, D. Wu, 1999 (a), "Modeling and simulation of an electrohydraulic steering system", ASAE/CSAE Annual International Meeting, Toronto, Ontario, Canada, paper no. 993076

Qiu, H. Q. Zhang, J.F. Reid, D. Wu, 1999 (b), "System identification of an electrohydraulic steering system", 1999 SAE International Off-Highway & Powerplant congress & Exposition, Indianapolis, Indiana, paper no. 1999-01-2854.

Recker, D., P. Kokotovic, and J. Winkelman, 1991, "Adaptive Nonlinear Control of Systems Containing a Dead-zone," *Proc. IEEE Conf. Decis. Control*, pp2111-2115.

Seraji, H., 1997, "A New Class of Nonlinear PID Controllers," *IFAC Robot Control, Nantes*, France.

Stombaugh, T., 1997, "Automatic Guidance of Agricultural Vehicles at Higher Speeds," *Ph.D. thesis*, University of Illinois at Urbana-Champaign Library, Urbana, IL.

Van Der Lely, C. 1985, "Tractor Having guidance System," *U. S. Patent No. 4515221.*

Wu, D., Q. Zhang, J. Reid, H. Qiu, and E. Benson, 1998, "Model Recognition and simulation of an E/H Steering Controller on Off-road Equipment," *Proc. '98 ASME Fluid Power Systems and Technology*, California.

FPST-Vol. 6, Fluid Power Systems and Technology — 1999
ASME 1999

A Unified Model of a Proportional Valve

Bora Eryilmaz* and Bruce H. Wilson†
Department of Mechanical, Industrial, and Manufacturing Engineering
Northeastern University
Boston, Massachusetts 02115 USA

ABSTRACT

Developments in nonlinear control theory have made it possible to design controllers for systems having non-smooth nonlinearities in their dynamics. Hydraulic systems that use inexpensive proportional valves are examples of such systems, where non-smooth nonlinearities arise due to valve geometry and spool imperfections. Without a proper valve model, however, nonlinear analysis and control of these hydraulic systems is not possible. We have developed nonlinear equations for a generic proportional valve model and have used them to obtain simplified flow rate expressions under generally accepted assumptions. These equations related a set of geometric spool properties and physical model variables to the flow rate through the valve ports. The development focuses on obtaining a *single set* of flow rate equations for the cases of critical center, overlapped, and underlapped proportional valves. The proportional valve model equations are useful for system simulation and for nonlinear controller design.

NOMENCLATURE

A_1	piston end cylinder area,	m^2
A_2	rod end cylinder area,	m^2
C_d	valve discharge coefficient,	–
F	piston force,	N
F_f	friction force,	N
K_q	critical center spool flow gain,	m^2/s
L	valve port opening,	m
P_i	Port i pressure,	N/m^2
P_R	return (tank) pressure,	N/m^2
P_S	supply pressure,	N/m^2
Q_i	Port i total flow rate,	m^3/s
Q_{1R}	Port 1 return side flow rate,	m^3/s
Q_{1S}	Port 1 supply side flow rate,	m^3/s
Q_{2R}	Port 2 return side flow rate,	m^3/s
Q_{2S}	Port 2 supply side flow rate,	m^3/s
Q_{ip}	leakage flow rate,	m^3/s
R_{ip}	leakage coefficient,	m^4s/kg
S	piston stroke,	m
V_{10}	the initial volume (chamber 1 side),	m^3
V_{20}	the initial volume (chamber 2 side),	m^3
x_p	hydraulic piston position,	m
x_v	valve spool position,	m
w	valve area gradient,	m
β	fluid bulk modulus,	N/m^2
ε_i	spool lapping parameters,	m
ρ	fluid mass density,	kg/m^3

*Email: eryilmaz@coe.neu.edu.
†Email: wilson@neu.edu. Corresponding author.

INTRODUCTION

Hydraulic systems are often employed in high performance applications that require fast response and high power. These applications include high bandwidth control of position and force [3, 9], active vibration isolation in vehicles [1], and control of multi-axis robotic manipulators [12]. The hydraulic actuator, usually a cylinder, provides the motion of the load attached to the hydraulic system. A control valve meters the fluid into and out of the actuator as a spool traverses within the valve body. The control valve is either a servovalve or a proportional valve. The less expensive proportional valve consists of a solenoid that is directly attached to the spool, and the spool displacement is, in principle, proportional to the input current. The performance of hydraulic systems strongly depends on the control valve and spool geometry and their manufacturing tolerances. The manufacturing precision distinguishes proportional valves from servovalves, in terms of both performance and cost.

In hydraulic control applications, proportional valves offer various advantages over servovalves. Proportional valves are much less expensive. They are more suitable for industrial environments because they are less prone to malfunction due to fluid contamination. In addition, since proportional valves do not contain sensitive, precision components, they are easier to handle and service. However, these advantages are somewhat offset by nonlinear response characteristics.

Since proportional valves have less precise manufacturing tolerances, they may suffer from performance degradation. The larger tolerances on spool geometry result in response nonlinearities, especially in the vicinity of neutral spool position. Proportional valves lack the smooth flow properties of "critical center" valves, a condition closely approximated by servovalves at the expense of high machining cost. Small changes in spool geometry (in terms of lapping) may have large effects on the hydraulic system dynamics. For example, a "closed center" (overlapped) spool may result in steady state error because of its deadzone characteristics in the flow gain [6]. An "open center" (underlapped) spool, on the other hand, may raise stability concerns due to the increased flow gain near neutral spool position. Hence, a critical center valve model is not sufficient for designing hydraulic control systems using proportional valves. A more comprehensive model of proportional valves is necessary in designing and analyzing hydraulic systems.

Various researchers have developed mathematical models for proportional control valves [6, 13]. However, these works usually assume a critical center valve spool whose motion is restricted to the vicinity of the neutral position for subsequent linearized analysis. Modeling and analysis of underlapped valves are briefly discussed under the restrictive assumption that the spool moves within the underlap region [6]. The overlapped valve models are, in general, ignored from the discussion. More general valve models have also been developed, however, they usually apply to expensive critical center servovalves [14, 15] or various configurations of flow control valves [5]. A generic proportional valve model applicable under all operating regimes and for all types of spool lapping is lacking.

Without the proper model, nonlinear analysis of hydraulic system performance is not possible. As a number of nonlinear control strategies have been developed in the last decade [10, 11], the availability of a general nonlinear model of a proportional valve would increase the likelihood that the former could be applied to the latter. In this paper, we develop a nonlinear mathematical model for a generic proportional control valve. This model relates a set of geometric spool properties and physical variables to the flow rate through the valve ports. The development focuses on obtaining a *single set* of flow rate equations for the cases of critical center, overlapped, and underlapped proportional valves. The flow rate is expressed as a continuous but nonlinear function of spool lapping parameters, as well as other conventional parameters. We derive simplified flow rate equations under certain generally accepted assumptions (such as equal cylinder chamber volumes and incompressible fluid), while keeping nonlinearities due to spool geometry. The variation of the flow gain and uncertainty bounds of the flow rates as a function of valve and lapping parameters are also analyzed.

This paper is organized as follows. We first develop the orifice flow equation for a flexible orifice model that describes various nonlinear effects. The orifice equation forms the basis of the hydraulic proportional valve model. In a typical hydraulic control application, a hydraulic cylinder serves as the actuator. In the next section, we present the mathematical model for such actuators. The proportional valve equations are then simplified under widely accepted assumptions, and flow properties of these valves as a function of the lapping parameters are analyzed. The paper closes with a brief discussion of conclusions and future research directions.

DEVELOPMENT OF PROPORTIONAL VALVE MODEL

Many fluid power control applications use a variable orifice for controlling the fluid flow. A sliding element, or spool, has various "lands" that modulate flow. The spool moves in a bore (sleeve) containing ports. The movement of the spool causes each land to vary the area of the corresponding ports exposed to the flow, thereby changing the flow-versus-pressure relationship of the orifices. An underlapped spool has an orifice port opening that is larger than the land; whereas an overlapped spool has an orifice port opening that is smaller than the land.

The variable orifice in Fig. 1 is the basic element for controlling hydraulic systems. It has three flow rates associated with the orifice ports (Q_{1S}, Q_{1R}, Q_1). The sup-

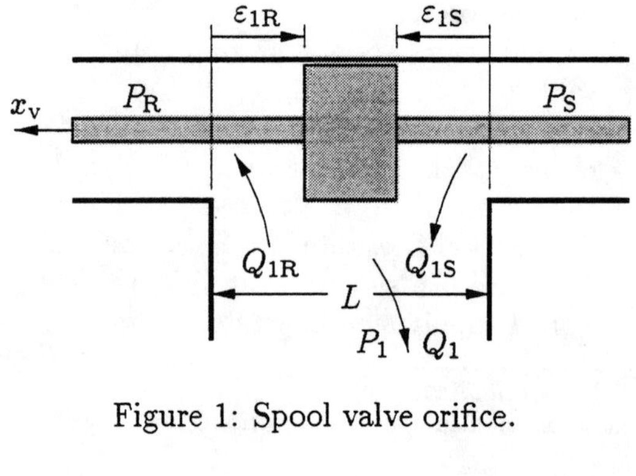

Figure 1: Spool valve orifice.

ply and return ports connect to the pump and tank lines respectively. The orifice output port provides the regulated flow and usually connects to an actuator such as a hydraulic cylinder. The parameters ε_{1S} and ε_{1R} represent the spool underlap or overlap (at neutral spool position) on the supply and return ports of the orifice, and they can be set independently. With respect to the neutral spool position, *positive* values of these parameters correspond to spool *underlap* (shown in Fig. 1). Similarly, *negative* parameter values define an *overlapped* spool.

Various simplifying assumptions and valve models are used in applications. Most hydraulic control systems employ critical center valves, where the spool lands perfectly match the valve ports so that $\varepsilon_{1S} = \varepsilon_{1R} = 0$. Typically, the supply pressure, P_S, is larger than the port pressures, (e.g. P_1), and the return (tank) pressure, P_R, is small. This assumption avoids the flow reversal so that the flow is either directed from the supply line to the output port, or from the output port to the return line. To avoid flow saturation and to use linearized analysis, the spool travel is in general restricted so that $|x_v| \ll L$. Under these conditions, the orifice equations simplify to

$$Q = \begin{cases} C_d w\sqrt{2/\rho}\sqrt{P_S - P_1}\, x_v, & x_v \geq 0, \\ -C_d w\sqrt{2/\rho}\sqrt{P_1 - P_R}\, x_v, & x_v < 0, \end{cases}$$

so that only the supply or the return orifice is open at a given time [6].

Although the above assumptions simplify the orifice model, they do not always hold in high performance applications. In hydraulic control systems rapid changes of fluid flow properties are associated with rough disturbance inputs and the motion of large dynamic loads. Under these conditions, various flow anomalies might occur. These include back flow into the supply line ($P_1 > P_S$) or from the tank ($P_1 < P_R$). Further, the spool may not be critically centered, either intentionally or due to manufacturing tolerances. Hence, both the supply and return orifices or none of them may be open at a given time. These considerations necessitate a more complete orifice model that will represent a wide variety of realistic operating conditions. The generalized equation, which includes various nonlinearities, for the supply side flow rate can be expressed as

$$Q_{1S} = \begin{cases} 0 \quad \text{for } x_v \leq -\varepsilon_{1S}, \\ \alpha \,\text{sgn}\,(P_S - P_1)\,|P_S - P_1|^{1/2}\,(x_v + \varepsilon_{1S}) \\ \qquad\qquad \text{for } -\varepsilon_{1S} < x_v < L - \varepsilon_{1S}, \\ \alpha \,\text{sgn}\,(P_S - P_1)\,|P_S - P_1|^{1/2}\,(L) \\ \qquad\qquad \text{for } L - \varepsilon_{1S} \leq x_v, \end{cases} \tag{1a}$$

where $\alpha = C_d w\sqrt{2/\rho}$. Similarly, the return side flow rate is given by

$$Q_{1R} = \begin{cases} 0 \quad \text{for } -x_v \leq -\varepsilon_{1R}, \\ \alpha \,\text{sgn}\,(P_1 - P_R)\,|P_1 - P_R|^{1/2}\,(-x_v + \varepsilon_{1R}) \\ \qquad\qquad \text{for } -\varepsilon_{1R} < -x_v < L - \varepsilon_{1R}, \\ \alpha \,\text{sgn}\,(P_1 - P_R)\,|P_1 - P_R|^{1/2}\,(L) \\ \qquad\qquad \text{for } L - \varepsilon_{1R} \leq -x_v. \end{cases} \tag{1b}$$

Note that while the ranges for the values of the parameter x_v remain similar for both equations in (1), the sign of x_v is reversed. This is because the orifice opening on one side of the output port becomes larger while the opening on the other side becomes smaller as the parameter x_v is varied. Further, the flow equations (1) assume turbulent flow through the orifices, no internal leakage, and no cavitation. The total flow rate, Q_1, through the output port is the difference between the supply and return side flow rates, so that

$$Q_1 = Q_{1S} - Q_{1R}. \tag{2}$$

These flow-rate equations are quite general and encompass the various nonlinear effects previously mentioned. In particular, the above equations model: valve underlap or overlap, asymmetric neutral land position, and flow reversal. Although the flow-rate equations in (1) appear difficult to analyze and interpret, they are simplified if we use a saturation function. The saturation function, shown in Fig. 2, is defined as follows

$$u(x, \varepsilon, L) = \begin{cases} 0 \quad \text{for } x \leq -\varepsilon, \\ x + \varepsilon \quad \text{for } -\varepsilon < x < L - \varepsilon, \\ L \quad \text{for } L - \varepsilon \leq x. \end{cases} \tag{3}$$

Using this function, the orifice equations can be expressed more simply as

$$\begin{aligned} Q_1 &= Q_{1S} - Q_{1R} \\ &= \alpha \Big[\text{sgn}\,(P_S - P_1)\,|P_S - P_1|^{1/2}\; u(x_v, \varepsilon_{1S}, L) \\ &\qquad - \text{sgn}\,(P_1 - P_R)\,|P_1 - P_R|^{1/2}\; u(-x_v, \varepsilon_{1R}, L)\Big]. \end{aligned} \tag{4}$$

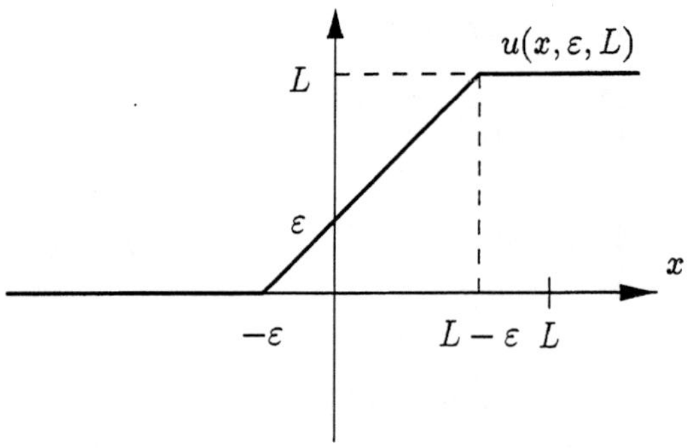

Figure 2: Saturation function (shown for $\varepsilon \geq 0$).

Proportional valve model

Having developed a concise relation for orifice flow, we will now incorporate this relation in the proportional valve shown in Fig. 3. The model results from com-

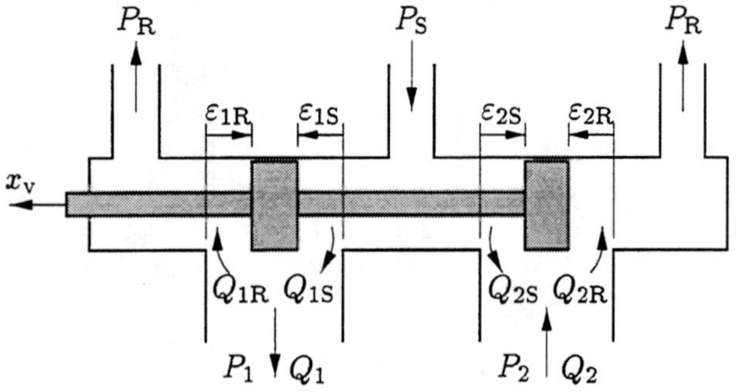

Figure 3: Hydraulic proportional valve.

bining two orifice equations so that the lands are rigidly connected and undergo identical displacements. The flow equation for Port 1 is identical to the orifice equation (4). The flow equation for Port 2 has a similar form, but the position of the supply and return lines are interchanged, resulting in a modified form in terms of the sign of the spool position variable, x_{v}:

$$\begin{aligned} Q_2 &= Q_{2\mathrm{R}} - Q_{2\mathrm{S}} \\ &= \alpha \Big[\operatorname{sgn}(P_2 - P_{\mathrm{R}}) \left| P_2 - P_{\mathrm{R}} \right|^{1/2} u(x_{\mathrm{v}}, \varepsilon_{2\mathrm{R}}, L) \\ &\quad - \operatorname{sgn}(P_{\mathrm{S}} - P_2) \left| P_{\mathrm{S}} - P_2 \right|^{1/2} u(-x_{\mathrm{v}}, \varepsilon_{2\mathrm{S}}, L) \Big] . \end{aligned} \tag{5}$$

The flow-rate equations (4) and (5) can be used to simulate *all* important nonlinear behavior of hydraulic proportional valves. In particular, valve land overlap or underlap and valve land asymmetry can be simulated by varying valve parameters. Port 1 characteristics are changed via the parameters $\varepsilon_{1\mathrm{S}}$ and $\varepsilon_{1\mathrm{R}}$. A symmetric overlapped land has $\varepsilon_{1\mathrm{S}} = \varepsilon_{1\mathrm{R}} < 0$, and a symmetric underlapped land has $\varepsilon_{1\mathrm{S}} = \varepsilon_{1\mathrm{R}} > 0$. Asymmetric spool properties and neutral position offset between valve lands can also be modeled using underlap or overlap values so that $\varepsilon_{1\mathrm{S}} \neq \varepsilon_{1\mathrm{R}}$. Similar considerations apply for the Port 2 parameters $\varepsilon_{2\mathrm{S}}$ and $\varepsilon_{2\mathrm{R}}$.

The overall intent of this paper is to develop a relationship between valve-spool displacement and fluid flow. The bond graph model of the proportional valve, shown in Fig. 4, will be used in the sequel to develop this relationship.

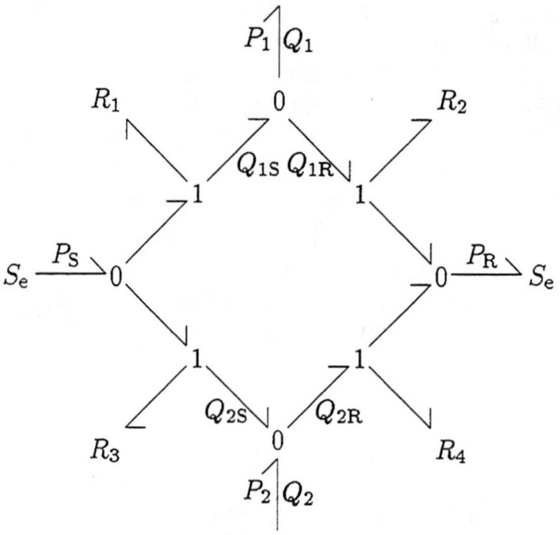

Figure 4: Proportional valve bond graph, modified from [2]

HYDRAULIC CYLINDER MODEL

In applications, the hydraulic actuator is typically a double-acting hydraulic cylinder. The cylinder ports are connected to a proportional valve, and piston motion is obtained by modulating the oil flow into and out of the cylinder chambers. A proportional valve provides this modulation.

The motion of the double-acting hydraulic actuator shown in Fig. 5 can be precisely controlled by regulating the flow rates Q_1 and Q_2. However, the relationship between the piston position, x_{p}, and the flow rates depends on the dynamic properties of the loads acting on the piston. The bond graph in Fig. 6, which represents the power flow in a hydraulic piston, facilitates analysis of this relationship.

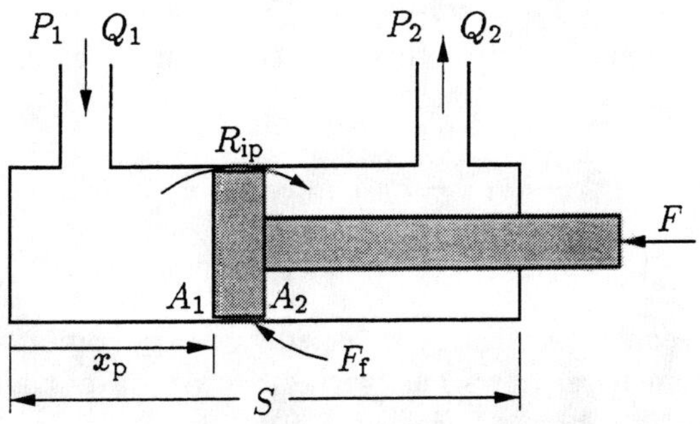

Figure 5: Double-acting hydraulic piston.

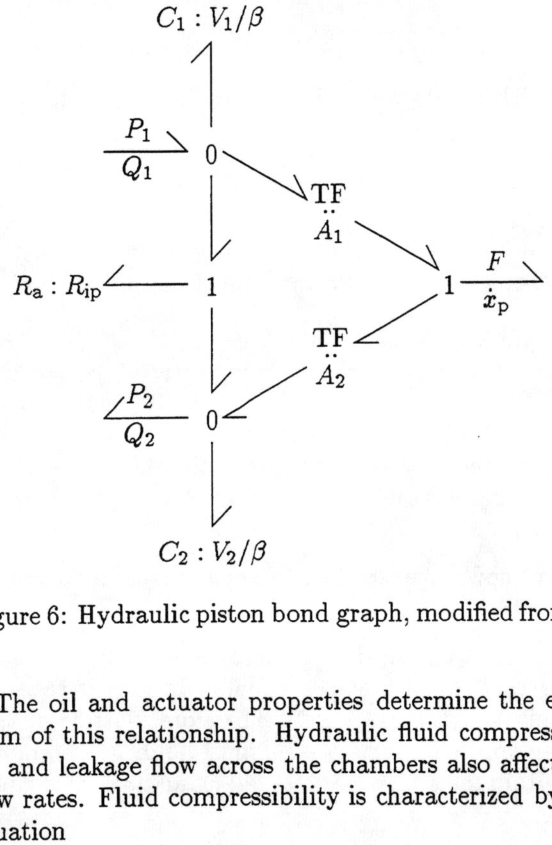

Figure 6: Hydraulic piston bond graph, modified from [2]

The oil and actuator properties determine the exact form of this relationship. Hydraulic fluid compressibility and leakage flow across the chambers also affect the flow rates. Fluid compressibility is characterized by the equation

$$dV = -\frac{V}{\beta}dP, \tag{6}$$

where β is the fluid bulk modulus, V is the chamber volume, and P is the fluid pressure. For hydraulic fluids, the bulk modulus has a nearly constant value. The leakage flow across the cylinder chambers is a linear function of the pressure differential across the piston [1], expressed as

$$Q_{ip} = R_{ip}(P_1 - P_2),$$

where R_{ip} is the coefficient of internal leakage.

The application of the continuity equation to the two sides of the cylinder yields

$$Q_1 = \dot{V}_1 + \frac{V_1}{\beta}\dot{P}_1 + R_{ip}(P_1 - P_2), \tag{7a}$$

$$Q_2 = -\dot{V}_2 - \frac{V_2}{\beta}\dot{P}_2 + R_{ip}(P_1 - P_2), \tag{7b}$$

where

$$V_1 = V_{10} + A_1 x_p, \tag{8a}$$

$$V_2 = V_{20} + A_2(S - x_p), \tag{8b}$$

are the total fluid volumes in the two sides of the cylinder [9]. Here, S is the piston stroke, and V_{10} and V_{20} are the fluid volumes in the lines and fittings on the two sides of the cylinder. The first right-hand term in (7) represents the flow rate as a function of volume change due to the piston motion. The second term expresses the effect of oil compressibility on the flow rate.

The net force, F, provided by the actuator is the difference between the hydraulic force and the frictional force, F_f, that is

$$F = A_1 P_1 - A_2 P_2 - F_f(x_p, \dot{x}_p). \tag{9}$$

The friction force may be modeled as a combination of Coulomb friction, which is approximated by the signum function, and viscous friction

$$F_f(x_p, \dot{x}_p) = \mu \operatorname{sgn} \dot{x}_p + b\dot{x}_p$$

where μ is the Coulomb friction coefficient and b is viscous friction coefficient. More general friction models can be incorporated into the friction force term, $F_f(x_p, \dot{x}_p)$, if necessary.

ANALYSIS OF THE HYDRAULIC SYSTEM PROPERTIES

The dynamics of a hydraulic system comprising a generic proportional valve and a cylinder is described by the model equations (4)-(5) and (7)-(9). Although various analyses have been performed using similar models, they are in general limited to systems with critical center valves. Using unified valve equations developed in previous section, however, we are able to extend analyses to all such systems, regardless of the valve type.

The general trend in hydraulic servo control systems is to eliminate underlapped and overlapped spools in proportional valves. Instead, precision manufactured and expensive critical center spools are used together with a matching valve body. However, advances in nonlinear control theory made it possible to use open and closed center valves to our advantage, or at least to cope with the nonlinearities they introduce [8, 10, 11].

In this section, we obtain various flow properties of generic proportional valves in a form suitable for nonlinear control applications. We assume the valve-cylinder configuration has the following characteristics:

1. a double-acting cylinder with equal chamber areas: $A = A_1 = A_2$,
2. an incompressible fluid: $\beta \rightarrow \infty$,
3. a matched spool with lapping parameters: $\varepsilon_1 = \varepsilon_{1S} = \varepsilon_{2R}$, and $\varepsilon_2 = \varepsilon_{1R} = \varepsilon_{2S}$,
4. no flow reversal or cavitation, i.e., $P_S \geq P_i \geq P_R \geq 0$, $i = 1, 2$.

Using Assumptions 1 and 2, the cylinder bond graph structure of Fig. 6 reduces to a simpler form as shown in Fig. 7. This allows simplified flow-rate equation to be

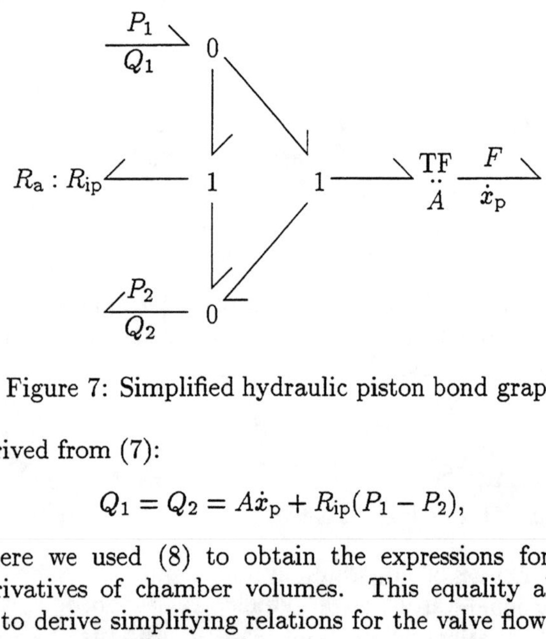

Figure 7: Simplified hydraulic piston bond graph

derived from (7):

$$Q_1 = Q_2 = A\dot{x}_p + R_{ip}(P_1 - P_2),$$

where we used (8) to obtain the expressions for the derivatives of chamber volumes. This equality allows us to derive simplifying relations for the valve flow rate equations in (4) and (5). From the proportional valve bond graph of Fig. 4, we obtain

$$Q_1 = Q_{1S} - Q_{1R}$$
$$Q_2 = Q_{2R} - Q_{2S}$$

so that

$$Q_{1S} - Q_{2R} = -(Q_{2S} - Q_{1R}),$$

since we have $Q_1 = Q_2$. Therefore, using (4)-(5) and Assumptions 3 and 4, this equation can be explicitely written in terms of systems variables as

$$\left(\sqrt{P_S - P_1} - \sqrt{P_2 - P_R}\right) u(x_v, \varepsilon_1, L) = - \left(\sqrt{P_S - P_2} - \sqrt{P_1 - P_R}\right) u(-x_v, \varepsilon_2, L). \quad (10)$$

This equation must be satisfied for every combination of system parameters and variables. We have available two possible means to satisfy (10):

1. The terms
$$\sqrt{P_S - P_1} - \sqrt{P_2 - P_R} = 0,$$
$$\sqrt{P_S - P_2} - \sqrt{P_1 - P_R} = 0,$$
which results when $P_S + P_R = P_1 + P_2$.

2. The terms
$$u(x_v, \varepsilon_1, L) = u(-x_v, \varepsilon_2, L) = 0 \quad (11)$$
which results when $P_S + P_R \neq P_1 + P_2$. To see this, we rearrange (10) in the form
$$u(x_v, \varepsilon_1, L) = - \left[\frac{\sqrt{P_S - P_2} - \sqrt{P_1 - P_R}}{\sqrt{P_S - P_1} - \sqrt{P_2 - P_R}}\right] u(-x_v, \varepsilon_2, L).$$
Since the bracketed term is positive for $P_S + P_R \neq P_1 + P_2$ and each $u(\cdot) \geq 0$, it is clear that (11) must hold.

Let us now consider these two cases vis-a-vis valve lapping:

(a) For an overlapped or critical-center spool: we have $\varepsilon_1, \varepsilon_2 \leq 0$, so that the equality in (11) is satisfied only for $x_v \in (\varepsilon_2, -\varepsilon_1)$. Therefore, outside of this region, $P_S + P_R = P_1 + P_2$ must hold. Furthermore, within the region $x_v \in (\varepsilon_2, -\varepsilon_1)$, all flow rates equal zero.

(b) For an underlapped spool: we have $\varepsilon_1, \varepsilon_2 > 0$, so that the equality in (11) is never satisfied for any value of x_v. Thus $P_S + P_R = P_1 + P_2$ must hold.

The consequences of (a) and (b) are the following: either $P_S + P_R = P_1 + P_2$ or all valve flow rates equal zero. Substituting this pressure equality into the flow rate equations, (4) and (5), and comparing terms, we obtain

$$Q_{1S} = Q_{2R}$$
$$Q_{1R} = Q_{2S},$$

which also holds for zero flow rates. Furthermore, since $Q_1 = Q_2$, the flow rate equations can be expressed as

$$Q_1 = Q_2 = m_1 u(x_v, \varepsilon_1, L) - m_2 u(-x_v, \varepsilon_2, L), \quad (12)$$

where

$$m_1 = \alpha\sqrt{\frac{P_S - P_R - P_L}{2}}, \quad m_2 = \alpha\sqrt{\frac{P_S - P_R + P_L}{2}}$$

are the slopes and $P_L \doteq P_1 - P_2$ is the load pressure.

Equation (12) provides a useful relation for flow-rate versus spool displacement. This relation is plotted in Fig. 8, where an overlapped valve is assumed. We see

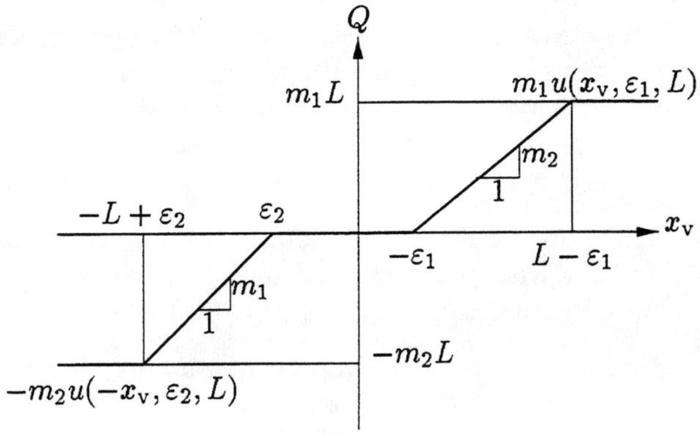

Figure 8: Plot of (12) for an overlapped valve.

five distinct regimes. Inside the deadzone, the flow rates are zero. This corresponds to the case $P_S + P_R \neq P_1 + P_2$, where the motion of the spool and the cylinder decouples, so that the equality $P_S + P_R = P_1 + P_2$ can no longer be maintained. Outside the deadzone, the motion of the piston is governed by (12), which provides the flow rate input to the cylinder.

For an underlapped valve, the analysis is slightly more complex. Outside the underlap region, $x_v \notin (-\varepsilon_1, \varepsilon_2)$, the sloped regions of Fig. 8, m_1 and m_2, as well as the saturations are all present as shown in Fig. 9. However, within the underlap region, we have

$$u(x_v, \varepsilon_1, L) = x_v + \varepsilon_1, \quad u(-x_v, \varepsilon_2, L) = -x_v + \varepsilon_2$$

so that

$$\begin{aligned} Q_1 = Q_2 &= m_1(x_v + \varepsilon_1) - m_2(-x_v + \varepsilon_2) \\ &= (m_1 + m_2)x_v + (m_1\varepsilon_1 - m_2\varepsilon_2), \end{aligned}$$

where the term $(m_1\varepsilon_1 - m_2\varepsilon_2)$ is the null flow rate due to the load pressure, P_L. This means that even for a

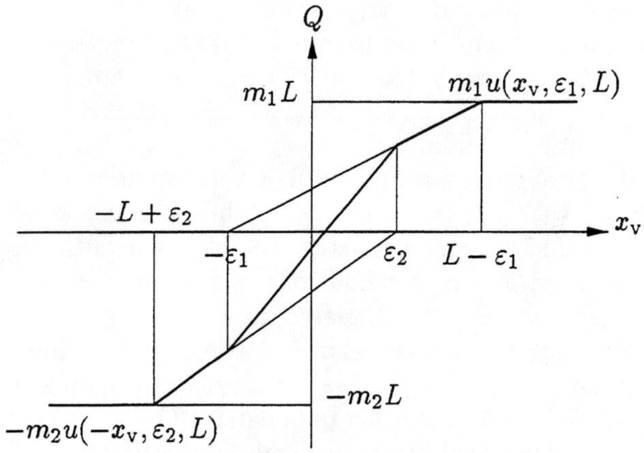

Figure 9: Plot of (12) for an underlapped valve.

centered spool, $x_v = 0$, we may have piston motion due to the null flow if $\varepsilon_1 \neq \varepsilon_2$ and $P_L \neq 0$.

On the other hand, the term $(m_1 + m_2)x_v$ represents the flow gain relation between spool position and flow rate. The flow gain, K_q, within the underlap region is an important number in designing hydraulic control systems, since it determines the open-loop plant gain. Large variations in the flow gain could lead to instability. Fortunately, its bounds can easily be characterized as a corollary of our analysis. The flow gain is defined as

$$\begin{aligned} K_q &\doteq \frac{\partial Q}{\partial x_v} = (m_1 + m_2) \\ &= \frac{\alpha}{\sqrt{2}}\left[\sqrt{P_S - P_R - P_L} + \sqrt{P_S - P_R + P_L}\right]. \end{aligned}$$

Since $|P_L| \leq P_S - P_R$ by our assumptions, it is easy to show that

$$\alpha\sqrt{P_S - P_R} \leq K_q \leq \sqrt{2}\alpha\sqrt{P_S - P_R}.$$

CONCLUSIONS

We have developed a unified model for proportional control valves and analyzed the effect of spool lapping on open-loop hydraulic system properties. The nonlinear mathematical equations relate the flow rates through the valve ports to the valve parameters. The flow rates are expressed as a continuous but nonlinear function of *lapping parameters*, as well as other conventional parameters. These equations are readily applicable to various types of proportional valves, and they unify the cases of critical center, overlapped, and underlapped valves.

The spool lands' geometry are individually controlled via model parameters. An offset between the lands and asymmetric neutral spool position can also be simulated using our model. The system model is fully implemented in Matlab's Simulink simulation package and the hydraulic system component models are combined in a library for easy reuse.

We also derived simplified flow rate equations under certain widely-used assumptions while keeping nonlinearities due to spool geometry. The variation of the flow gain and uncertainty bounds of the flow rates of an underlapped valve are also analyzed.

Although this paper assumes a constant bulk modulus, changes in temperature, pressure, and percent air in the fluid will alter this parameter. The assumption facilitated the analysis, in particular the piecewise-linear characterization of flow versus spool displacement in Figures 8–9. If operating conditions are such that this assumption is invalid, adaptive [7] and singular perturbation [4] control approaches are available to accommodate the variation in bulk modulus.

Our future work includes developing internal leakage flow models for proportional valves to provide improved servo performance. The design of nonlinear controllers for overlapped and underlapped valve spool is also a promising research area. Simple nonlinear compensators together with less expensive proportional valves offers substantial cost savings in hydraulic control systems design.

References

[1] Andrew Alleyne and J. Karl Hedrick. Nonlinear adaptive control of active suspensions. *IEEE Transactions on Control Systems Technology*, 3(1):94–101, March 1995.

[2] B. W. Barnard and P. Dransfield. Predicting response of a proposed hydraulic control system using bond graphs. *ASME Journal of Dynamic Systems, Measurement, and Control*, pages 1–8, March 1977.

[3] J. E. Bobrow and K. Lum. Adaptive, high bandwidth control of a hydraulic actuator. In *Proceedings of the 1995 American Control Conference*, pages 71–75, Seattle, WA, June 1995.

[4] Bora Eryilmaz and Bruce H. Wilson. Improved control of hydraulic systems using singular perturbation theory. *ASME Journal of Dynamic Systems, Measurement, and Control*, 1999. In review.

[5] H. M. Handroos and M. J. Vilenius. Flexible semi-emprical models for hydraulic flow control valves. *ASME Journal of Mechanical Design*, 113:232–238, September 1991.

[6] Herbert E. Merritt. *Hydraulic Control Systems.* John Wiley & Sons, New York, NY, 1967.

[7] H. J. Park, H. S. Cho, and B. S. Hyun. An adaptive control of nonlinear time-varying hydraulic servo systems. In *Proceedings of the 1989 American Control Conference*, pages 1894–1898, Pittsburgh, PA, June 1989.

[8] Darrel A. Recker. *Adaptive Control of Systems Containing Piecewise Linear Nonlinearities.* PhD thesis, University of Illinois at Urbana-Champaign, Urbana, IL, 1993.

[9] Garett A. Sohl and James E. Bobrow. Experiments and simulations on the nonlinear control of a hydraulic servosystem. In *Proceedings of the 1997 American Control Conference*, pages 631–635, Albuquerque, New Mexico, June 1997.

[10] Gang Tao and Petar V. Kokotović. Adaptive control of plants with unknown dead-zones. *IEEE Transactions on Automatic Control*, AC-39(1):59–68, 1994.

[11] Gang Tao and Petar V. Kokotović. *Adaptive Control of Systems with Actuator and Sensor Nonlinearities.* John Wiley & Sons, New York, NY, 1996.

[12] Gert-Wim van der Linden and Peter Valk. Digital control of an experimental hydraulic manipulator. In *Proceedings of the 1994 American Control Conference*, pages 2455–2459, Baltimore, MD, June 1994.

[13] R. B. Walters. *Hydraulic and Electro-Hydraulic Control Systems.* Elsevier Science, New York, NY, 1991.

[14] D. Wang, R. Dolid, M. Donath, and J. Albright. Development and verification of two-stage flow control servovalve model. In *Proceedings of the ASME Fluid Power Systems and Technology Division*, volume FPST-Vol. 2, pages 121–129, San Francisco, CA, November 1995. 1995 ASME International Mechanical Engineering Congress and Exposition.

[15] Heather Wright, Andrew Alleyne, and Rui Liu. On the stability and performance of two-stage hydraulic servovalves. In *Proceedings of the ASME Fluid Power Systems and Technology Division*, volume FPST-Vol. 4, pages 215–222, November 1997. 1997 ASME International Mechanical Engineering Congress and Exposition.

FPST-Vol. 6, Fluid Power Systems and Technology — 1999
ASME 1999

INJECTION MOULDING SIMULATION FOR HYDRAULIC VALVE-ACTUATOR SYSTEM CONTROLLER DESIGN

Paul K Guerrier and Kevin A Edge
Department of Mechanical Engineering
University of Bath, UK.
e-mail: P.K.Guerrier@bath.ac.uk

Abstract

In the development of advanced injection controllers and more effective control valves, full scale testing is often too expensive. This paper is concerned with virtual testing using simulation and experimental testing using a 'hardware-in-the-loop' technique in which a servovalve-actuator system emulates the moulding process. Simulation results for the purely simulated load and a 'hardware-in-the-loop' emulated load are compared with experimental data. Good agreement is achieved. Although the particular case of an hydraulic injection moulding machine has been considered, the model of the moulding process would be equally valid for studies of an all-electric machine.

Nomenclature

k	power law index viscosity
l	length
m	mass
n	power law index
P	pressure
Q	flow rate
R	radius
r_1	radius at the beginning of a section
r_2	radius at the end of a section
λ_T	extensional viscosity constant
μ	viscosity
η_N	zero shear Newtonian viscosity
$\dot{\gamma}$	shear rate
τ	shear stress
τ_y	yield shear stress
σ_E	extensional stress
θ	convergence angle

Introduction

For plastic and rubber part production, injection moulding is one of the most important manufacturing processes available. In hydraulically driven machines, injection actuators are typically velocity controlled until the polymer being injected has filled the mould; once the filling phase is over pressure-controlled packing begins. Many existing controllers are relatively simple but nonetheless can be difficult to commission. A particular problem is to achieve a smooth transition from the velocity control phase to the pressure control phase. With the advent of digital control, advanced flowrate and pressure filling and packing controllers are now becoming available on commercial moulding machines (Zhang et al., 1996). However high performance control is far from being an industry standard and is currently an area of research interest. This paper is concerned with the development of simulation models to aid in the process of control system design and assessment

Whilst it would be advantageous in many ways to test new control strategies on a full size injection moulding machine, this would be very costly. Simulation offers a cheap and effective way of filtering out unsuitable strategies, with only the most promising being evaluated using full scale testing. However, as a result of simplifications and assumptions some effects are not adequately modelled. For this reason a more detailed evaluation following on from simulation is required. Ideally this should be cheaper and more flexible than through full scale testing. Such a requirement can be met using a 'hardware-in-the-loop' approach (Ramden et al., 1997). In this application the moulding machine can be considered as the injection cylinder 'load' and can be emulated by a computer-controlled servo-hydraulic actuator. This will allow the evaluation of suitable control schemes over a wide range of carefully-controlled moulding situations.

Polymer Structure & Flow Characteristics

To model the injection moulding process it is essential to have an appreciation of the non-Newtonian properties of polymeric liquids.

Polymeric fluids are often called viscoelastic fluids. Such materials have elasticity and after deformation will attempt to return to the original shape but will not usually reach it because of the viscous properties. Full mathematical modelling of viscoelastic behaviour is complicated, usually containing temperature-dependent relaxation constants which describe how the fluid body reacts to forces acting on it (Bird et al.,

1987). This level of detail is unnecessary for the study reported here.

Newtonian viscosity is of little or no use for describing polymer flow. There are 5 recognised relationships for predicting shear rate of non-Newtonian fluids. These are as follows:

- Pseudoplastic (shear thinning), where n<1 and where
  $\tau = \eta_N \dot{\gamma}^n$ (1)
- Dilatant (shear thickening), which again follows equation (1) but with n>1
- Bingham, where
  $\tau - \tau_y = \eta \dot{\gamma}$ (2)
- Plastic, which is a combination of the Bingham and Pseudoplastic relationships.
- Ostwald which according to (Lenk, 1978) is yet to be fully proven as existing.

Graphical illustrations of the relationships between shear stress and shear rate for these 5 flow phenotypes together with Newtonian Behaviour are shown in Figure 1. Of these the Ostwald and Bingham types are not relevant here as they do not describe polymer flow. Pseudoplastic (shear thinning) and dilatant (shear thickening) are of greatest interest with good examples being polypropylene and a blend of polycarbonate and ABS respectively. The plastic relationship is also relevant to polymer although, in practice the yield stress needed for polymer flow to start is so low that there is no need for the inclusion of the Bingham relationship in practice.

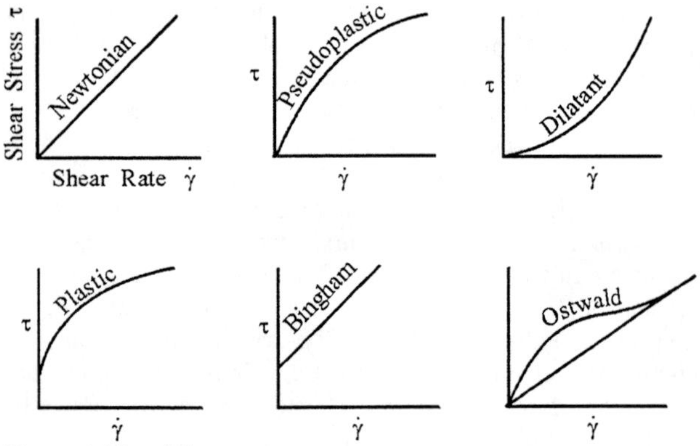

Figure 1 Flow Phenotypes

For both the pseudoplastic and dilatant relationships described by equation (1) η_N is the zero shear (Newtonian) viscosity. If n=1 then the equation (1) describes the Newtonian relationship.

Polymer Flow Governing Equations

For simulation purposes a relationship between polymer flow and injection actuator pressure is required. Incorporating the power law for viscosity into the Hagen-Poiseuille equation for flow through pipes gives equation (3) where k is both material and temperature dependent

$$\Delta P = \frac{2\left(\frac{4}{\pi}\right)^n klQ^n}{R^{3n+1}} \tag{3}$$

(if $n = 1$ then the power law index equation becomes the Hagen-Poiseuille equation)
With some manipulation, equations based on equation (2) can be derived to describe pressure requirements for diverging and converging sections. However, in such sections, a further flow elongation effect occurs that needs to be taken into account. Figure 2 shows an example of the behaviour in a converging section. The corresponding differential pressure over the converging section is described by equations (4) and (5) where λ_T is both material and temperature dependent.

$$\sigma_E = \lambda_T \frac{\dot{\gamma}}{2} \tan\theta \tag{4}$$

$$\Delta P_E = \frac{2}{3}\sigma_E \left[1 - \left(\frac{r_2}{r_1}\right)^3\right] \tag{5}$$

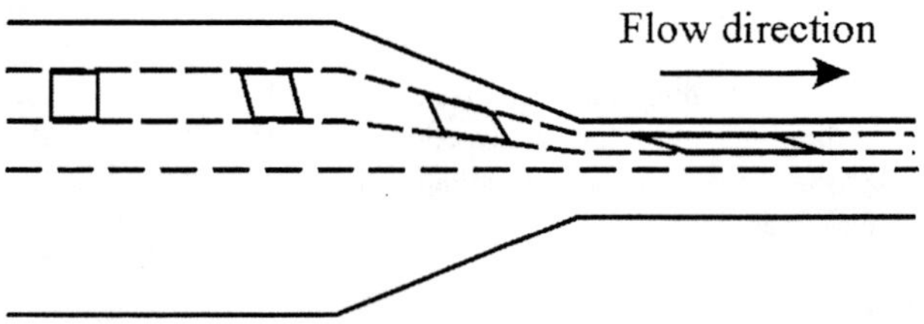

Figure 2 Flow Extension

Injection Moulding Process Modelling & Simulation

At the onset of packing the polymeric material being injected is still molten but since the mould temperature is well below the melting temperature of the polymer it soon cools and freezes. Subsequently there is a point in time called gate freeze after which material can no longer flow in or out of the mould (the gate is the opening to the mould). According to Gao et al. (1996) the only feasible way to effect the mould cavity pressure after gate freeze is by changing the temperature of the mould itself. This can be achieved by passing a cooling fluid (typically water or oil) through channels in the dies. Since the work in this paper is intended to aid the study of the control of the filling and packing phases, the cooling phase after gate freeze is not addressed here.

The Bath*fp* simulation package (Tilley & Richards, 1991) was employed for the simulation study. This package is based on a multi-step multi-order numerical integrator appropriate for the solution of fluid power systems. The mathematical model of the injection moulding process is represented in Bath*fp* as a 'load'. This can be linked to an actuator and will from this point on be referred to as the load model.

Load Model Description

The load model was developed for the case of a real mould (comprised of two materials testing specimens). An image of the part produced by this mould can be seen in Figure 3. In order to model the mould the following assumptions were made:

- Flow is one-dimensional and laminar.
- Flow is incompressible during the filling stage. This assumption is adopted in the majority of simulations detailed in previously-published work. If the flow were to be modelled as compressible during the filling phase this would complicate the model considerably and finite elements would have to be used.
- Flow is compressible as soon as the mould is full.
- Polymer flow is perfectly balanced between the 2 flow paths.
- Viscous filling forces include both the power law index and extensional viscosity.
- The process is isothermal. This is another simplification that has been used in the majority of simulations detailed in previously-published works. To include the effect of different temperatures throughout the part would add complexity and require a finite element approach. The current model can however be run at different polymer temperatures.
- The mass of the polymer being injected, about 20g for the materials testing specimen mould, is considered negligibly small in comparison to the mass of the moving parts of the machine, 60kg. The mass of the polymer is not included in any calculation.

These assumptions are generally in agreement with those adopted by previous researchers, in particular (Shankar & Paul, 1982) (Cheng-Ping Chiu et al, 1991) (Jong-Hwei Wei et al, 1994) (Rafizadeh et al, 1996) and 2 companion papers (Chiang et al 1991a) (Chiang et al 1991b)

The Bath*fp* load model has been developed such that a force is the output and injection actuator position and velocity are required as inputs. To determine the equations required to calculate the load force, the mould melt front is tracked during mould filling. The mould melt front position is calculated from the volume of material that has entered the mould, which is then compared with the total volume of the mould. This calculation is generally straightforward except in the case of converging and diverging sections when there are cubic equations to be solved. A Newton-Raphson approximation was included in the code to deal with this requirement. While this method of solution usually takes 7 iterations to converge this did not slow down computation significantly.

The Bath*fp* model treats the part shown in Figure 3 as being comprised of the following series of shapes in series; a diagram showing this representation is presented in Figure 4:

At the extreme left of the figure is a converging section, equivalent to the injection moulding machine nozzle (this has no region number as it is permanently full). This is followed by a short, constant radius, section, corresponding to the mould entry gate (region 1) and then a diverging section, corresponding to the section joining the larger diameter runner to the small cross section gate (region 2). Region 3 corresponds to a constant radius section, equivalent to the runner and region 4 compromises two constant radius sections in parallel, equivalent to the runners serving each materials testing part. The two short converging sections at the end of each runner (region 5) join the larger radius runners with the small second gates at entry to the moulded parts. This is followed by two short constant radius sections, equivalent to the gates before the moulded parts (region 6) and finally two constant area rectangular sections, equivalent to the parts to be moulded (region 7).

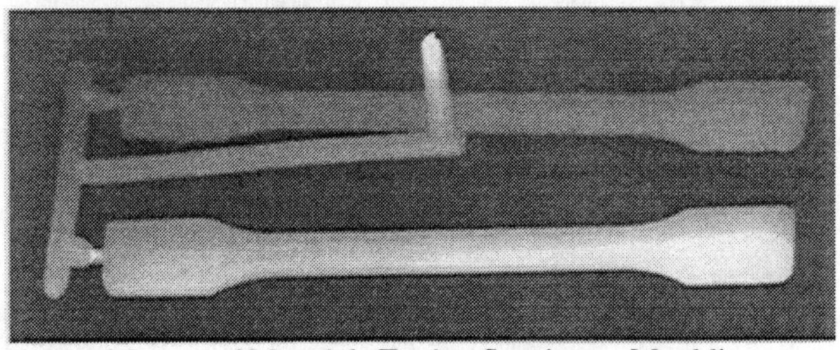

Figure 3 Image of Materials Testing Specimens Moulding

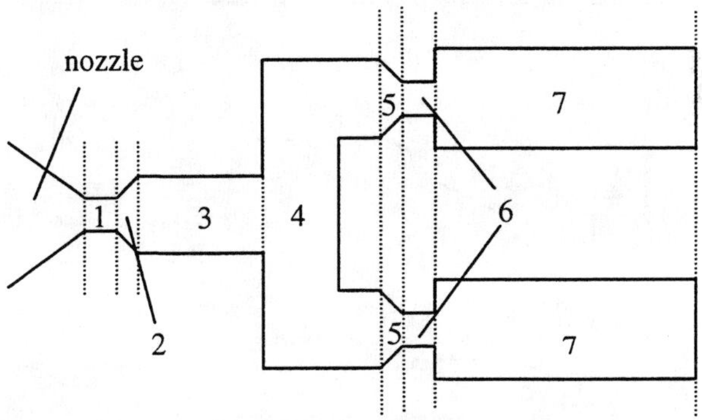

Figure 4 Simulated Mould Region Diagram

This model readily allows the simulation of other similar moulds since the dimensions of each section can be changed.

Simulation Study

The first simulation runs were done using the load model connected directly to an hydraulic actuator. The electrohydraulic circuit diagram of this is shown in Figure 5 . Except for the actuator all the hydraulic components and the signal input are standard models (Tilley & Richards, 1997). The standard actuator model was modified to have different annular areas.

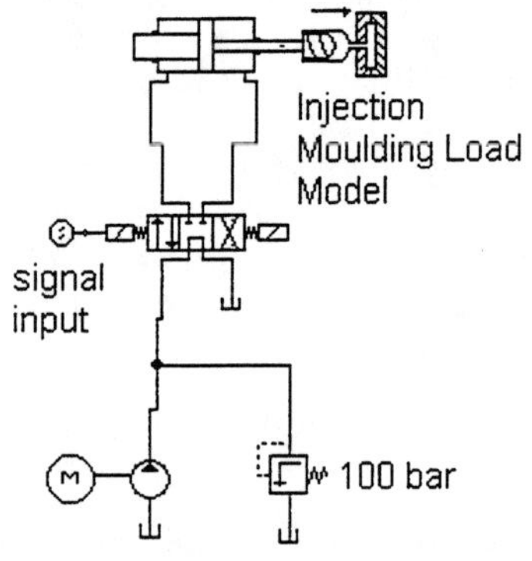

Figure 5 Hydraulic Load Simulation Circuit Diagram

Results from the simulation are shown in Figure 6. During the initial increase in pressure up to 0.18s mould regions 1-5 are filled. The subsequent short sharp rise in pressure at 0.2s occurs as regions 5 and 6 are filled. This is because flow elongation effects are significant in these regions and because region 6 is a small constriction.

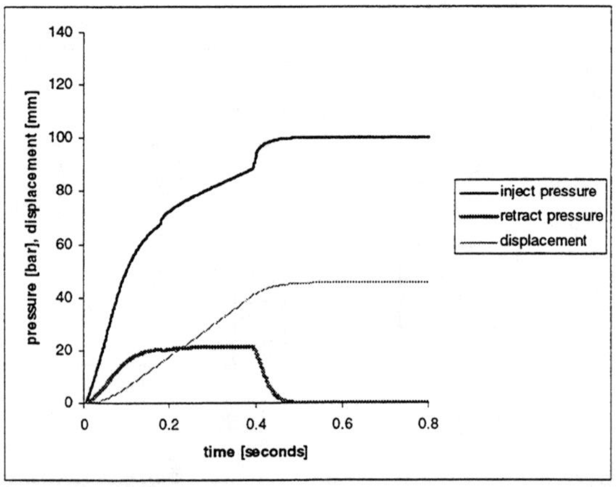

Figure 6 Load Model Simulation Results

After the initial load model simulations, a hardware-in-the-loop loading system was designed. Load emulation can be achieved hydraulically (Nimegeers et al., 1997), using a circuit consisting of a servo-valve-controlled actuator. The load emulation circuit adopted here was simulated as part of the design and evaluation process prior to implementation in the laboratory. As previously mentioned, such an arrangement would allow promising control schemes developed in simulation to be evaluated experimentally under well-controlled conditions and at a fraction of the cost of tests on a real machine. The electrohydraulic circuit diagram of this arrangement is shown in Figure 7.

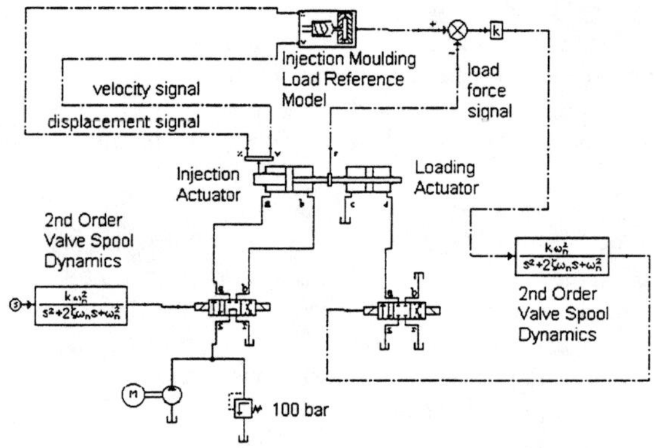

Figure 7 Hydraulic Load Emulator Circuit Diagram

The whole circuit is based around two actuators connected together with a force transducer between them and with load position and velocity transducers. The actuator on the left is the injection actuator, which is controlled by an electrohydraulic valve with a supply-to-tank central position. Second order valve spool dynamics are separate from the valve because the valve model itself was instantaneous. Position and velocity signals are fed into the injection moulding load model, which then in turn acts as a reference model, producing demanded force output. The actual force, measured by the force transducer is subtracted from the demanded force and the resultant force error is amplified and fed to the load actuator servo-valve. This forms a closed loop proportional force controller. The spool dynamics are again introduced separately. The load actuator valve meters flow out of the actuator during extension (movement to the right in Figure 7). It was found during simulation that metering out was adequate and simpler to control than having a pressure source connected via a spool valve to the load actuator.

Simulation results for the arrangement are shown in Figure 8. Again the slight sharp rise in pressure during filling, in this case just after 0.2 s, is due to the filling of mould regions 5 and 6. For the case of an injection actuator valve with a 10Hz bandwidth and a step input it was found that a 20Hz load actuator valve was sufficiently fast.

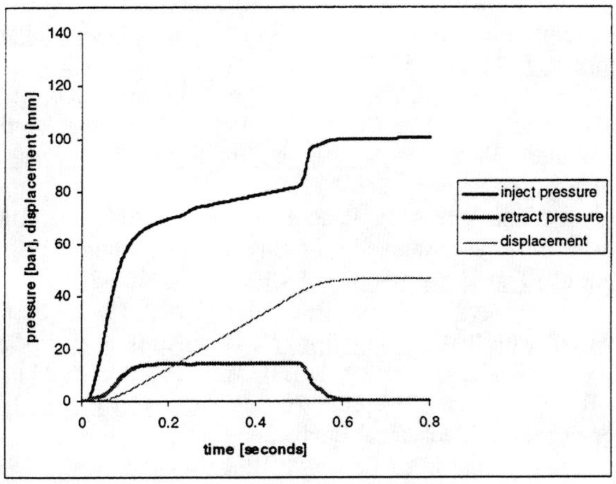

Figure 8 Load Emulator Simulation Using 20Hz Load Control Valve

Experimental Work

To achieve confidence in simulation results experimental tests were undertaken on a 25 tonne clamping force injection moulding machine using the same mould that was used in simulation. Two injection actuators, connected in parallel and of equivalent area to the single actuator used in simulation, were instrumented with pressure transducers and a displacement transducer. The injection moulding machine was set up with just enough material in the barrel to fill the mould and the injection controller was set up to give the maximum injection velocity with no switch to pressure control. The results obtained experimentally are presented in Figure 9. The rise in pressure up to 0.2s corresponds to the mould fill up to the moulded part section. The injection pressure peak at the end of injection (0.5s) is at the point where filling is complete the material is being compressed. However, this compression only accounts for the first 5 bar of the pressure transient with the rest due to the injection actuators hitting their endstops.

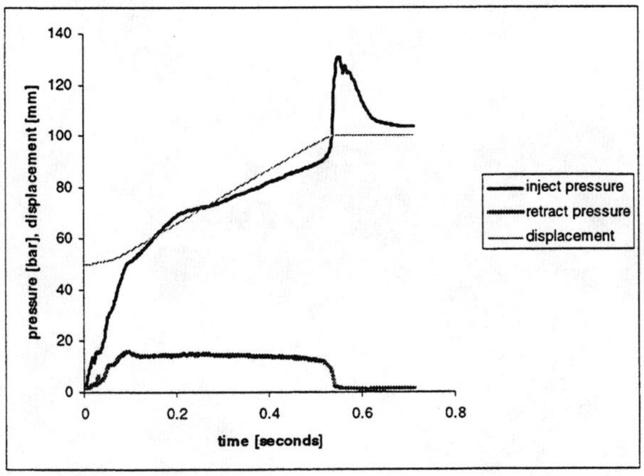

Figure 9 Experimental Results

Comparison of Experimental & Simulation Results

Experimental and load model injection pressures are compared in Figure 10. Mould fill can be seen to occur in both sets of results with the pressure increasing sharply at 0.4s in simulation and at 0.5s in the experimental results. The reason for this discrepancy is that polymer was able to leak internally in the injection barrel of the moulding machine used. This leakage was not modelled in simulation.

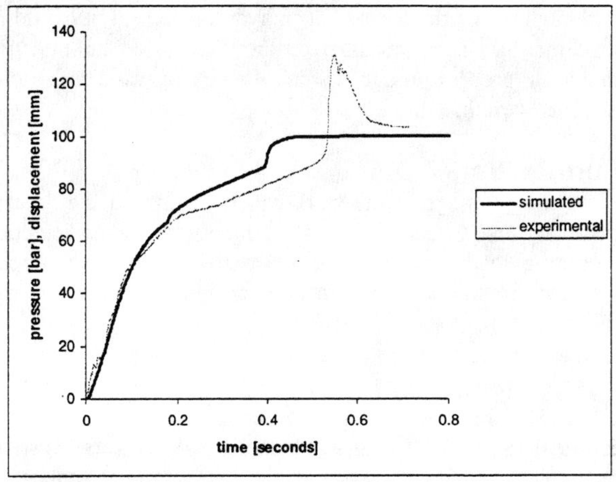

Figure 10 Experimental & load Model

A comparison of the experimental and simulated injection pressure results for the case of the load emulation circuit is shown in Figure 11. The time at which the mould is completely full is now the same in both the real and simulated results. This was achieved by slowing down the simulated injection speed to compensate for the injection barrel leakage. As with the previous simulation results the two pressure time traces are very similar.

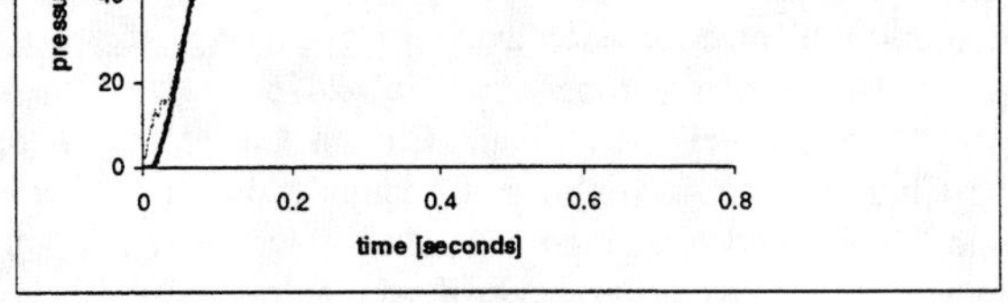

Figure 11 Experimental & Load Emulation

Conclusions

This paper has examined the feasibility of using simulation to model the injection moulding process. An injection moulding load model has been implemented in the Bath*fp* simulation package. For the case of a pseudoplastic material (polypropylene) good agreement has been achieved between simulated and experimental results. Following on from this a further simulation study was carried out to investigate the feasibility of emulating the hydraulic injection moulding process using a 'hardware-in-the-loop' approach. Again good agreement was found between simulated and experimental results. The simulated injection moulding load produced by this work will provide an effective means of evaluating both novel control schemes and new designs of control valve.

Although the particular case of an hydraulic injection moulding machine has been considered, the model of the injection moulding process produced would be equally valid for studies an all-electric machine.

Further Work

The next stage of the work will be to investigate dilatant materials and to set up a test rig incorporating an injection moulding load emulation circuit, based on the load circuit tested in simulation. New controller design and evaluation will follow.

Acknowledgements

The authors would like to thank the Engineering and Physical Science Research Council and Vickers Systems Division Aeroquip-Vickers Ltd for their financial support of this project under the CASE postgraduate studentship scheme.

References

Bird, R. B., Armstrong, R. C. & Hassager, O., 1987, 'Dynamics of Polymeric Liquids', John Wiley & Sons, 2nd edition

Chiang, H. H., Hieber, C.A. & Wang, K.K., 1991a, 'A Unified Simulation of the Filling and Postfilling Stages in Injection Molding. Part 1: Formulation', Polymer Engineering and Science, **31**(2, January), pp. 117-124

Chiang, H.H, Hieber, C. A. & Wang, K.K., 1991b, 'A Unified Simulation of the Filling and Postfilling Stages in Injection Molding. Part 2: Experimental Verification', Polymer Engineering and Science, **31** (2, January), pp. 125-139

Cheng-Ping Chiu, Jong-Hwei Wei & Ming-Chang Shih 1991, 'Adaptive Model Following Control of the Mold Filling Process in an Injection Molding Machine', Polymer Engineering and Science, **31** (15, Mid-August), pp. 1123-1129

Gao F., Patterson, W.I. & Kamal, M.R., 1996, 'Cavity Pressure Control During the Cooling Stage in Thermoplastic Injection Moulding', Polymer Engineering and Science, **36** (19, mid-October), pp. 2467-2476

Jong-Hwei Wei, Chiann-Chiunn Chang, & Cheng-Ping Chiu, 1994, 'A Nonlinear Dynamic Model of a Servo-Pump Controlled Injection Molding Machine', Polymer Engineering and Science, **34** (11, mid-June), pp. 881-887

Lenk, R. S., 1978, 'Polymer Rheology', Applied Science Publishers Ltd London

Nimegeers, C.R., Burton, R. & Schoenau (1997), 'A Computer Controlled Hydraulic Loading System for Simulating Loads', 9th Bath International Workshop, University of Bath, pp. 10-23

Rafizadeh, M., Patterson, W.I. & Kamal M.R., 1996, 'Physically–Based Model of Thermoplastics Injection Moulding for Control Applications', International Polymer Processing, **11**, pp. 352-362

Ramden, C., Jansson, A. & Palmberg, J-O., 1997, 'Design and Analysis of a Load Simulator for Testing Hydraulic Valves', 10th Bath International Workshop, University of Bath, pp. 260-277

Shankar, A. & Paul, F.W., 1982, 'A Mathematical Model for the Evaluation of Injection Molding Machine Control', Transactions of the ASME, **104**, pp. 86-92

Tilley, D.G. & Richards, C.W., 1991 'System Analysis with Bath*fp*' Fluid Power, March/April, **3**, pp.50-53

Tilley, D.G & Richards, C.W., 1997 'Bath*fp* Manual 2 Bath*fp* Model Reference Guide' University of Bath, Department of Mechanical Engineering

Zhang, C.Y, Leonard, J. & Speight, R.G., 1996, 'Adaptive Controller Performance Used For Ram Control During Filling Phase', Society of Plastics Engineers proceedings, ANTEC, pp. 593-597

RESEARCH ON DIGITAL CONTROL OF METER-IN AND METER-OUT INDEPENDENT REGULATING FOR HIGH INERTIA LOAD

Wang Qingfeng Gu Linyi (The State Key Laboratory of Fluid Power Transmission and Control, Zhejiang University, Hangzhou, P.R.China. E-mail: qfwang@sfp.zju.edu.cn, lygu@sfp.zju.edu.cn)

Lu Yongxiang (Chinese Academy of Science, Beijing, P.R.China)

[ABSTRACT] The smoothness of acceleration and deceleration process is a serious problem in valve control system with high inertia load, especially in the hydraulic systems in construction machines.

In this paper, a meter-in and meter-out independent regulating method, in which the two sides of actuator are controlled by a meter-in valve and a meter-out valve respectively, is put forward. In one hand, the meter-out valve could control the actuator's outlet pressure to avoid the ultra-high outlet pressure when actuator decelerates or brakes suddenly. On the other hand, the dynamic damping ratio of valve control system could be raised through calculated flow feedback control algorithm.

Secondly, a grading control algorithm in dynamic process of high inertia load is adopted. When the actuator's velocity is far from its command value, the actuator's inlet and outlet pressure are controlled. After the velocity error decrease to a threshold, a state feedback control algorithm based on parameters on line estimating is employed to realize both its velocity accuracy and the smoothness of dynamic process.

Experiments show that the actuator's velocity could increase or decrease to its command value accurately, smoothly and rapidly after the above method and algorithm are applied.

[KEY WORDS] Hydraulics, High inertia load, Meter-in and meter-out independent regulating, Acceleration and deceleration process, Grading control

1. Introduction

The smoothness of acceleration and deceleration process is a serious problem in valve control system with high inertia load (Aoki et al., 1994)(Ichiryu et al., 1993), especially in the hydraulic systems in construction machines.

There are three main reasons that lead to this problem:

1. Interference between meter-in and meter-out.

2. The high inertia load leads to large time delay and low damping ratio in valve control system.

3. High brake torque in deceleration or brake process.

Accordingly, corresponding means are applied to improve its performance, such as:

1. Adopt a special valve control unit whose meter-out is not in the same spool of meter-in throttle valve, but in the same spool of pressure compensator (Masuda, 1993).

2. Introduce electronic control into pressure compensator to raise the damping ratio of valve control system (Aoki et al., 1994)(Ichiryu, 1995).

3. Apply a flow-pressure compounded control method (Tamura et al., 1993).

4. Use load sensing modulated valve (Ichiryu, 1995).

In this paper, a meter-in and meter-out independent regulating method based on pressure difference sensing is advanced. In this method, the two sides of actuator are controlled by a meter-in valve and a meter-out valve respectively, and the pressure compensator is replaced by electronic control.

2. Structure and principle

Fig.1 shows the schematic structure of meter-in and meter-out independent regulated control method based on pressure difference sensing.

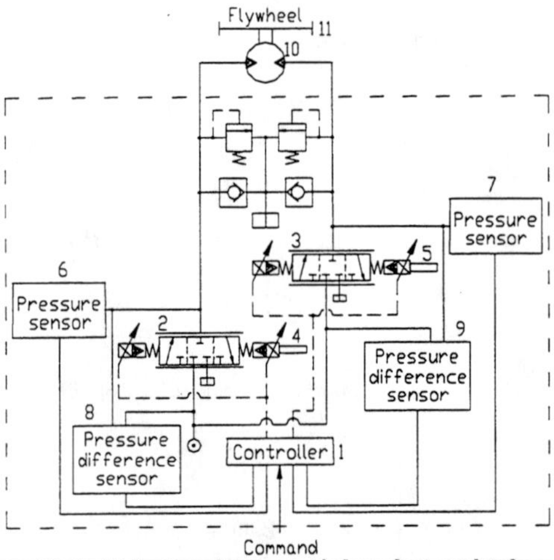

Fig.1 Meter-in and meter-out independent regulated control method based on pressure difference sensing

Valve 2 and valve 3 are meter-in and meter-out valve. The meter-in valve could control the inlet pressure or flow of actuator and the meter-out valve control its outlet pressure. The inlet and outlet pressure could be detected by pressure sensors (6 and 7). The meter-in valve's flow could be calculated through its spool's displacement and pressure decrease, which could be detected by its spool displacement sensor (4 or 5) and pressure difference sensor (8 or 9) respectively.

Thank to the independently regulated meter-out valve, the outlet pressure of actuator could be controlled freely. In acceleration process, the meter-out valve is fully opened to shorten the transitional process; in deceleration process, it could control the outlet pressure to its permissible limit to avoid ultra-high brake torque.

Take advantage of pressure difference sensing, the dynamic damping ratio could be changed freely through calculated flow feedback control algorithm, a close loop control of meter-in valve's flow whose feedback is calculated from meter-in valve spool's displacement and pressure decrease.

Table 1 shows a relative measure of both the outlet pressure in acceleration/deceleration processes of high inertia load, and the dynamic damping ratio for each of the three controllers listed below:

1. Traditional method (meter-out is in the same spool of meter-in valve, and pressure compensator is applied).

2. Pressure difference sensing method (meter-out is still in the same spool of meter-in valve, but calculated flow feedback control algorithm is applied instead of pressure compensator).

3. Meter-in and meter-out independent regulating method based on pressure difference sensing respectively.

Table 1 Dynamic performance of valve control system with high inertia load

	Outlet pressure	Damping ratio
1	High	Small
2	Ultra-high	Could be adjusted
3	Could be controlled	Could be adjusted

As can be seen from the table, of the three controllers tested the only one that was able to provide control over both the outlet pressure and the damping ratio was meter-in and meter-out independent regulating method based on pressure difference sensing.

Although the high brake torque, low dynamic damping ratio and interference between meter-in and meter-out could be avoided when the above-mentioned method is applied, the large time delay would still leads to large overstroke. It should be considered in control algorithm.

In this paper, a grading control algorithm is introduced in the dynamic process of high inertia load.

3. Acceleration process control

To avoid the large time delay between the actuator's velocity and its command value, the acceleration process is divided into two grades: full accelerating grade and controlling grade. When the actuator's velocity is much slower than its command value, the meter-in valve and meter-out valve are all fully opened to shorten the acceleration process. After the velocity error decreased to a threshold, a state feedback control algorithm based on parameters on line estimating is employed to realize both its velocity accuracy and the smoothness of acceleration process. The schematic structure of grading control is shown as fig.2.

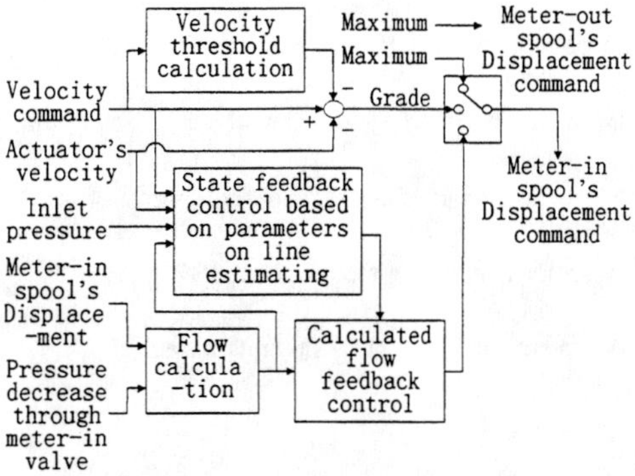

Fig.2 Grading control of acceleration process

There are three main problems that should be considered in the control of acceleration process: Firstly, overstroke should be avoided in controlling grade because the fully opened meter-out valve could not supply enough brake torque. Therefore, the new damping ratio of the system should be about 1.0 after the state feedback control to obtain a critically no-overstroke acceleration process. Secondly, some parameters such as the inertia, viscous damping and leakage change constantly in practical systems. These parameters should be estimated on line in controlling grade. Thirdly, the threshold of the two grades is essential to the grading control algorithm. Too small threshold would leads to long acceleration time but too large threshold would leads to overstroke. An optimum control could be applied to calculate the best threshold. The object function is:

$$\int_0^\infty (\omega_{m0} - \omega_m)^2 dt = \min \tag{1}$$

Where ω_m and ω_{m0} means the motor's shaft-speed and its command respectively.

The experimental setup of high inertia load is also shown as fig.1. The actuator is a high-speed hydraulic motor 10. A flywheel 11, which is connected with the motor shaft directly, is applied as the high inertia load.

The main parameters of this setup are:

Moment of inertia for the flywheel and the rotational part of motor: about 5.1kg·m^2

Viscous damping on motor: about 0.37kg·m^2·s^{-1}

Motor volume/fluid bulk-modulus: about 1.58ml/Mpa

Motor volumetric-displacement: 107ml/r

Maximum pump discharge pressure: 15Mpa

Maximum pump discharge flow: 95L/min

Velocity command: 600r/min.

Fig.3 and fig.4 show the acceleration process of high inertia load under direct close loop of motor's shaft-speed and the above mentioned grading control algorithm respectively.

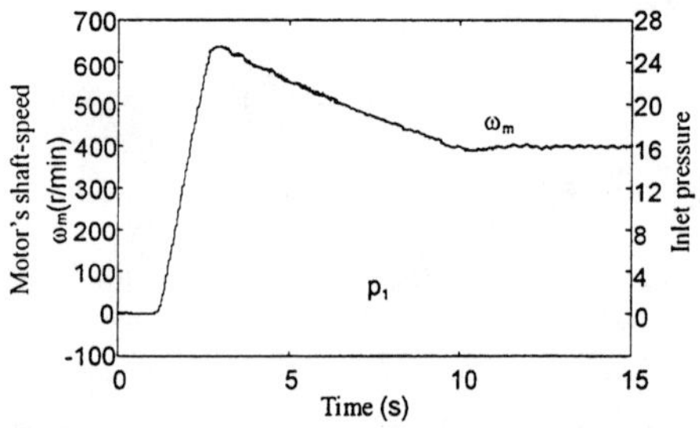

Fig.3 Acceleration process under direct close loop of speed

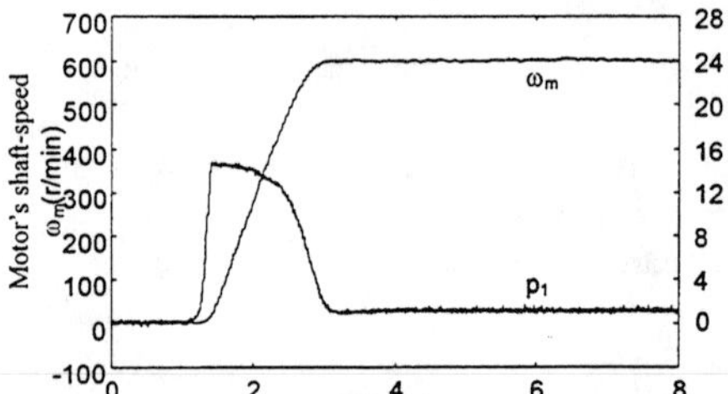

Fig.4 Acceleration process under grading control

The motor's shaft-speed ω_m increased to its command value accurately, smoothly and rapidly when grading control algorithm is introduced.

In CLSS (Close Center Load Sensing System), the pressure of oil supply should adapt to its load pressure. In full accelerating grade of acceleration process, if the meter-in valve were still fully opened, the pump's discharge pressure would increase continuously together with the actuator's inlet pressure until the relief valve opened. Therefore, the inlet pressure should be controlled to its permissible limit in full accelerating grade of acceleration process in CLSS. The schematic structure of modified grading control is shown as fig.5.

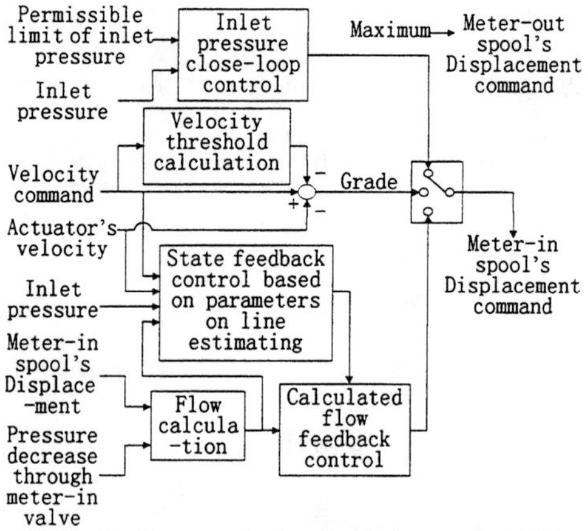

Fig.5 Grading control of acceleration process in CLSS

Fig.6 shows the acceleration process of CLSS when the modified grading control algorithm is applied.

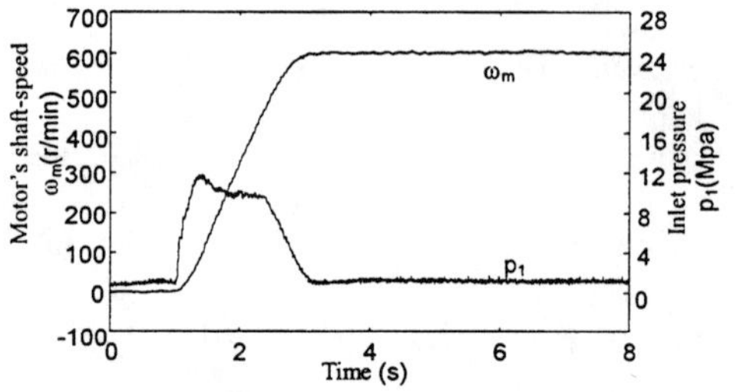

Fig.6 Acceleration process of CLSS

The inlet pressure is controlled to its permissible limit (10Mpa) accurately in full accelerating grade and decreased gradually in controlling grade, while the motor's shaft-speed increased to its command value (600r/min) accurately, smoothly and rapidly.

4. Deceleration and brake process control

The deceleration process is also divided into full decelerating grade and controlling grade. When the velocity of actuator is much faster than its command value, the meter-in valve is fully closed and the actuator's outlet pressure is controlled to its permissible limit through meter-out valve to shorten the deceleration process. After the velocity error decreased to a threshold, a state feedback control algorithm based on parameters on line estimating is also employed to realize both its velocity accuracy and the smoothness of deceleration process. The schematic structure of grading control is shown as fig.7.

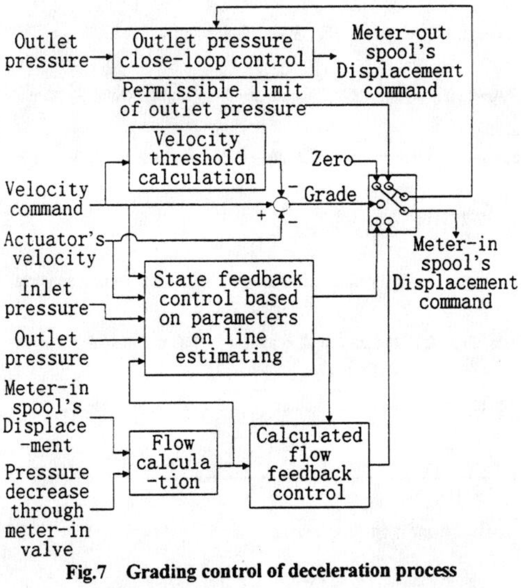

Fig.7 Grading control of deceleration process

There are two additional problems that should be considered in the control of deceleration process: Firstly, the meter-out valve should be fully opened at the end of deceleration process to reduce energy consume. It would become a restrictive factor in state feedback controlling. Secondly, due to the strong nonlinear in the outlet pressure control system through meter-out valve, it would be difficult to control the outlet pressure accurately with a simple close loop control, which would decrease the smoothness of stage switching. Therefore, nonlinear compensation algorithm should be applied to raise the accuracy of outlet pressure.

A simple nonlinear compensation algorithm is to control $\omega_m / \sqrt{p_2}$ instead of the outlet pressure p_2. The main purpose of this compensation is to take the nonlinear out of close loop. Without nonlinear, close loop gain could be improved to raise the accuracy of outlet pressure.

Fig.8 shows the deceleration process of high inertia load when grading control algorithm is introduced.

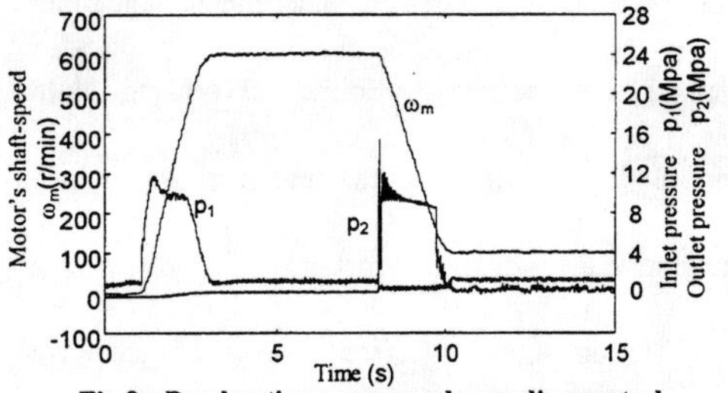

Fig.8 Deceleration process under grading control

The outlet pressure is controlled to its permissible limit (10Mpa) accurately in full decelerating grade and decreased to zero gradually in controlling grade, while the motor's shaft-speed ω_m decreased to its command value (100r/min) accurately, smoothly and rapidly.

In brake process, unlike deceleration process, the accuracy of outlet pressure is no longer needs to be considered, and the meter-out valve also needn't to be fully opened at the end of the process. Therefore, the control algorithm could be simplified: the controlling grade of deceleration process could be canceled in brake process, and the nonlinear compensation algorithm based on close loop control of $\omega_m / \sqrt{p_2}$ is also useless.

Fig.9 shows the brake process of high inertia load when the simplified control algorithm is applied.

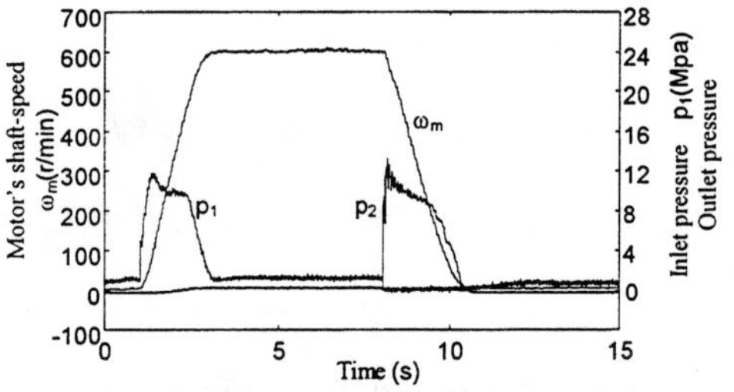

Fig.9 Brake process of high inertia load

The motor's shaft-speed decreased to zero smoothly and rapidly.

5. Conclusions

1. Meter-in and meter-out independent regulating method based on pressure difference sensing could control the actuator's outlet pressure and adjust damping ratio in dynamic process.

2. Through the above mentioned grading control algorithm, the actuator's velocity could increase /decrease to its command value accurately, smoothly and rapidly.

3. Through an additional control of inlet pressure, the grading control of acceleration process could be used in CLSS to improve its performance.

4. A simple nonlinear compensation algorithm, in which $\omega_m / \sqrt{p_2}$ is close-loop controlled instead of the outlet pressure p_2, could be applied in the grading control of deceleration process to improve the smoothness of stage switching.

5. Based on meter-in and meter-out independent regulating method, the brake process's performance could be improved conveniently.

References

Masuda, K., 1993, "Inertial Load Driven by the Pressure Compensated Flow Control Circuit", Journal of the Japan Hydraulics & Pneumatics Society, Vol.24, No.2, pp.196-200.

Tamura, M., et al., 1993, "Development of electric control system for large-sized mining shovel," Proceedings, 2nd JHPS International Symposium on Fluid Power, Maeda, T., et al., ed., E&FN Spon, Tokyo, pp.63-68.

Ichiryu, K., Ochiai, M., and Toyooka, T., 1993, "Simulation Study of Load Sensing Dynamics on Hydraulic Excavator," Proceedings, 2nd JHPS International Symposium on Fluid Power, Maeda, T., et al., ed., E&FN Spon, Tokyo, pp.477-482.

Aoki, Y., et al., 1994, "Load Sensing Fluid Power System," Journal of Commercial Vehicle, SAE Transactions, Vol.103, pp.139-153.

Ichiryu, K., 1995, "Simulation Study of Load-sensing Dynamics for Hydraulic Excavator," Journal of the Japan Hydraulics & Pneumatics Society, Vol.26, No.4, pp.496-503.

Symbols used in this paper

ω, Greek l.c. omega

A NEW APPROACH OF DIRECT DIGITAL CONTROL FOR SPOOL VALVES

J. Ruan
Department of Mechanical Engineering
Zhejiang University of Technology
Hangzhou 310014, P. R. china
Email: wxmin@public.hz.zj.cn

R. Burton
Department of Mechanical Engineering
University of Saskatchewan, 57 Campus Drive
Saskatoon, Saskatchewan, Canada, S7N 5A9
Email: burton@engr.usask.ca

ABSTRACT

In many applications, digital valves driven from stepping motors are often characterized by quantitative errors and in some cases, slow response. A new means of direct digital control is introduced for a spool valve actuated by a stepper motor. With this control strategy, both excellent speed of response and accuracy are simultaneously sustained for the valve. By way of illustration, the characteristics of a digital spool valve are theoretically and experimentally investigated. This paper also deals with the design of the controller and some concepts concerning the digital control of a valve, such as initialization, false protection, etc. An example is given to demonstrate the effectiveness of this digital control approach for a practical electro-hydraulic control system.

INTRODUCTION

In the past two decades, the wide application of computers in the hydraulic industry has seen fundamental changes to fluid power systems; one such change can be labeled 'digitalization'. At present, the most common way to achieved digital control of a fluid power system is to interface a proportional or a servo-valve to a digital controller via a D/A converter. Digital control of this type may be labeled "indirect" for it contains analogue signals and requires A/D conversion. Apart from the need of an additional D/A converter, an indirect digital control system is dependent upon fluid temperature changes and hysterisis in proportional or servo valves. Feedback control of the spool's displacement can effectively reduce the hysterisis, but greatly increase the cost of the valve. In comparison to indirect digital control, there are also "direct" methods of digital control — systems that do not require A/D conversion — for a fluid power system. Two particular methods are PWM (Pulse-Width Modulation) and stepper motor actuation.

In a PWM control mode, an on-off pulse of a finite and variable duration is applied to an on-off valve. The valve responds to the time average of the signal; hence, flow can be modulated by varying the width of the pulse. Some of the first studies reported on PWM techniques were by Murtaugh in 1959. Since then, many studies which have focused on the development and improvement of PWM control both for hydraulic and pneumatic systems, have been initiated. Researchers, such as Tsai and Ukrainetz (1979), El-Ibiary, Ukrainetz and Nikiforuk (1978); Taft and Harned (1980), Lee and Wormaley (1980), Taft, Herrickand and Burke (1981), Usman and Parker (1984), and Tanaka (1986) amongst many others, have been instrumental in advancing PWM control.

Direct digital control using PWM has the advantage of simple construction, and of insensitivity to oil contamination. However, the physical "on/off" states of the valve can induce significant pressure pulsation. Performance of PWM control systems is also limited to a large extent by the mechanical switching speed of the on-off mechanism. Use in open loop applications is also limited.

A second means to achieve direct digital control is to introduce in the system a digital valve which uses a stepper motor to translate the input signal into a stepping motion of a spool. Since the input signals to digital valves are, in fact, a step (pulse) number, digital valves are also called discrete proportional valves. Ramchandran et al (1985) developed a stepper motor driven two-stage digital valve with mechanical feedback. One of the authors, in cooperation with several other researchers, has also developed digital valves for both pressure and flow rate control purposes, by utilizing the rotary and linear motions of a single spool (Xu and Ruan, 1989; Ruan and Xu, 1990; Jiang and Ruan, 1996). A discrete proportional valve suitably interfaced with a stepper motor has been shown to display excellent performance characteristics, such as high repetitive accuracy and low hysteresis (Henke, 1989). However, some digital valve designs tend to display slow responses (EL-Ibiary, etc., Oct.1978; Xu and Ruan, 1989). This is in part because transforming the input signal into a discrete step number can cause a quantitative error. To compensate, the gain of the valve (output per step) should be kept small enough in order to maintain the error rate to be within an acceptable level in practical application. Unfortunately, this small gain corresponds to a large number of steps to achieve the final value and can result in a slow response of the digital valve and hence, system.

This paper considers a new means of direct digital control which controls the angular displacement of the stepper motor continually. In this way the quantitative error for the conventional digital valve actuated by the stepper motor is eliminated and thus both excellent speed of response and accuracy are simultaneously sustained. The characteristics of this digital spool valve are theoretically and experimentally investigated. This paper also deals with the design of the controller and some concepts concerning the digital control of the valve, such as initialization, false protection, etc.

DIRECT DIGITAL CONTROL OF A SPOOL VALVE

Consider the digital spool valve illustrated in Figure 1. The valve uses a two-phase Slo-syn stepper motor as its electro-mechanical transformer to actuate and position the spool. A transmission device designed for rotary to linear motion conversion is used. Metal string is wound once on the driving wheel of the stepper motor. Part of the metal string is inlaid in a narrow groove on the driving wheel to prevent it from sliding. The two ends of the string are fixed to an elastic bow which is rigidly connected with the spool; tension in the string is maintained by the elastic bow. Using this arrangement, the spool displays precise linear motion for an angular displacement in the stepper motor. This valve will now be used to illustrate the control algorithm proposed in this paper.

Mathematical Model of the Valve

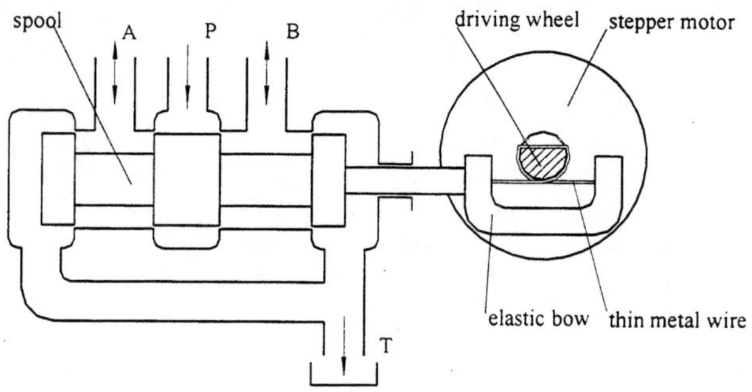

Figure 1 Schematic of Digital Spool Valve

Voltage Balance Equation of the Stepper Motor. When the stepper motor is driven with a phase current not beyond the rated value, the magnetic circuit does not saturate. And because the magnetic circuit is designed to have a large resistance and good magnetic conductivity, both the eddy current and the magnetic hysteresis are negligible. Thus the voltage balance equation for phase A and B of the stepper motor can be express (Takashi, 1986):

$$u_A = Ri_A + L\frac{di_A}{dt} - KZ_i\sin(Z_i\theta)\frac{d\theta}{dt} \quad (1)$$

$$u_B = Ri_B + L\frac{di_B}{dt} + KZ_i\cos(Z_i\theta)\frac{d\theta}{dt} \quad (2)$$

Equation (1) and (2) provide the voltage balance of the phase A and B in discrete form. From them a model coordinating the continual control of the angular displacement is further established. It is assumed that:

$$\left.\begin{aligned} i_A &= i_m\cos(Z_r\theta_m) \\ i_B &= i_m\sin(Z_r\theta_m) \end{aligned}\right\} \quad (3)$$

where $i_m = \sqrt{i_a^2 + i_b^2}$ and $\theta_m = \frac{1}{Z_i}\arctan\frac{i_B}{i_A}$.

Substituting Equation (3) in Equations (1) and (2) and considering the case in which the stepper motor has a round-shape magnetic field, (i.e., i_m is constant), yields:

$$u_A = Ri_m\cos(Z_i\theta_m) - LZ_ii_m\cos(Z_i\theta_m)\frac{d\theta_m}{dt} - KZ_i\sin(Z_i\theta_m)\frac{d\theta_m}{dt} \quad (4)$$

$$u_B = Ri_m\sin(Z_i\theta_m) + LZ_ii_m\sin(Z_i\theta_m)\frac{d\theta_m}{dt} + KZ_i\cos(Z_i\theta_m)\frac{d\theta_m}{dt} \quad (5)$$

Multiplying Equation (4) by $\sin\theta$ and Equation (5) by $\cos\theta$ and subtracting yields:

$$u_m\sin Z_i(\theta_{mc} - \theta_m) = L_1Z_ii_m\frac{d\theta_m}{dt} + KZ_i\cos Z_i(\theta_m - \theta)\frac{d\theta}{dt} \quad (6)$$

In Equation (6), $u_m = \sqrt{u_A^2 + u_B^2}$ and $\theta_{mc} = \frac{1}{Z_i}\arctan\frac{u_b}{u_a}$. In this expression θ_{mc} is the control signal of the angular displacement of the rotary magnetic field θ_m.

Torque Equations. The torque of the stepper motor is:

$$T_e = -K_{TP}(i_A\sin(Z_r\theta) + i_B\cos(Z_r\theta)) \quad (7)$$

which simplifies to:

$$T_e = T_m\sin Z_r(\theta_m - \theta) \quad (8)$$

In Equation (8) $T_m = K_{TP}\sqrt{i_A{}^2 + i_B{}^2}$ and K_{TP} is the coefficient of torque.

Force Balance Equation of the Spool. For the spool illustrated in Figure 1,

$$\frac{T_e}{r} = m_g\frac{d^2x_v}{dt^2} + B_c\frac{dx_v}{dt} + F_b + F_f \quad (9)$$

Where $m_g = m_v + J/r$, $x_v = r\theta$, and $F_b = 2C_dC_Vx_vW(p_s - p_L)$; From Equations (6) and (8) it can be observed that the angular displacement of the rotary magnetic field θ_m has a phase shift with respect to the control signal θ_{mc} due to the effect of the inductance; in addition, the angular displacement of the rotor θ experiences a phase shift with respect to θ_m because of load. The phase relationship is illustrated in Figure 2. It can be noted that θ lags θ_m and θ_{mc} to some extent. The phase shift between θ_m and θ_{mc} can be attributed to the type of the power supplied to the stepper motor. For a stepper motor operating under a constant-current power supply, the phase shift between θ_m and θ_{mc} will be greatly reduced; in the limit, θ_m is in phase with θ_{mc}.

Continual Control of a Stepper Motor

The spool's displacement (and hence the flow rate through the valve) is proportional to the angular displacement of the stepper motor, $x_v = r\theta$. Because the stepper motor can only provide discrete angular displacement increments, $N\Delta\theta$ ($N = 1,2,\cdots,N_0$; $\Delta\theta$ is the step length) the resolution of the spool displacement (and hence the flowrate) is very dependent upon the size of the angular

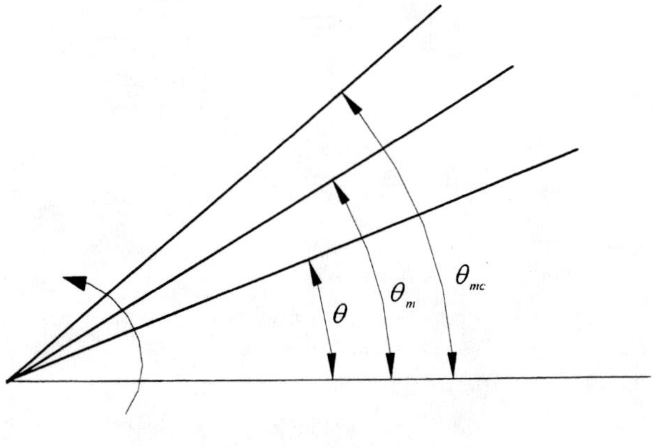

Figure 2 Relation between θ, θ_m and θ_{mc}

steps. It thus becomes a problem of finding a compromise between fast response (large $\Delta\theta$) and accuracy (small $\Delta\theta$). Ideally, it would be highly desirable to control the angular displacement of the stepper motor continually.

There are several possible methods to obtain continual control of the stepper motor. The first is by controlling the phase currents. An arbitrary angular displacement $N\Delta\theta+\alpha$ (where α is smaller than $\Delta\theta$) can be reached by adjusting the phase currents i_A and i_B of the stepper motor to meet following condition (Takashi 1986):

$$\alpha = \frac{1}{Z_i}\arctan\frac{i_B}{i_A} \tag{10}$$

or:

$$\begin{cases} i_A = i_m \cos Z_i\alpha \\ i_B = i_m \sin Z_i\alpha \end{cases} \tag{11}$$

A second method is by introducing PWM control to the phase currents and creating energizing intervals T_A and T_B to correspond to the phases A and B using the following relationships:

$$\alpha = \frac{1}{Z_i} a\tan\frac{T_B}{T_A} \tag{12}$$

or:

$$\begin{cases} T_A = T\dfrac{\cos Z_i\alpha}{\cos Z_i\alpha + \sin Z_i\alpha} \\ T_B = T\dfrac{\sin Z_i\alpha}{\sin Z_i\alpha + \cos Z_i\alpha} \end{cases} \tag{13}$$

The continual control of the digital valve is physically realized using a "one-chip" computer system. A special algorithm has been

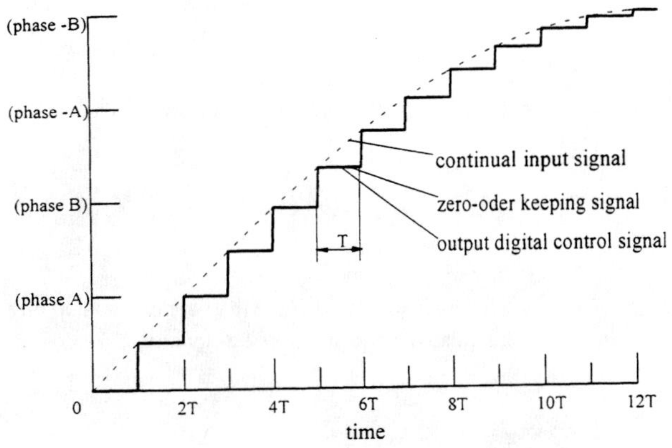

(a) phase current control

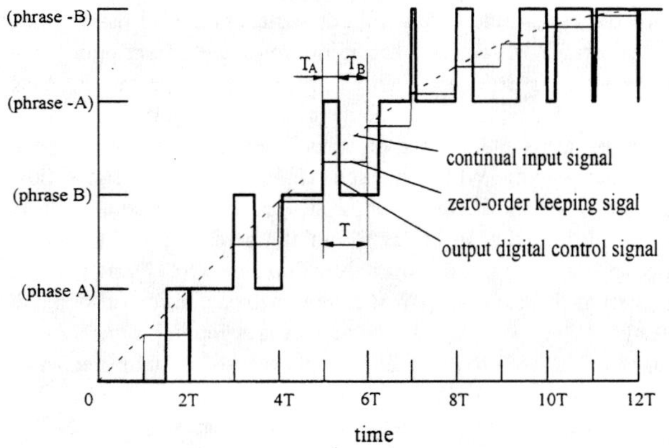

(b) PWM control

Figure 3 Digital Control Signals

developed for this purpose. Within each sampling cycle, *T*, the one-chip computer reads the discrete value of the input digital signal, *u(k)*, and converts it into a special digital control signal to the stepper motor, θ_{mc}. This process is conducted in three steps. For the first step, an initial control signal for the stepper motor is obtained by dividing the input signal by the electrical gain of the motor, K_e, i.e.,

$$\theta_{mc}(k) = u(k)/K_e \tag{14}$$

In the second step, a limiting value, θ_v, is set to the maximum increment of the digital control signal. This keeps the stepper motor operating in a state of a perfect tracking to the input control signal without from losing steps. This is accomplished in the algorithm program using the following logic:

If

$$|\ \theta_{mc}(k) - \theta_{mc}(k-1)\ | < \theta_v \tag{15}$$

then:

$$\theta_{mc}(k) = \theta_{mc}(k) \quad (16)$$

If conditions specified in Equation (15) is not true, then:

$$\theta_{mc}(k) = \theta_{mc}(k-1) + \mathrm{sign}(\theta_{mc}(k) - \theta_{mc}((k-1)) \cdot \theta_v \quad (17)$$

In the last step, the distance of $\theta_{mc}(k)$ to the nearest stepping point $N\Delta\theta$, α, is calculated through following equation:

$$\alpha = \theta_{mc} - \Delta\theta \times \mathrm{integer}(\theta_{mc} / \Delta\theta) \quad (18)$$

The phase currents, i_A and i_B, (or their equivalent energizing intervals, T_A and T_B) are established by Equation (11) (or (13)). Based above discussion, the digital control signals produced using phase control or PWM control according to above descriptions are illustrated in Figure 3a and Figure 3b respectively. The pattern of the control signal produced with the method of phase current control is essentially the same as that of an ordinary sampling signal with zero-order keeping. When Equation (15) is satisfied, the signal flow diagrams of continual control of the stepper motor for both methods can be illustrated as in Figure 4. In this figure, *V(s)* is the transfer function which can be developed from Equations (6), (7) and (9) with the linear analysis. For the PWM control method, the control signal of the stepper motor is a composition of a sampling signal and a "fragment" signal, *f(s),* which exists because of high frequency components of the PWM signal (Ruan and Burton, 1999). The fragment signal will stimulate addition phase currents, $\Delta i_A, \Delta i_B$ in the form of:

$$\Delta i_A = \begin{cases} i_m(1-\cos Z_r\alpha)\,(kT \le t < (k+\cos Z_r\alpha)T) \\ -i_m \cos Z_r\alpha\,((k+\cos Z_r\alpha)T \le t \le (k+1)T) \end{cases} \quad (19)$$

$$\Delta i_B = \begin{cases} -i_m \sin Z_r\alpha\,(kT \le t < (k+1-\sin Z_r\alpha)T) \\ i_m(1-\sin Z_r\alpha)\,((k+1-\sin Z_r\alpha)T \le t \le (k+1)T) \end{cases} \quad (20)$$

$\Delta i_A, \Delta i_B$ must satisfy the following integration equations:

$$\begin{cases} \int_0^{kT} \Delta i_A(t)dt = 0 \\ \int_0^{kT} \Delta i_B(t)dt = 0 \end{cases} \quad (k \in N) \quad (21)$$

Equation (21) implies that the time-mean values of $\Delta i_A, \Delta i_B$ are both equal to zero. This means that under steady state conditions, the control signal produced by the PWM control method is equivalent to that produced by the phase current method. A spectral analysis of Δi_A,and Δi_B for a typical application would show that the energy is concentrated in the 500 to 4000 Hz. range. The fragment signal, *f(t),* will be filtered to a large degree by the digital valve and the loading

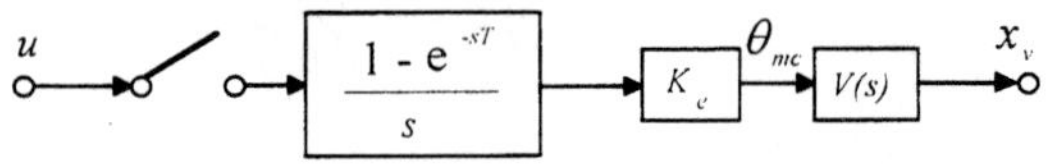

(a) phase current control

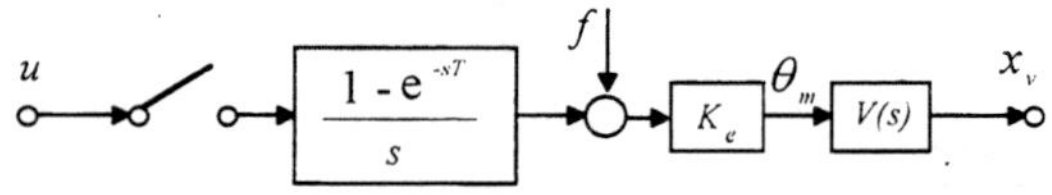

(b) PWM control

Figure 4 Signal Flow Diagrams

actuator. Therefore, the fragment signal can be considered as having little effect on either the static or the dynamic characteristics of the driving system; it does, however acts as a dither in the valve and is favorable in reducing the frictional force of the spool.

Control of a Digital Spool Valve

The algorithm is just one consideration in the control of a digital spool valve. In the complete control program of the digital valve, several other considerations concerning the digital valve have been taken into account. They include initialization, power monitoring and energy saving modes.

Initialization. When the power supply of the controller is turned on, the stepper motor is "zeroed" to some starting position. This process is defined as the initialization of the stepper motor. There are two ways to carry this out; physical initialization and logical initialization.

In physical initialization the spool is positioned (centered) physically to coincide to the zero position of the stepper motor. This requires the setting of a "mechanical dead-point". The stepper motor is first moved to this point to establish a reference. Then, the motor is stepped back a known number of steps to its zero point.

In logical initialization, when the power supply is turned off after use, the stepper motor is returned to zero position through a power monitoring control (to be discussed). At this point, the stepper motor is connected as a generator and the rotor is locked. The stepper motor is kept in the zero position by the action of the permanent magnetic torque of the stepper motor. When establishing a logical initialization, the controller directly controls the phases of the stepper motor which correspond to the zero position.

Power Monitoring. When the power supply turns off or fails due to an accident, the stepper motor is forced to its zero position using a large capacitor.

Energy Saving Working Mode. If the input signal is unchanged for a number of sampling cycles, the phase current of the stepper motor is decreased to a lower percentage of the rated current. When the input signal changes, the phase current is returned to the rated current. This type of control is often found in electro-hydraulic open loop applications.

ANALYSIS OF FREQUENCY RESPONSE OF DIGITAL CONTROL

Analysis of a non-linearity in the Digital Control

Because there is a limitation to the rate of θ_{mc} (given by the Equation (17)), the digital control signal waveform will show distortion if the rate of input signal, *u(t)*, exceeds this limitation. To demonstrate this assume that the sampling cycle T is small enough that the digital control signal, *u(t)*, can be regarded as a smooth wave. For a sinusoidal wave signal input, i.e., $u(t)=A_m \sin(2\pi f_s\ t)$, the increment between two sample points, k and $k-1$ (that is $u(k)-u(k-1)$), of the input signal can be approximately expressed as:

$$u(k)-u(k-1)\approx \dot{u}(t)T = 2\pi f_s A_m T\cos 2\pi f_s t$$
$$= \lambda \cos 2\pi f_s t \qquad (22)$$

where $\lambda = 2\pi f_s A_m T/\theta_v$ (the non-dimensional frequency). If the digital control signal θ_{mc} can smoothly restore the input signal, *u(t)*, then according to Equation (15), the following relationship should exist (here, the electrical gain K_e is assumed to be equal to 1):

$$\lambda\left|\cos 2\pi f_s t\right| < 1 \qquad (23)$$

When $\lambda \leq 1$, the digital control signal, θ_{mc}, will linearly restore the input sinusoidal wave, as illustrated in Figure 5a. This is because the maximum slope of the stepper motor angular displacement (with respect to time) is always greater than the instantaneous slope of the input signal. Thus, it can always "catch up" to the input signal at every step. With an increase in the value of frequency, f_s, or amplitude, A_s, and with $\lambda > 1$, the digital control signal, θ_{mc}, will be unable to restore the input sinusoidal wave completely and shows a distortion, see Figure 5b. In this case, the maximum slope of the stepper motor angular displacement (with respect to time) is less than the slope of the input signal and hence cannot catch up to the desired input signal. Thus the slope will stay constant until it finally catches up to the input signal as shown in Figure 5(b). In the regions BC and DE, the digital control signal can still restore the input sinusoidal wave; while in regions AB, CD and EF the digital control signal diverges from the sinusoid wave.

As λ increases further, the digital control signal will diverge completely from the sinusoidal wave and become a triangular signal with a phase delay to the input sinusoidal wave, see Figure 5c. Between the cases illustrated in Figures 5a and 5c, there is a critical value for λ, at which the linear areas BC and DE contract to zero. Then points C and D coincide with points B and E respectively and the distance from point A to axis ωt is equal to that from point B or point D to axis ωt. Moreover, B and D are the points of tangency of lines CD and EF to the sinusoidal curve. From above geometric relations it can be calculated that the critical value of λ is 1.618.

Based on the geometrical relationships shown in Figure 5 and the discussion above, the Digital control signal wave for different ranges of λ value are developed to be:

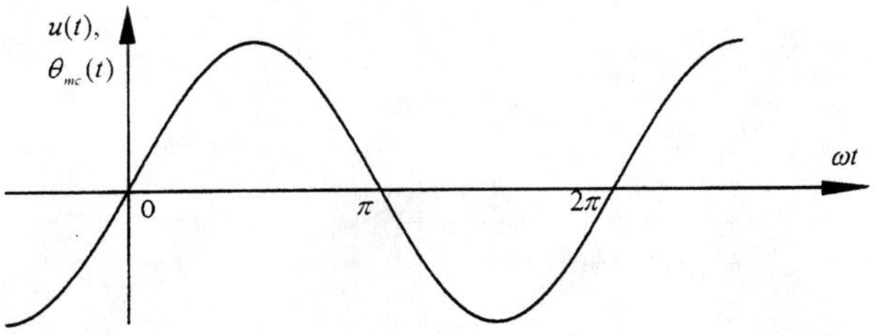

(a) $\lambda \leq 1$

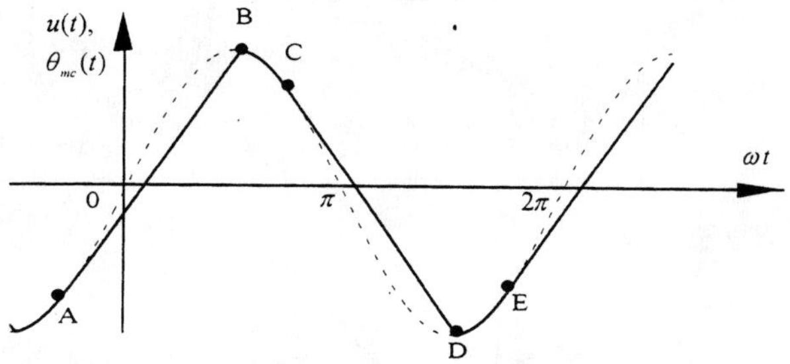

(b) λ=1.4

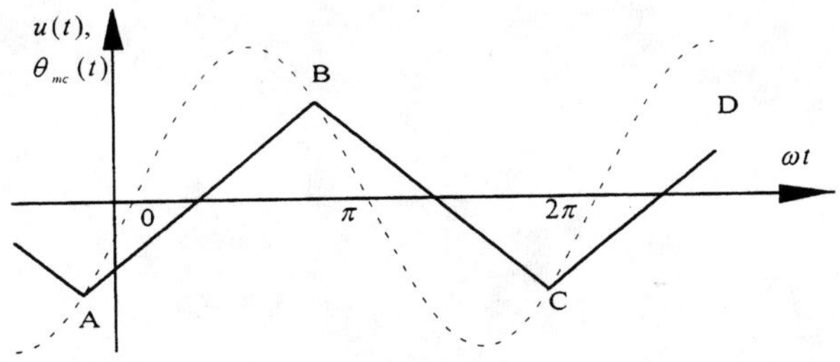

(c) λ=2.5

Figure 5 Digital Control Signal Wave

(a) $\lambda < 1$,

$$\theta_{mc}(\alpha_t) = A\sin\alpha_t \ \text{ where } \ \alpha_t = 2\pi f_s t \qquad (24)$$

(b) $1.168 > \lambda \geq 1$,

$$\theta_{mc}(\alpha_t) = \begin{cases} \dfrac{\theta_v}{2\pi f_s T}\alpha_t + A_s\sin\alpha_t - A_s\alpha_A\cos\alpha_A \\ \quad (\alpha_A \leq \alpha_t < \alpha_B) \\ A_s\sin\alpha_t(\alpha_B \leq \alpha_t < \pi+\alpha_A) \\ \dfrac{\theta_v}{2\pi f_s T}\alpha_t - A_s\sin\alpha_A - A_s(\pi+\alpha_A)\cos\alpha_A \\ \quad (\pi+\alpha_A \leq \alpha_t < \pi+\alpha_B) \\ A_s\sin\alpha_t(\pi+\alpha_B \leq \alpha_t < 2\pi+\alpha_A) \\ \dfrac{\theta_v}{2\pi f_s T}\alpha_t + A_s\sin\alpha_A - A_s(2\pi+\alpha_A)\cos\alpha_A \\ \quad (2\pi+\alpha_A \leq \alpha_t < 2\pi+\alpha_B) \end{cases} \qquad (25)$$

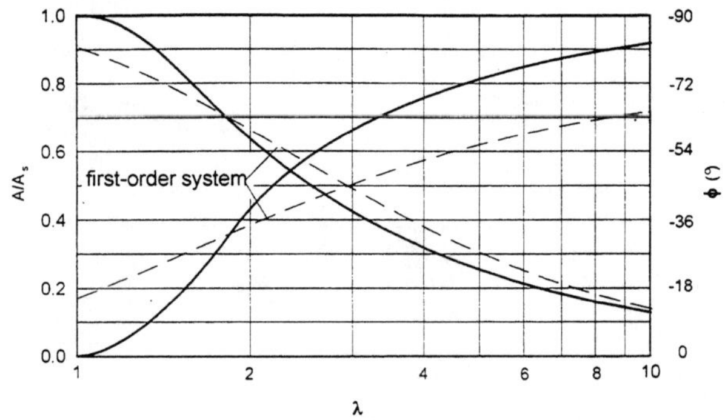

Figure 6 Frequency Response of the Digital Control Signal

where $\alpha_A = -\arccos\frac{1}{\lambda}$. α_B is given by the following implicit equation:

$$\sin\alpha_B = \alpha_B\cos\alpha_A + \sin\alpha_A - \alpha_A\cos\alpha_A \tag{26}$$

(c) $\lambda \geq 1.618$, then:

$$\theta_{mc}(\alpha_t) = \begin{cases} \frac{\theta_v}{2\pi f_s T}(\alpha_t - \alpha_A) + A_s\sin\alpha_t \\ (\alpha_A \leq \alpha_t < \pi + \alpha_A) \\ \frac{\theta_v}{2\pi f_s T}(\pi + \alpha_a - \alpha_t) - A_s\sin\alpha_A \\ (\pi + \alpha_A \leq \alpha_t < 2\pi + \alpha_A) \\ \frac{\theta_v}{2\pi f_s T}(\alpha_t - 2\pi - \alpha_A) + A_s\sin\alpha_A \\ (2\pi + \alpha_A \leq \alpha_t < 3\pi + \alpha_A) \end{cases} \tag{27}$$

where $\alpha_A = -\arcsin\frac{\pi}{2\lambda}$.

Frequency Response of the Digital Control

The digital control of the stepper motor will display a non-linearity when the value of λ is larger than 1. In order to do a frequency response for the nonlinear digital control signal, the describing function method is used. The fundamental component of the digital control signal can be expressed as:

$$\theta_{mc0}(\alpha_t) = A_0\sin(\alpha_t + \phi) \tag{28}$$

where $A_0 = \sqrt{a^2 + b^2}$, $\phi = a\tan\frac{a}{b}$. The values of a and b are

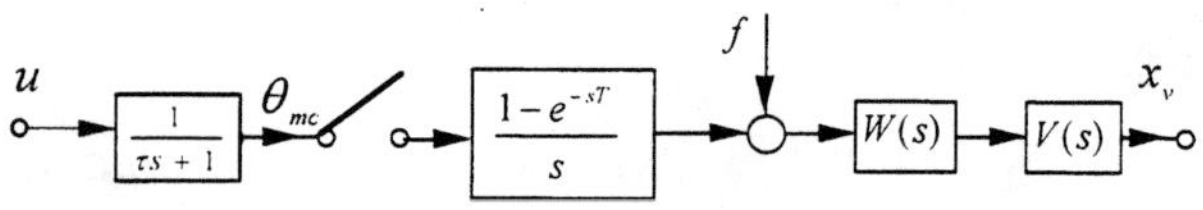

Figure 7 Linear Model of the Digital Valve

established using the Fourier integration equations:

$$a = \frac{1}{\pi}\int_0^{2\pi}\theta_{mc}(\alpha_t)\cos\alpha_t d\alpha_t \tag{29}$$

$$b = \frac{1}{\pi}\int_0^{2\pi}\theta_{mc}(\alpha_t)\sin\alpha_t d\alpha_t \tag{30}$$

Substituting Equations (25) and (27) into Equations (29) and (30), the values of a and b can be calculated. The amplitude ratio, A_0 / A_s, and the phase angle, ϕ, versus the nondimensionalized frequency, λ, can then be evaluated. The frequency response for typical parameter values is shown in Figure 6. It is noted that $\phi \to -90$ as $\lambda \to \infty$ and at the critical point i.e., λ =1.618, $A_0 / A_s \approx 0.7$. For this reason the digital control of the stepper motor may be approximated as a first order system as illustrated in Figure 6 for $\lambda > 1$. The transfer-function between $\theta_{mc}(s)$ and $u(s)$ is thus:

$$\frac{\theta_{mc}(s)}{u(s)} = \frac{1}{\tau s + 1} \tag{31}$$

The time constant τ is evaluated at the point λ =1.618, $A_0 / A_s \approx 0.7$, which means that the bandwith of the digital control is the same as that of a first-order system:

$$\tau = \frac{3.88TA_s}{\theta_v} \tag{32}$$

The time constant, τ, will decrease with a decrease in the amplitude of the input signal, A_s. This means that the digital spool valve will have a faster response when operating near the neutral position (where A_s is small). In practical applications, the valve usually operates near the neutral position in order to avoid flow rate saturation.An approximate linear model of the digital valve which reflects the various linear and nonlinear regions using direct digital control is illustrated in Figure 7.

Convergence of the Digital Control Signal

The above analysis is based on a periodic steady Digital control signal which is symmetrical about the zero-axis and has a zero DC-component. From a static condition the digital control signal can experience a transient when a sinusoid or other periodic signal is the input. A question arises as to whether the digital control signal will converge to the steady-state waveform after the transient period. When $\lambda < 1.618$, the digital control signal converges to a steady

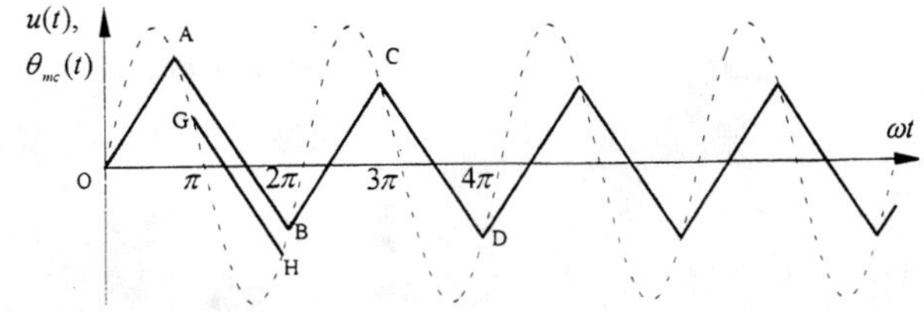

Figure 8 Dynamic Response of the Digital Control Signal

state waveform as shown in Figures 5a or 5b. To address the convergence concern, it is necessary only to consider $\lambda \geq 1.618$. The dynamic response of the digital control signal to a sinusoidal input wave is shown in Figure 8. Lines OA, AB, BC and CD illustrate the digital control signal for this condition. In an equivalent *x-y* coordinate system, A, B, C and D can expressed as $A(x_A, y_A)$, $B(x_B, y_B)$, $C(x_C, y_C)$ and $D(x_D, y_D)$ respectively. If convergence occurs, then the following geometric considerations exist:

(1) If $|y_A| > |y_B|$, then $|y_A| > |y_C|$;

(2) If $|y_A| > |y_B|$, then $|y_C| > |y_B|$;

(3) If $|y_A| > |y_B|$, then $|y_D| > |y_B|$;

(4) If $|y_A| > |y_B|$; then $|y_C| > |y_D|$.

It is interesting to note that conditions (2) and (3) ensure that the control signal will converge to a steady value whilst conditions (1) and (4) mean that the convergence is asymptotical.

EXPERIMENTAL CONSIDERATIONS

An experimental system was set up to determine the static and dynamic characteristics of the digital servo valve illustrated in Figure 1. A schematic of the circuit is shown in Figure 9. The circuit consisted of a hydraulic power supply and loading sub-system, a digital valve control sub-system and a performance monitoring and recording sub-system. The hydraulic power supply and loading sub-system included the pump, the digital valve to be tested and a throttle valve (which served as a loading valve). The digital valve control sub-system included a Personal Computer, which was utilized as a signal generator, and a digital valve controller. The controller contained a signal conversion board and a power amplification board. The signal converter was used to convert the input signal to the digital control signal (via the rotary magnetic field of the stepper motor). Within each sampling cycle, *T*, the controller received the signal sent from the signal generator, executed the digital algorithm program and outputing the appropriate signal (via a power amplifier) to the stepper motor. In this experiment, the PWM method is used to control the angular displacement in a continuous sense.

The performance monitoring and recording subsystem included a non-contact eddy-current displacement transducer, a flow transducer and a storage oscilloscope. The measured data from the displacement transducer was first stored in the oscilloscope and later transferred to a PC. By utilizing a special curve-drawing software called WINGRAF, the measured data was converted into graphical transient and frequency plots. Some of the physical parameters of the valve and stepper motor are summarized in Table 1.

To illustrate the response of the digital valve to the proposed algorithm, several types of inputs were considered: ramp, step and sinusoidal.

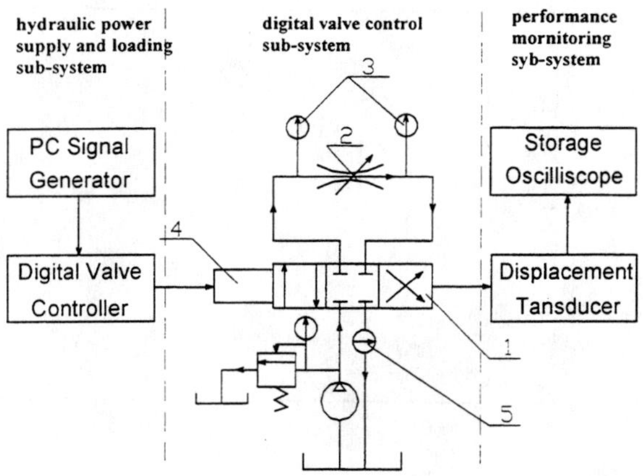

1---digital spool valve; 2---throttle valve; 3---pressure gauge
4---stepper motor; 5---flow rate meter

Figure 9 Experimental System

Table 1 Physical Parameters

Valve			
Maximum step number	**spool stroke (mm)**	**pressure (MPa)**	**maximum flow rate (l/min, p_A-p_B=3.5 MPa)**
±4	0.34	21	12
Stepper Motor			Controller
Phase current(A)	**static torque (NM)**	**Stepping angle**	**sampling cycle T (ms)**
2.4	0.67	0.9°	0.5

RESPONSE TO A TRIANGULAR WAVE

Figures 10 (a) and (b) illustrate the simulated and experimental spool response to a triangular input signal. Because of the nature of this kind of input, the output waveform can be used to assess the static characteristic of the digital valve. From Figure 10(b) it can be observe that there is no hysteresis in the valve. The output waveform is composed of a fundamental triangular wave and a barely visible high-frequency perturbation which introduced by the fragment signal. This perturbation acts as a dither and is particularly significant in the reducing coulomb frictional force at low motion speed of the spool and sustains the high accuracy of the valve (in the spool digital valve coulomb frictional is a most important factor affecting the repetitive accuracy of the valve).

Response to a Rectangular Wave

The spool's response to rectangular waveform (step) is presented in Figure 11. For the step response of the -95% - +95% full-stroke of the spool, the dynamic response time, which is defined as the time interval in which the angular displacement reaches 98% of its steady value, is about 10 milliseconds.

Response to a Sinusoidal Wave

The spool's responses to sinusoidal inputs at different frequencies are shown in Figure 12. When the non-dimensional frequency, λ, is smaller than 1, the output waveform shows no

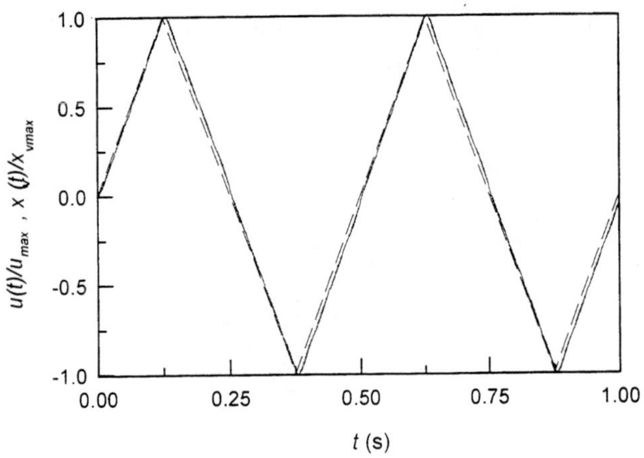

(a) Simulated results

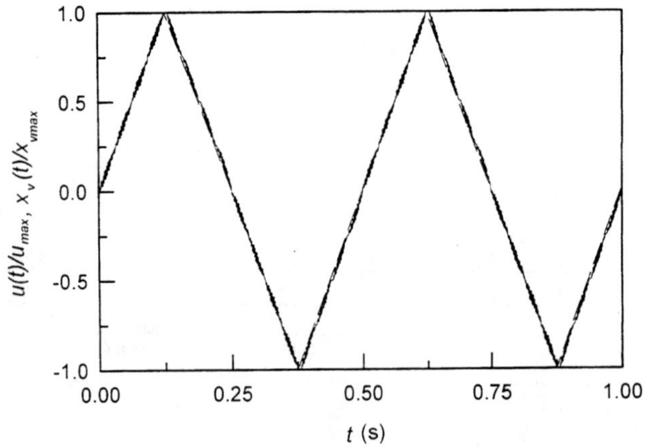

(b) Experimental results

Figure 10 Response to a Triangular Wave

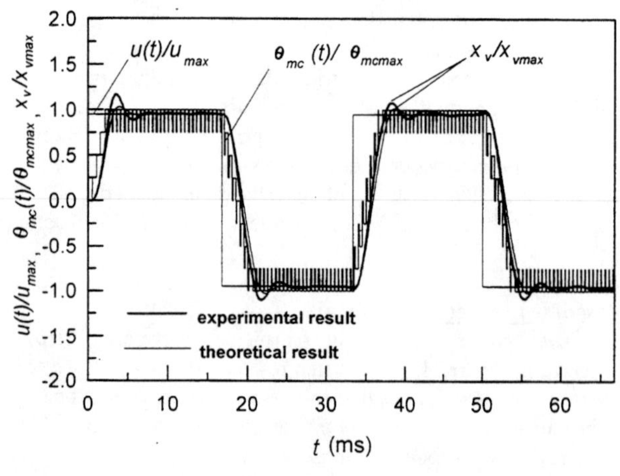

Figure 11 Response to Rectangular Wave

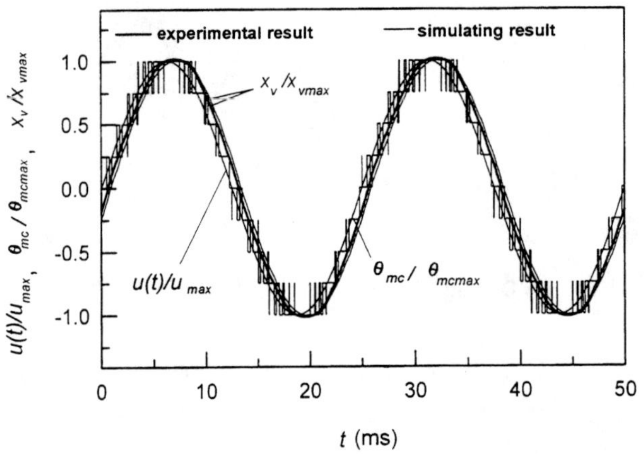

(a) λ=0.5

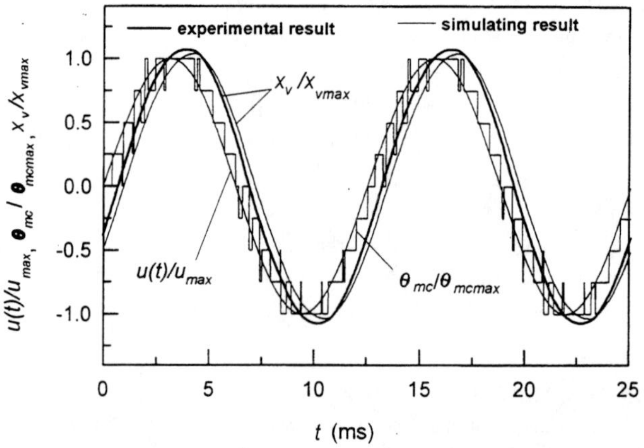

(b) λ=1

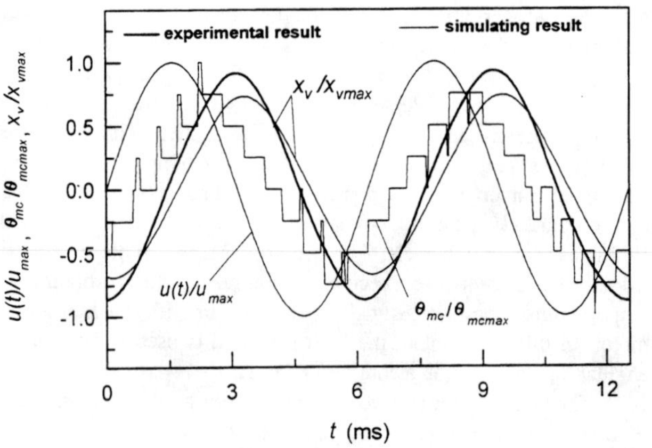

(c) λ = 2

Figure 12 Spool's Response to Sinusoidal Wave

observable distortion (see Figures 12(a) and (b)). When λ is larger that 1, significant distortion in the output waveform in the form of a phase shift, is noted (Figure 12(c)).

For the different range of inputs, the characteristics of the spool's frequency response are given in Figure 13. There is a change of the bandwidth of the valve for the input sinusoidal signals of different amplitude. For an input sinusoidal signal with an amplitude

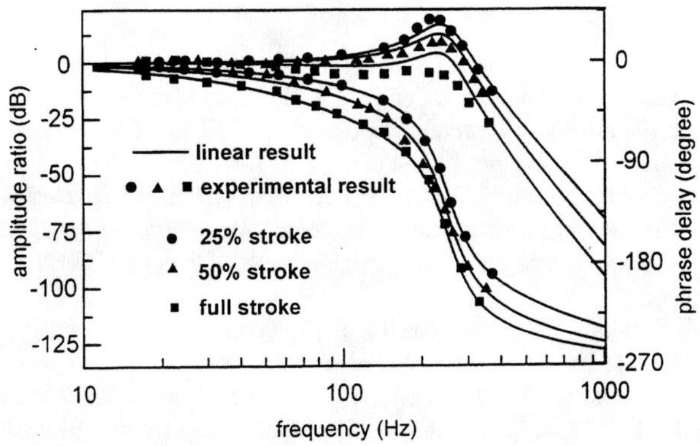

Figure 13 Frequency Response of the Valve

corresponding to 25% stroke, the bandwidth is about 250 Hz. This value drops to nearly 200 Hz for an amplitude corresponding to full spool's stroke. The change is caused by the non-linearity of proposed algorithm.

From Figures (10), (11), (12) and (13), it can be observed that the theoretical results diverge somewhat from the experimental ones. The reason for these differences is that some nonlinear factors, (eddy current and hysteresis, for example) have been neglected in the analysis. In addition, it is difficult to precisely assign a realistic value to the spool's viscous friction

APPLICATION EXAMPLE

The digital control approach and digital valve discussed in this paper has been applied to the electro-hydraulic control system of a cement strength-testing device. In this application, the digital valve output to be controlled was pressure. Consequently, the signal driving the digital valve was, in fact, a pressure error signal. A schematic of the system is shown in Figure 14. In accordance to cement testing standards, the cement sample must be loaded at a constant rate of 2.4 KN/s, until it breaks (the maximum load is 300 KN). In this process, the load on the cement sample must be very smooth. This is accomplished using an electro-hydraulic closed-loop pressure control system (Figure 14). The digital valve consisted of a spool valve digitally operated and a pressure compensator placed in parallel to the valve. The pressure compensator was used to reduce the heat of the system by forcing the pressure drop across the valve to be very small and to the absorb some of the pressure ripple introduced by the piston pump. The PC (Figure 14) was used as the system controller. The measured pressure was compared to the desired input loading ramp signal. Within the PC, the error was then processed by the PID algorithm and sent through a parallel port (printing port) directly to the digital valve controller to operate the digital valve.

The experimental system showed that the system could load the cement sample smoothly at 2.4 KN/s as well as 0.3 KN/s. Under normal control conditions, the cement strength-testing device tends to "creep" at lower rates. The two loading process are presented in Figures 15 (a) and (b).

CONCLUSIONS

For a more conventional step operating mode, a stepper motor can only be controlled to position at finite discrete points. Thus, when it is applied as an electro-mechanical interface to a valve, it is very difficult to maintain high accuracy and fast response. In this paper a different tracking algorithm is introduced which controls the angular

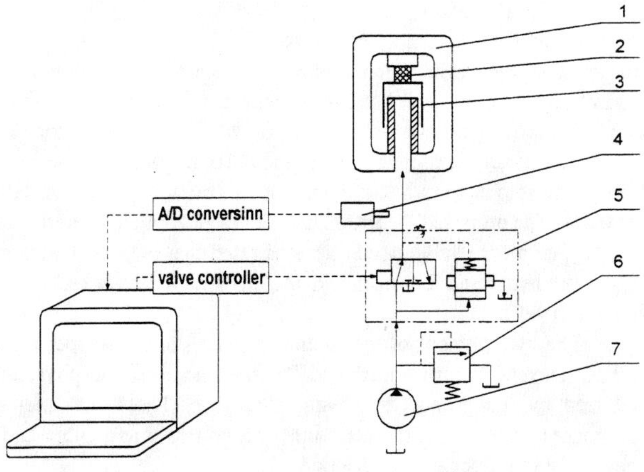

1---frame; 2---cement sample; 3---cylinder; 4---pressure sensor; 5---digital flow rate valve; 6---relief valve; 7---piston pump

Figure 14 Schematic of the Cement Testing Device

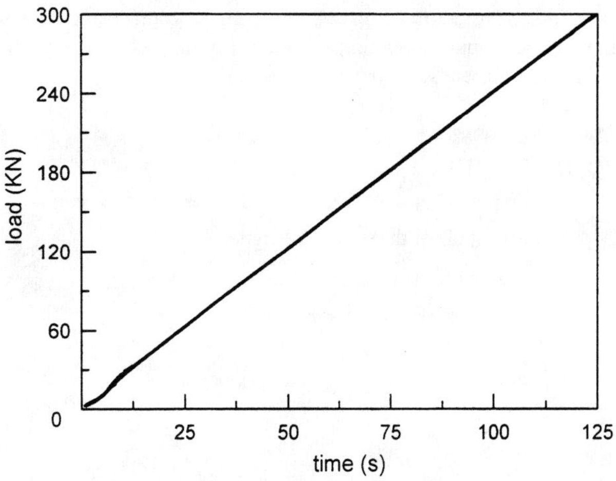

(a) loading rate equal to 2.4 KN/s

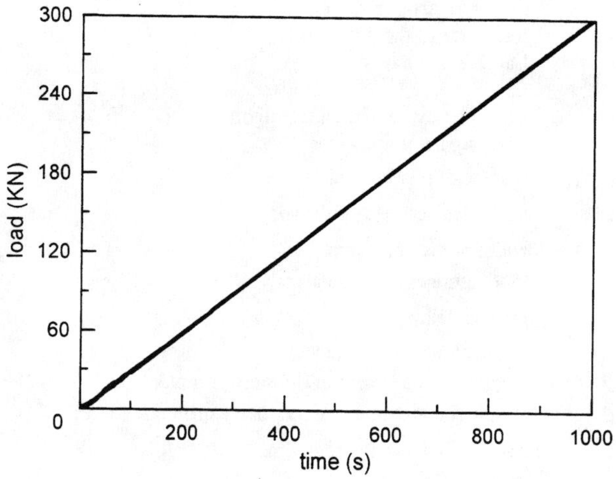

(b) loading rate equal to 0.3 KN/s

Figure 15 Experimental Loading Process

displacement of the stepper motor in a continuous PWM mode. The response of the valve is maintained by selecting small number of steps to achieve full spool stroke whilst the accuracy of the valve is kept through continuous PWM control to the stepper motor. The response time of the digital valve is about 10 ms for an −95%~+95% full stroke step input; while the repetitive of the valve is estimated to be 0.4%. The proposed algorithm has also been applied to another 2D digital directional valve which demonstrated that both accuracy and fast response could be sustained at fairly high levels (Ruan and Burton, 1999).

The two methods of the continual control of the stepper motor (phase current control and the PWM approach) can be physically realized by the proposed tracking algorithm. The PWM method produces an additional fragment signal which serves as a dither and is beneficial in reducing static friction.

In order to prevent the stepper motor from losing steps, the rotary speed of the stepper motor is limited in the tracking algorithm; this introduces a non-linearity in the valve's operation ($\lambda > 1$). It has been shown that the non-linearity of the valve can be approximated as a first-order linear system for these conditions.

In conclusion, the proposed direct digital control algorithm is effective for fluid power control. Indeed, it has been successfully applied to a cement strength testing device.

NOMENCLATURE

A_0	amplitude of the fundamental sinusoidal component
A_m	amplitude of the sinusoidal signal
f_s	frequency of sinusoid signal
i_A, i_B	current of phase A and phase B
J	rotary inertia
K	coefficient of the inductive of E. M. F.
K_e	gain of electrical part
L	impedence of phase
m_v	mass of spool and attached elements
N_0	stepping number corresponding to the spool's full stroke
p_L	pressure of load
p_s	pressure of oil supply
r	radius of driving wheel
R	resistance of the phase coil
s	Laplace operator
t	time
T	executing cycle of tracking algorithm
$u(t)$	input signal of the valve
u_A, u_B	voltage of phase A and B
x_v	axial displacement of the spool
W	width of the valve port
Z_i	tooth number of the rotator
ϕ	phrasal delay
λ	non-dimensional frequency
θ	angular displacement of the stepper motor
θ_m	angular displacement of the rotary magnetic field
θ_{mc}	control signal of the rotary magnetic field
$\Delta\theta$	stepping angle
τ	time constant of track control
ρ	density of oil

REFERENCE

Araki, K., Tanahasi, T. and Ehana, M., Sept., 1985, "Freguency Response of Pressure Control with an Electro-Pneumatic Proportional Valve", *Inter.Symposuim on Fluid Control and Measurement, Tokyo,*pp. 132-141

Cui, P., Burton, R.T. and Ukraintz, P.R., March, 1992, "*Design and* Assessment of a Digital On/off V*alve*", *Proceeding of the 1992 International Fluid Power Exposition and Conference* (IFPE'92), Chicago.

Franklin, G.D. and Powell J .D., "Digital Control of Dynamic System", 1980, *Addison Wesley Pub. Co. Ltd.*

EL-Ibiary, Y., Ukrainez, P. R., and Nikiforuk, P. N., Oct.1978, "Design and Performance of Some New Degital Electro-hydraulic Valves", *Proc. of the 33rd National Conference on Fluid Power.*

Henke, R., Sept. 1989, "Fluid Power for the '90s: State-of-the-Art, Trends and Developments", *Proceedings of the 2nd Inter. Conference on Fluid Power*, Hangzhou, China, pp235-240.

Hesse, H. and Msymbol "Kinoller, H., 1972, "Pulsdauermodulierte Steuerueng von Magnetvetilen", pp. 16-11.

Jiang, W. and Ruan, J., "Pneumatic Servo Control System Utilizing 2D Digital Servo Valve", 1996, *Hydraulics and Pneumatics*, No.3, pp. 34-40.

Lee, C.S., and Wormaley, G., Sept., 1980, "On the Development of a PWM Control Based Pneumatic Servomechanism", *Int. Symp. on fluid Control & Measurement*, Tokyo, pp.2-6.

Merritt, H.E., "Hydraulic Control Systems", 1967, *John Wiley & Sons.*

Murtaugh, J., 1959, "On-off Characteristics of an Electro-servo Control System", *Trans. ASME, J. Basic Eng.*, pp254-270.

Parker, G. A., Usman, T. and Yukel, I., April, 1981, "A Novel Electro-hydraulic Switching Valve", *6th Inter. Fluid Power Symposium.*

Ramechandran, S., Ukrainetz, P.R., and Nikiforuk P.N., "Digital Flow Control Valve-an Evaluation", 1985, *Proceeding of International Conference on Fluid Power*, pp. 508-512, Hangzhou, China.

Ruan, J., Xu, Y.M. and Yu, C.Y., "Technique of Pressure Regulation with a Capillary Damper Tube", 1990, *Proceedings of International Conference on Fluid Power Control and Robotics*, pp.324-329, Chendu.

Ruan J. and Burton R.T., 1999, "Direct Digital Control of a 2D Directional Valve", *Proceedings of Sixth Scandinavian Conference on Fluid Power,* Tampere, Finland, May, Vol. 2, pp. 135- 148.

Takashi, K., "Stepper Motors and Their Microprocessor Control", 1986, *Clareudon Press*, Oxford.

Taft, C.K., and Harned, T.J., 1980, "Electro-fluid Four-way NO-OFF Pneumatic Valve", *Trans. ASME*, Vol. 102.9, PP174-179.

Taft C. K., Herrick, B. M. and Burke, C. K., 1981, "A Proportional Piezoelectric Electro-pneumatic Servo-valve Design", *Joint Automatic Control Conference.*

Tanaka, H., 1981, "Electro-hydraulic PCM Control", *J. Fluid Control*, Vol. 17. 1, 33-36.

Tsai, S. C. and Ukrainetz, P. R., 1970, "Response Characteristics of a Pulse-width modulated Electro-hydraulic Servo", *Trans. ASME, J. Basic Eng.*, June, pp204-214.

WINGRAFT , Golden Software Inc., Golden, CO, 80401-1866, USA.

Xu, Y.M. and Ruan, J., "Study of a New Type of Electro-Pneumatic Digital Pressure Valve", 1989, *Proceedings of First Japanese Conference on Fluid Power Control and Measurement*, pp.107-112, Tokyo.

FPST-Vol. 6, Fluid Power Systems and Technology — 1999
ASME 1999

A MODIFIED TUSTIN TRANSFORMATION FOR IMPROVED CONTROLLER PERFORMANCE WITH APPLICATION TO THE ELECTROHYDRAULIC ACTUATOR

Dean H. Kim
Shawn J. Moulton

Department of Mechanical Engineering
Bradley University
Peoria, IL 61615

Tsu-Chin Tsao

Department of Mechanical and Aerospace Engineering
University of California – Los Angeles
Los Angeles, CA 90095-1597

ABSTRACT

This paper presents a modified Tustin transformation which can be used in the design process to improve controller performance significantly by reducing control input fluctuations caused by unfavorable controller pole locations. An important application is the design of robust performance controllers, which can exhibit control input fluctuations due to the rigidity of the known optimal solutions. This modified Tustin transformation utilizes a scaling factor to map the stability region of the continuous domain to a region in the discrete domain which excludes the undesirable points near $z = -1$. In comparison to the standard Tustin transformation, this scaling factor affects the mapping of only the points in the continuous domain corresponding to high frequencies. Hence, a critical feature of this modified transformation is that it preserves the mapping from the origin in the continuous domain to point $z = 1$ in the discrete domain, because this pole location corresponds to integral control. This modified Tustin transformation is applied to the robust performance controller designs for the electrohydraulic actuator. The objective is to perform a number of controller designs with different scaling factors and then determine the design which provides the best compromise between improved controller performance and achieved system performance. The simulation results and experimental data demonstrate that a scaling factor equal to 1.2 improves controller performance and maintains system performance when compared to the controller design using the standard Tustin transformation. Specifically, the control input signal energy is reduced by 53 %, the system overshoot is reduced by 47 %, and the settling time is reduced by 42 %.

INTRODUCTION

The control of physical systems using discrete methodologies has become very common, because of features such as better flexibility in control strategy and more reliable implementation. However, design engineers have been able to implement only a small percentage of the recent advances in modern control theory. This gap between control theory and application must be bridged so that system performance can be improved to the fullest extent. For example, efforts to implement the results of robust control theory often encounter problems during controller implementation. The optimal robust controller typically is obtained from pre-determined solution methods, and the resulting controller may exhibit undesirable characteristics such as large controller order and oscillatory controller performance due to unfavorable controller pole location. A better approach is needed where the design engineer can choose the most suitable controller for the given application (Nwokah, 1998).

The specific issue of enhancing controller performance by improving the controller pole locations is addressed in this paper. Discrete controllers are well-known to exhibit oscillatory fluctuations in control input if any controller pole is located near the point $z = -1$ in the discrete domain (Franklin, et al., 1990). This type of controller performance is undesirable for numerous reasons, especially for electrohydraulic systems: wasted input energy, unnecessary wear on mechanical and/or electrical components, and excessive noise in the working environment. One clear objective is to design controllers which do not have poles in this region. However, robust performance controllers are obtained using a pre-determined optimal solution which depends on the chosen weighting filters (Glover and Doyle, 1988), where the only constraint on this optimal controller is that it must be stable. Because the unit circle represents the limit of the stability region for discrete systems, the discretized controller may have poles near the undesired location $z = -1$. The key to improving controller performance, while maintaining system performance, is to modify the robust performance controller design process itself.

Typically, the robust performance control designs are performed in the w-plane, as the standard Tustin transformation is performed on the zero-order-hold plant model $P_o(z)$ in order to obtain the plant model $P_o(w)$ (Kim and Tsao, 1997, Tsao and Kim, 1997). The designs are performed in the w-plane because each robust performance controller can be mapped directly to the discrete controller without any approximation. In this paper, a modified version of the standard Tustin transformation is presented which improves controller performance while maintaining system performance. This new transformation utilizes a scaling factor greater than unity to establish a modified continuous domain, i.e. the $\hat{w}$-plane, whose stability region is mapped to a constricted region in the discrete domain, i.e. the z-plane, which excludes the undesired points near z = -1. Larger scaling factors correspond to discrete stability regions which are driven further away from the point z = -1. A crucial feature of this modified Tustin transformation is that it preserves the mapping from $\hat{w} = 0$ to $z = 1$, because a controller pole at this location indicates integral control.

The modified Tustin transformation is used to improve the robust performance controller designs for the electrohydraulic actuator. The design software which incorporates the pre-determined optimal solution is used to design the feedback controller corresponding to the particular scaling factor. The design objective is to perform a number of controller designs with different scaling factors and then determine the design which best improves controller performance while maintaining the system performance originally achieved using the standard Tustin transformation. The correct controller design procedure must be determined to transform the plant models and weighting functions to the modified continuous domain, and then to evaluate the resulting performance. A particular matter of concern is to what extent the high frequency dynamics are affected by the modified Tustin transformation for different scaling factors.

The remainder of the paper is organized as follows. The robust performance control design methodology is described briefly. The modified Tustin transformation is presented, and its effects on mapping from the continuous plane to the discrete plane are discussed. The electrohydraulic actuator is described, in addition to the system models and weighting functions. The robust performance controller designs are performed using scaling factors from 1.0 to 1.35, in increments of 0.05. The controller design results demonstrate that as the scaling factor increases, the discrete controller poles are shifted progressively further away from the undesired z = -1 location. The effects of the scaling factor on the plant dynamics and the weighting functions are discussed. The most suitable scaling factor is chosen for this set of controller designs. The step response data, both from simulations and experiments, confirm that the correct usage of this modified Tustin transformation improves the controller performance while maintaining the system performance.

ROBUST PERFORMANCE CONTROL DESIGN METHODOLOGY

The control system block diagram is shown in figure 1. Controllers are said to provide robust performance for a system if they maintain stability and achieve performance in the presence of model uncertainty. The system dynamics are modeled with multiplicative uncertainty in the form of

$$P(s) = [1 + \Delta(s)W_r(s)]\, P_o(s), \tag{1}$$

where P(s) represents the experimental frequency response data, $P_o(s)$ is the nominal model, and $W_r(s)$ is a fixed stable transfer function which bounds the model uncertainty. The function $\Delta(s)$ is a variable stable transfer function which satisfies

$$\| \Delta \|_\infty := \sup_\omega | \Delta(j\omega) | \leq 1. \tag{2}$$

A necessary and sufficient condition for robust performance with the feedback controller $K_{fb}(s)$ is (Doyle, 1982)

$$\| \, | W_p S_o | + | W_r T_o | \, \|_\infty < 1, \tag{3}$$

where

$$S_o := \frac{1}{1 + P_o K_{fb}}, \tag{4}$$

$$T_o := \frac{P_o K_{fb}}{1 + P_o K_{fb}}. \tag{5}$$

The performance weight W_p typically has larger magnitudes at lower frequencies, signifying desired tracking at these frequencies. The robust weight W_r typically has larger magnitudes at higher frequencies, indicating greater model uncertainty at these frequencies.

There is no known procedure to find directly the minimum achievable value of (3), and the corresponding controller K_{fb}. Instead, a modified robust performance problem known as the mixed sensitivity problem is solved (Kwakernaak, 1985) where the objective is to minimize

$$\sup_\omega \sqrt{|W_p(j\omega)S_o(j\omega)|^2 + |W_r(j\omega)T_o(j\omega)|^2}. \tag{6}$$

The software to synthesize the controller which obtains the minimum achievable value of (3) is used (Balas, et al., 1991). For the sake of simplicity, the controller synthesis performed in this paper is referred to as robust performance control design.

MODIFIED TUSTIN TRANSFORMATION

The modified Tustin transformation has been developed to reduce the control input fluctuations of the robust performance controller for the electrohydraulic actuator, in comparison to the controller obtained using the standard Tustin transformation. Typically, the discrete nominal plant $P_o(z)$ is computed from the continuous-time plant model $P_o(s)$ using a zero-order-hold transformation, and then the Tustin transformation is performed via the relation

$$z = \frac{1 + \frac{T}{2}w}{1 - \frac{T}{2}w}, \tag{7}$$

resulting in the plant $P_o(w)$ where "T" is the sampling interval (Kim and Tsao, 1997, Tsao and Kim, 1997). A key advantage to designing controllers in the w-plane is that each feedback controller $K_{fb}(w)$ can be mapped directly to its discrete equivalent $K_{fb}(z)$ without any approximation. However, a potential problem from designing robust performance controllers using the standard Tustin transformation is due to the fact that the stable region in the w-plane maps to the interior of the unit circle in the z-plane. Specifically, large values along the imaginary axis in the w-plane (i.e. as $w \to j\infty$) are mapped to points near $z = -1$, as can be seen by (7). The only constraint on the design software is to provide robust performance controllers $K_{fb}(w)$ with stable poles (Balas, et al., 1991). As a result, the discrete controllers can have poles near $z = -1$, which indicates oscillatory controller performance.

In this paper, a modified version of the standard Tustin transformation is presented which improves controller performance while maintaining system performance. This modified Tustin transformation establishes a modified continuous domain, hereafter referred to as the $\hat{w}$-plane:

$$z = \frac{1 + (\frac{T}{2}\hat{w})}{1 - f(\frac{T}{2}\hat{w})}. \qquad (8)$$

The design engineer performs a number of controller designs with different scaling factors "f" and then determines which one provides the best compromise between controller performance and system performance. Setting the scaling factor to unity results in the standard Tustin transformation. Figure 2 shows how the stable region in the continuous $\hat{w}$-plane is mapped to the discrete z-plane for f = 1.0, 1.1, 1.2, and 1.3. The boundary of each discrete stability region corresponds to $\hat{w}$ values on the imaginary axis, i.e. $\hat{w} = j\omega$. This modified Tustin transformation preserves the mapping from $\hat{w} = 0$ to $z = 1$ for every scaling factor, as this controller pole indicates integral control. The discrete stability region is moved further away from the undesired $z = -1$ location as the scaling factor increases. Specifically, large values along the imaginary axis in the $\hat{w}$-plane (as $\hat{w} \to j\infty$) are mapped to points near $z = -1/f$, as can be seen by (8). The robust performance controller designs in this paper will demonstrate that although the modified Tustin transformation affects the high frequency system dynamics, the improved controller performance does not compromise the achieved system performance. The state-space mapping relations for the modified Tustin transformation have been derived.

SYSTEM DESCRIPTION AND MODELS

The usage of the modified Tustin transformation is demonstrated with the robust performance controller designs for the electrohydraulic actuator. The electrohydraulic system consists of a two-stage servovalve utilizing a cantilever feedback spring and double-ended actuator. The supply pressure is 18.6 MPa (2700 psi) and Mobil DTE Light oil is used at an average temperature of 80 degrees F. For the open-loop system, the system input is a voltage corresponding to current supplied to the servovalve torque motor, and the output is the actuator displacement "x_{am}." An analog proportional control loop, established with an electronic servocontroller card, creates a stabilized plant with input signal "v_{3m}," as indicated in figure 1. A laser-driven linear encoder with a resolution of 0.6 microns per count is used for the digital control feedback loop.

The plant model corresponding to small perturbations near the null actuator position has the form of

$$P_o(s) = \frac{x_{am}(s)}{v_{3m}(s)} = \frac{-f_0s^5 + f_1s^4 + f_2s^3 + f_3s^2 + f_4s + f_5}{s^8 + g_1s^7 + g_2s^6 + g_3s^5 + g_4s^4 + g_5s^3 + g_6s^2 + g_7s + g_8}, \qquad (9)$$

which incorporates the actuator model and servovalve model (Kim, 1997). The model uncertainty bound $W_r(s)$ has been determined from the modeled plant data and the experimental plant data (Kim 1997):

$$W_r(s) = \frac{1.25s^2 + 5.06e^2s + 2.50e^3}{s^2 + 5.05e^3s + 2.50e^5}. \qquad (10)$$

ROBUST PERFORMANCE CONTROL DESIGN PROCEDURE USING MODIFIED TUSTIN TRANSFORMATION

The following describes the general procedure for the robust performance controller designs using the modified Tustin transformation (8). The sampling period is fixed for each set of controller designs. The design objective is to perform a number of controller designs with different scaling factors and then to determine the design which best improves controller performance while maintaining the system performance originally achieved using the standard Tustin transformation. Specifically, the reduction in control input fluctuations will result in smaller overshoot, faster settling time, and comparable rise time.

The robust performance controller design using the standard Tustin transformation is performed first. The plant model $P_o(w)$ is obtained by applying the standard Tustin transformation to the zero-order-hold model $P_o(z)$. The robustness weight $W_r(w)$ is assumed to be approximately equal to $W_r(s)$. The performance weight has the low-pass form

$$W_p(w) = \frac{K_{wp}}{\tau w + 1}, \qquad (11)$$

with a fixed low frequency magnitude ($K_{wp} = 3200$) to ensure steady state performance. The break frequency of $W_p(w)$ ($= 1/\tau$) is increased until the limit of the robust performance condition (3) is reached. The robust performance controller $K_{fb}(w)$ is synthesized using the available software (Balas, et al., 1991) which determines the optimal solution for the robust performance controller problem (Glover and Doyle, 1988). Lastly, the discrete controller $K_{fb}(z)$ is obtained from $K_{fb}(w)$ by applying the standard Tustin transformation (7).

For each subsequent controller design which uses the modified Tustin transformation (8), the plant model P_o, the performance weight W_p, and the robustness weight W_r must be converted correctly to the $\hat{w}$-plane. The plant model $P_o(\hat{w})$ is obtained from the zero-order-hold model $P_o(z)$ by applying the modified Tustin transformation with the specified scaling factor. A two-step process is needed to obtain the weighting functions $W_p(\hat{w})$ and $W_r(\hat{w})$. First, $W_p(w)$ and $W_r(w)$ are transformed temporarily to artificial functions $W_p(z)$ and $W_r(z)$, respectively, by applying the standard Tustin transformation. Next, $W_p(\hat{w})$

and $W_r(\hat{w})$ are obtained by applying the modified Tustin transformation (8) with the specified scaling factor.

The robust performance controller $K_{fb}(\hat{w})$ is synthesized, with the criterion for limit performance being the case where the robust performance condition limit (3) is reached in the standard w-plane for each scaling factor. This criterion is used because condition (3) must be satisfied in the standard w-plane for the achievement of robust performance. To check this condition, the modified Tustin transformation is applied to each $K_{fb}(\hat{w})$ to obtain the corresponding discrete controller $K_{fb}(z)$, and then the standard Tustin transformation is applied to obtain $K_{fb}(w)$. The simulated step response is obtained by implementing each discrete controller $K_{fb}(z)$ on the continuous plant $P_o(s)$ (9).

RESULTS OF ROBUST PERFORMANCE CONTROL DESIGN

The robust performance controller designs are performed using the modified Tustin transformation with scaling factors of f = 1.0, 1.05, 1.1, 1.15, 1.2, 1.25, 1.3, and 1.35. A sampling interval of T = 0.5 msec is chosen, which is sufficient because the plant bandwidth ≈ 50 Hz (Kim, 1997). The results are presented in Table 1 and Table 2. The controller design first is performed for the standard Tustin transformation, i.e. f = 1.0: the limit condition (3) is reached using the performance weight $W_p(w)$ (11) with τ = 6.3 sec, and the robustness weight $W_r(w)$ (10).

Each subsequent robust performance controller design is performed using the appropriate transfer functions as described previously. Figure 3 shows the frequency responses of the plant models $P_o(\hat{w})$ for scaling factors of f = 1.0, 1.1, 1.2, and 1.3. The frequency response plots for plant model $P_o(\hat{w})$ are almost identical for frequencies smaller than 3,000 rad/sec (≈ 480 Hz). The plots are shown on the $\hat{w}$ frequency axis because these transfer functions are the inputs for the design software which determines the controller $K_{fb}(\hat{w})$. The functions $W_p(\hat{w})$ and $W_r(\hat{w})$ are almost identical for different scaling factors. In fact, for each scaling factor the equivalent $W_p(w)$ has the same limit parameter, i.e. τ = 6.3 sec. Only the high frequency system dynamics are affected from the usage of the modified Tustin transformation. Hence, the robust performance controller and resulting system performance will be very similar in each case.

The effect of the scaling factor on the designed robust performance controller is discussed next. The frequency response plots for the controllers $K_{fb}(\hat{w})$, obtained for scaling factors of f = 1.0, 1.1, 1.2, and 1.3, are almost identical for frequencies smaller than 2,000 rad/sec (≈ 320 Hz). These results prove further that only the high frequency system dynamics are affected from the usage of the modified Tustin transformation. The poles of each discretized controller $K_{fb}(z)$ are provided in Table 1. As the scaling factor increases, the undesirable discrete controller poles obtained with the standard Tustin transformation are moved further away from the undesired z = -1 location. In addition, the integrator pole (z = 1) is preserved for each scaling factor.

The effect of the scaling factor on the system performance is discussed next. The frequency response plots for the nominal return difference functions $T_o(\hat{w})$ (5) resulting from each robust performance controller $K_{fb}(\hat{w})$ are almost identical for frequencies smaller than 650 rad/sec (100 Hz). The fact that the $T_o(\hat{w})$ frequency responses plot are very similar for the different scaling factors further indicates that the achieved performance will be very similar.

The simulated step responses confirm that the controllers designed with the modified Tustin transformation improve controller performance, due to the more favorable location of the controller poles. Figure 4 shows the simulated output responses and input signals using the discrete controllers for f = 1.0, f = 1.1, and f = 1.3. These performance characteristics obtained with the different scaling factors are provided in Table 2: rise time, percent overshoot, settling time (2 %), and input signal energy. The energy of the input signal "v_{3m}" for each step response is defined as

$$V_{3m_{en}} := \sqrt{\int v_{3m}^2(t)\, dt} = \sqrt{\sum_{k=1}^{k=N_{step}} [v_{3m}^2(k)\, T]}\,. \quad (12)$$

The input signal energy is reduced as the scaling factor increases because the discrete controller poles are driven away from the z = -1 location. Significant reductions are realized for scaling factors f ≥ 1.2. The settling time is reduced as the scaling factor increases, again due to the reduced input fluctuations. The percent overshoot is reduced and the rise time is increased slightly as the scaling factor increases, because of smaller initial control input signals.

The design objective is to determine the scaling factor which improves the controller performance while best maintaining the system performance originally obtained with the standard Tustin transformation (i.e. with f = 1.0). The scaling factor f = 1.2 is chosen based on the data of Table 2. Specifically, although the rise time is increased by 14 %, the overall system response is improved because the input signal energy is reduced by 53 %, the system overshoot is reduced by 47 %, and the settling time is reduced by 42 %.

Lastly, the digital controller obtained from converting this $K_{fb}(\hat{w})$ to the z-domain is implemented with a TMS320C30 digital signal processor which utilizes floating-point arithmetic. The experimental step response results with scaling factor = 1.2 are shown in figure 5. A reference amplitude of 0.1 mm is used because the plant model $P_o(s)$ (9) corresponds to small perturbations about the null actuator position. The experimental results match the simulations, confirming that the modified Tustin transformation with this scaling factor improves the controller performance by reducing control input fluctuations and the corresponding input energy. Additional experimental results confirm that the use of smaller scaling factors (f = 1.1, f = 1.0) results in worse controller performance and worse system performance due to the increased control input fluctuations.

CONCLUSIONS

A modified version of the standard Tustin transformation has been presented which improves controller performance while maintaining system performance. This modified transformation utilizes a scaling factor which constricts the stability region in the discrete domain such that the points near z = -1 are excluded, in order to eliminate undesirable fluctuations in the control input signal. Only the high frequency system dynamics are affected significantly by this modified transformation, so a good compromise between improved controller performance and achieved system performance can be realized. The usage of the modified Tustin transformation is demonstrated with the robust performance controller designs for the electrohydraulic actuator. The simulation results and experimental data demonstrate that the choice of the suitable scaling factor results in improved controller performance while maintaining the system performance obtained originally with the standard Tustin transformation.

REFERENCES

Balas, G., Doyle, J., Glover, K., Packard, A., and Smith, R., 1991, μ-Analysis and Synthesis Toolbox, Natick, Massachusetts: The MathWorks, Inc.

Doyle, J., 1982, "Analysis of Feedback Systems with Structured Uncertainties," Proceedings of the Institution of Electrical Engineers, Part D, Vol. 129, pp. 242-250.

Doyle, J., Francis, B., and Tannenbaum, A., 1992, Feedback Control Theory, MacMillan Publishing Company, New York, pp. 88-91.

Francis, B.A., 1987, A Course in Hinf Control Theory, Springer-Verlag Inc., Berlin, pp. 15-22, 75-83.

Franklin, G., Powell, J., and Workman, M. Digital Control of Dynamic Systems, Addison-Wesley, Reading, MA, 1990.

Glover, K., and Doyle, J., 1988, "State-Space Formulae for All Stabilizing Controllers That Satisfy an Hinf Norm Bound and Relations to Risk Sensitivity," System and Control Letters, Vol. 11, pp. 167-172.

Kim, Dean H., 1997, "Precision Motion Control of Electrohydraulic Actuators," Ph.D. Dissertation, Department of Mechanical and Industrial Engineering, University of Illinois.

Kim, D., and Tsao, T.C., "Robust Performance Control of Electrohydraulic Actuators for Camshaft Machining," Proceedings of the 1997 ASME International Mechanical Engineering Congress and Exposition, November 1997, pp. 1801-1806.

Kwakernaak, H., 1985, "Minimax Frequency Domain Performance and Robustness Optimization of Linear Feedback Systems," IEEE Transactions for Automatic Control, Vol. AC-30, No. 10, pp. 994-1004.

Nwokah, O.D.I, "The Future of Robust Control Engineering," Proceedings of the 1998 ASME International Mechanical Engineering Congress and Exposition, November 1998, pp. 325-330.

Tsao, T.C., and Kim, D., "An Improved Linearized Model of Electrohydraulic Servovalves and Its Usage for Robust Performance Control System Design," Proceedings of the 1997 American Control Conference, June 1997, pp. 3807 – 3808.

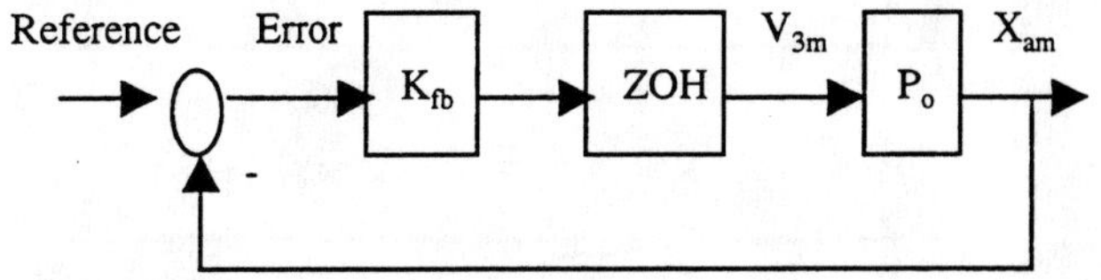

Fig. 1 Control System Block Diagram

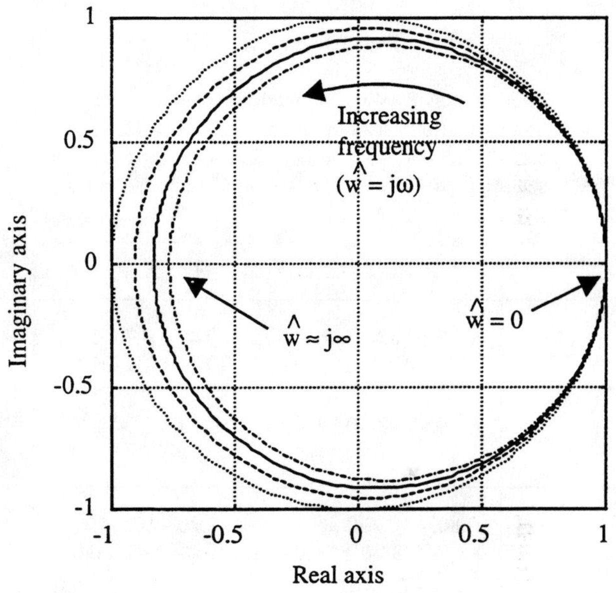

Fig. 2 Mapping of continuous w-plane stability region to discrete z-plane using modified Tustin transformation: scaling factor = 1.0 (dot), scaling factor = 1.1 (dash), scaling factor = 1.2 (solid), scaling factor = 1.3 (dash-dot).

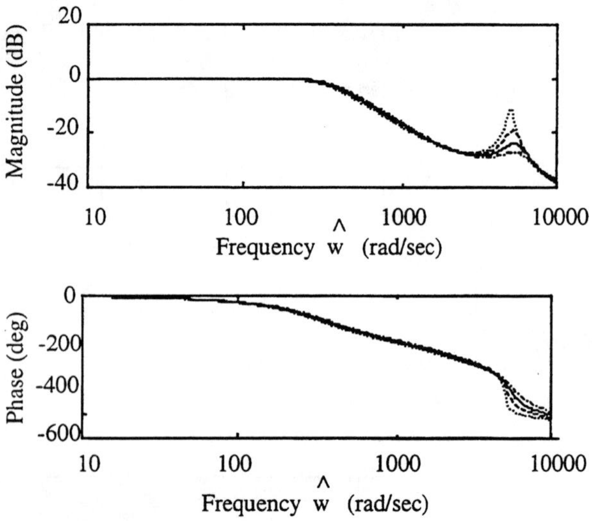

Fig. 3 Plant models $P_o(\hat{w})$ using modified Tustin transformation: scaling factor = 1.0 (dash-dot), scaling factor = 1.1 (solid), scaling factor = 1.2 (dash), scaling factor = 1.3 (dot).

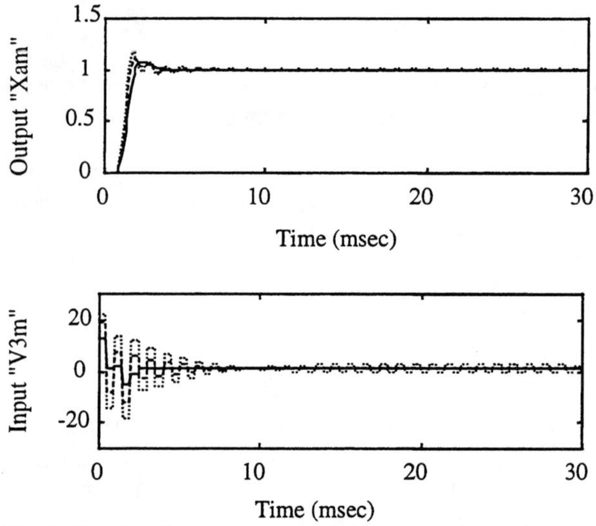

Fig. 4 Simulated steps responses (with unity reference) using robust performance controllers obtained from modified Tustin transformation: scaling factor = 1.0 (dot), scaling factor = 1.1 (dash), scaling factor = 1.3 (solid).

Scaling factor	Poles of Discrete Controller $K_{fb}(z)$
1.0	-0.99, -0.88, -0.46 + 0.74 j, -0.46 – 0.74 j, 1.0, -0.43, -0.13 + 0.09 j, -0.13 – 0.09 j, 0.03
1.1	-0.86, -0.88, -0.57 + 0.74 j, -0.57 – 0.74 j, 1.0, -0.43, -0.13 + 0.09 j, -0.13 – 0.09 j, 0.03
1.2	-0.80, -0.82, -0.26 + 0.74 j, -0.26 – 0.74 j, 1.0, -0.43, -0.13 + 0.09 j, -0.13 – 0.09 j, 0.03
1.3	-0.71, -0.73, -0.19 + 0.75 j, -0.19 – 0.75 j, 1.0, -0.43, -0.13 + 0.09 j, -0.13 – 0.09 j, 0.03

Table 1 Results of robust performance controller design: poles of discrete controllers

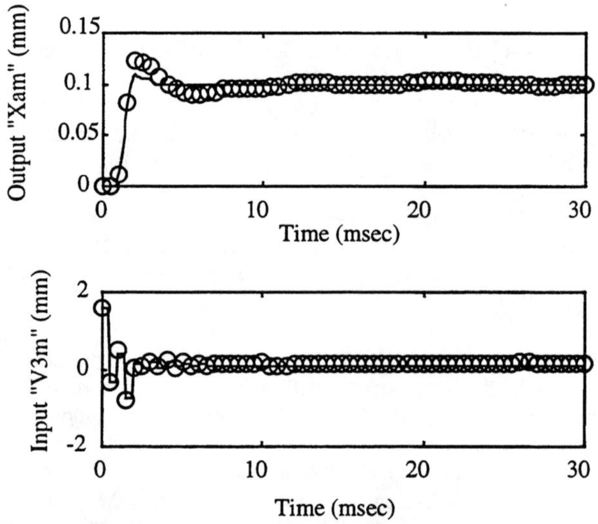

Fig. 5 Step responses (with reference = 0.1 mm) using robust performance controllers obtained from modified Tustin transformation with scaling factor = 1.2: simulated response (solid), experimental data (circles).

Scaling factor	Rise time (msec)	Percent Overshoot	Settling time (msec)	Input energy
1.0	1.50	16.2	5.32	0.92
1.1	1.60	11.2	3.42	0.60
1.2	1.71	8.6	3.17	0.43
1.3	1.83	7.6	3.09	0.36

Table 2 Results of robust performance controller design: step response characteristics (from simulations)

FPST-Vol. 6, Fluid Power Systems and Technology — 1999
ASME 1999

ADAPTIVE ROBUST PRECISION MOTION CONTROL OF SINGLE-ROD HYDRAULIC ACTUATORS WITH TIME-VARYING UNKNOWN INERTIA: A CASE STUDY

Fanping Bu and Bin Yao +

Ray W. Herrick Laboratory
School of Mechanical Engineering
Purdue University, West Lafayette, IN 47907
+ Email: *byao@ecn.purdue.edu*

ABSTRACT

This paper studies the precision motion control of single-rod hydraulic actuators with time-varying unknown inertia loads. The swing motion control of a robot arm driven by a single-rod hydraulic actuator (a scaled down version of industrial backhoe loader arm) is used as a case study. During the swing motion, the effective mass acting on the hydraulic actuator varies greatly with the swing angle and thus time-varying. Furthermore, due to the change of the load of the robot arm in operations, the effective mass is also unknown. A discontinuous projection based adaptive robust controller is constructed to take into account this strong dependence of the effective mass on the fast changing swing angle and the unknown payload of the arm. The resulting ARC controller is able to take into account not only the effect of parametric uncertainties coming from the payload and various hydraulic parameters but also the effect of uncertain nonlinearities such as uncompensated friction forces and external disturbances. The ARC controller guarantees a prescribed output tracking transient performance and final tracking accuracy while achieving asymptotic output tracking in the presence of parametric uncertainties. In addition, the zero output tracking error dynamics for tracking a large class of time-varying trajectories is shown to be globally uniformly asymptotically stable. Comparative experimental results are presented to show the effectiveness and the high performance nature of the proposed control algorithm.

1 INTRODUCTION

The dynamics of hydraulic systems are highly nonlinear [1]. The system may be subjected to non-smooth and discontinuous nonlinearities due to control input saturation, directional change of valve opening, friction, and valve overlap. Aside from the nonlinear nature of hydraulic dynamics, hydraulic servosystems also have large extent of model uncertainties such as the large changes in load and hydraulic parameters, external disturbances, leakage, and friction. In the past, much of the work in the control of hydraulic systems uses linear control theory [2; 3; 4; 5] and feedback linearization techniques [6; 7; 8]. In [8], Alleyne and Hedrick applied the nonlinear adaptive control to the force control of an active suspension driven by a double-rod cylinder, in which only parametric uncertainties of the cylinder are considered.

In [9], the adaptive robust control (ARC) approach proposed by Yao and Tomizuka in [10; 11; 12] was generalized to provide a rigorous theoretic framework for the high performance robust motion control of a double-rod hydraulic actuator by taking into account the particular nonlinearities and model uncertainties of the electro-hydraulic servo-systems. The presented ARC scheme uses smooth projections [11] to solve the design conflicts between adaptive control technique and robust control technique, which may not be convenient for practical implementation. In [13], the recently proposed discontinuous projection method based ARC design [12] is generalized to solve the practical problems associated with smooth projections [9]. In [14], the discontinuous projection based ARC Scheme [13] is extended to the precision control of a single-rod hydraulic actuator, in which a constant inertia load is considered.

This paper continues the work done in [14] and will extend the results to the precision control of single-rod hydraulic actuators with time-varying and unknown inertia load. To make the development more understandable and application specific, the swing motion control of a robot arm driven by single-rod hydraulic actuators will be used as a case study. During the swing motion, the effective mass acting on the hydraulic actuator varies greatly with the swing angle and thus time-varying. Furthermore, due to the change of the payload of the robot arm, the effective mass is also unknown. This strong dependence of the effective mass on the changing swing angle and the unknown payload of the robot arm makes the swing motion control a perfect case study for the precision motion control of single-rod actuators with time-varying unknown inertia load. A discontinuous projection based ARC controller will be constructed to take into account the strong dependence of the effective mass on the fast changing swing angle and the unknown payload of the arm. Comparative experimental results will be presented to show the effectiveness and the high performance nature of the proposed control algorithm.

As pointed out in [14], a particular phenomenon associated with the motion control of single-rod hydraulic actuators is that there exists a one-dimensional internal dynamics, which is due to the fact that the two chambers of a single-rod actuator have different areas. It is shown in this paper that when tracking a fairly general class of time-varying desired trajectories, the resulting zero output tracking error dynamics is globally uniformly asymptotically stable, which is the first theoretical result available in the literature on the stability of zero output tracking error dynamics when tracking a non-constant velocity trajectory.

2 PROBLEM FORMULATION AND DYNAMIC MODELS

The system under consideration is depicted in Fig. 1, which represents the swing motion of a 3 degree-of-freedom (DOF) hydraulic robot arm with the other two joints fixed. The goal is to have the cylinder displacement x_L track any feasible desired motion trajectory as closely as possible for precision maneuver of the inertia load of the hydraulic cylinder.

Using the joint angle q as the coordinate, the kinetic energy of the swing motion can be expressed as $T = \frac{1}{2}(J_0+J_1)\dot{q}^2$ where J_0 is the swing inertia of the robotic arm and J_1 is the swing inertia due to the payload mounted at the end of the robot arm [1]. Since the other two links of the robot arm are fixed, J_0 and J_1 are constant; J_1 is unknown since it depends on the unknown payload of the robot arm. In terms of the cylinder displacement x_L, the kinetic energy of the swing motion can be expressed as $T = \frac{1}{2}(J_0+J_1)(\frac{\partial q}{\partial x_L})^2 \dot{x_L}^2$, where the fact that q is uniquely related to x_L has been used, i.e., $q(x_L)$. Using Euler-Lagrange equation, The dynamics of the swing motion can be described by

[1]For simplicity, the swing inertia due to the swing cylinder is neglected since the mass of the swing cylinder is quite small comparing to the inertia of the whole robotic arm

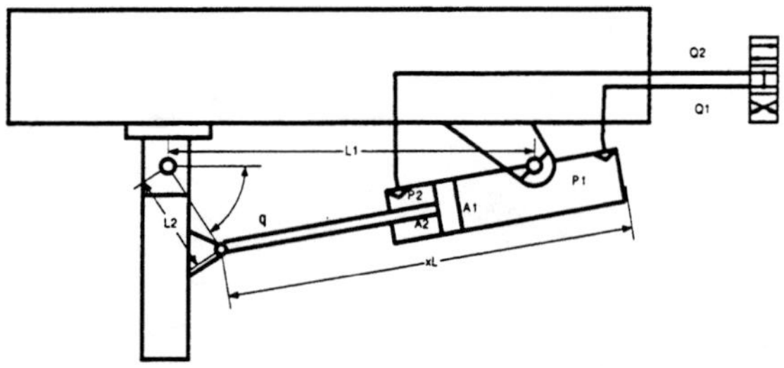

Figure 1. Swing Motion of a Robot Arm Driven by a Single-Rod Hydraulic Cylinder

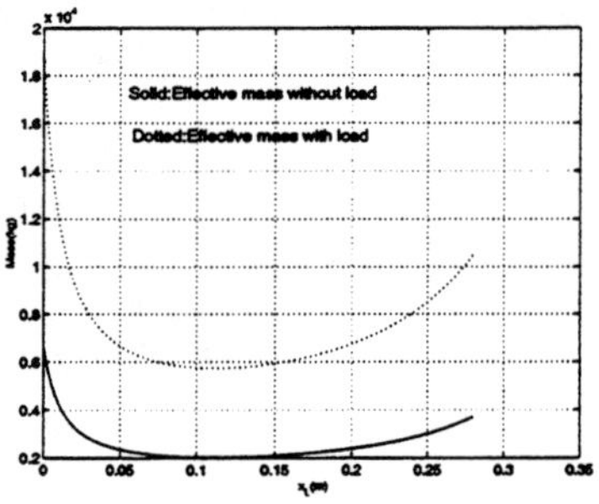

Figure 2. Effective Mass versus Cylinder Displacement

$$\begin{aligned}(J_0+J_1)(\tfrac{\partial q}{\partial x_L})^2\ddot{x}_L &= P_1A_1 - P_2A_2 - b\dot{x}_L - F_{fc}(\dot{x}_L)\\ &-(J_0+J_1)\tfrac{\partial q}{\partial x_L}\tfrac{\partial^2 q}{\partial x_L^2}\dot{x}_L^2 + \tilde{f}(t,x_L,\dot{x}_L)\end{aligned} \quad (1)$$

where P_1 and P_2 are the pressures of the two cylinder chambers respectively, A_1 and A_2 are the ram areas of the two cylinder chambers respectively, b represents the combined coefficient of the modeled damping and viscous friction forces on the load and the cylinder rod, F_{fc} represents the modeled Coulomb friction force, and $\tilde{f}(t,x_L,\dot{x}_L)$ represents the lumped modeling error including external disturbances and terms like the unmodeled friction forces.

As seen in (1), the effective mass driven by the cylinder is $(J_0+J_1)(\frac{\partial q}{\partial x_L})^2$, which could change greatly during operations since $(\frac{\partial q}{\partial x_L})$ is a nonlinear function of the cylinder position x_L and varies significantly; for the system used in the experiments in section 4, the dependence of the effective mass on the cylinder position with and without a payload of $45kg$ is shown in Fig.2, in which the effective mass varies from 2.030×10^3kg to 6.656×10^3kg for no-load situation and 5.746×10^3kg to 1.884×10^4kg for loaded situation. Furthermore, it is unknown due to the unknown swing inertia J_1 caused by the payload.

The actuator (or the cylinder) dynamics can be written as [1]:

$$\begin{aligned}\tfrac{V_1}{\beta_e}\dot{P}_1 &= -A_1\dot{x}_L - C_{tm}(P_1-P_2) - C_{em1}(P_1-P_r) + Q_1\\ \tfrac{V_2}{\beta_e}\dot{P}_2 &= A_2\dot{x}_L + C_{tm}(P_1-P_2) - C_{em2}(P_2-P_r) - Q_2\end{aligned} \quad (2)$$

where $V_1 = V_{h_1} + A_1x_L$ and $V_2 = V_{h_2} - A_2x_L$ are the total control volumes of the two chambers respectively, V_{h1} and V_{h2} are the chamber volumes when $x_L = 0$, β_e is the effective bulk modulus, C_{tm} is the coefficient of the internal leakage of the cylinder, and C_{em1} and C_{em2} are the coefficients of the external leakage for the forward and return chamber respectively. In (2), Q_1 and Q_2 are the forward flow and return flow to the two chambers of the cylinder respectively, which are related to the spool valve displacement of the servovalve, x_v, by [1]

$$\begin{aligned} Q_1 = k_{q1} x_v \sqrt{\Delta P_1}, \quad \Delta P_1 = \begin{cases} P_s - P_1 & \text{for } x_v \geq 0 \\ P_1 - P_r & x_v < 0 \end{cases} \\ Q_2 = k_{q2} x_v \sqrt{\Delta P_2}, \quad \Delta P_2 = \begin{cases} P_2 - P_r & \text{for } x_v \geq 0 \\ P_s - P_2 & x_v < 0 \end{cases} \end{aligned} \tag{3}$$

where k_{q1} and k_{q2} are the flow gain coefficients of the forward and return loop respectively, and P_s and P_r are the supply pressure and the tank pressure of the fluid respectively.

As in [14], to minimize the numerical error and facilitating the gain-tuning process, constant scaling factors are introduced to the pressures and the valve opening respectively; the scaled pressures and valve opening are $\bar{P}_1 = \frac{1}{S_{c3}} P_1$, $\bar{P}_2 = \frac{1}{S_{c3}} P_2$, $\bar{P}_s = \frac{1}{S_{c3}} P_s$, $\bar{P}_r = \frac{1}{S_{c3}} P_r$, and $\bar{x}_v = \frac{1}{S_{c4}} x_v$ respectively. Define a set of state variables as $x = [x_1, x_2, x_3, x_4]^T \triangleq [x_L, \dot{x}_L, \bar{P}_1, \bar{P}_2]^T$, the entire system, Eqs.(1)-(3) with control input $u = \bar{x}_v$ can be expressed in state space form as:

$$\begin{aligned} \dot{x}_1 &= x_2 \\ \dot{x}_2 &= \frac{1}{\bar{m}_0} \frac{1}{1+\frac{J_1}{J_0}} (x_3 - \bar{A}_c x_4 - \bar{b} x_2 - \bar{F}_{fc}) - p_q(x_1) {x_2}^2 + \frac{d}{m_0} \\ \dot{x}_3 &= h_1(x_1) [\frac{\beta_e S_{c4} k_{q1}}{\sqrt{S_{c3}} V_{h1}} (-\bar{A}_1 x_2 + g_3 u) - \frac{C_{tm} \beta_e}{V_{h1}} (x_3 - x_4) \\ &\quad - \frac{C_{em1} \beta_e}{V_{h1}} (x_3 - \bar{P}_r)] \\ \dot{x}_4 &= h_2(x_1) [\frac{\beta_e S_{c4} k_{q1}}{\sqrt{S_{c3}} V_{h1}} (\bar{A}_2 x_2 - g_4 u) + \frac{C_{tm} \beta_e}{V_{h1}} (x_3 - x_4) \\ &\quad \frac{C_{em2} \beta_e}{V_{h1}} (x_4 - \bar{P}_r)] \end{aligned} \tag{4}$$

where $m_0(x_1) = J_0 (\frac{\partial q}{\partial x_1})^2$, $\bar{m}_0(x_1) = \frac{m_0(x_1)}{s_{c3} A_1}$, $\bar{A}_c = \frac{A_2}{A_1}$, $\bar{b} = \frac{b}{S_{c3} A_1}$, $\bar{F}_{fc} = \frac{F_{fc}(x_2)}{S_{c3} A_1}$, $p_q(x_1) = \frac{\partial^2 q}{\partial x_1^2} / \frac{\partial q}{\partial x_1}$, $d = \frac{1}{1+\frac{J_1}{J_0}} \tilde{f}(t, x_1, x_2)$, $h_1(x_1) = \frac{1}{1+\bar{A}_{h1} x_1}$, $h_2(x_1) = \frac{1}{k_v - \bar{A}_{h2} x_1}$, $k_v = \frac{V_{h2}}{V_{h1}}$, $\bar{A}_{h1} = \frac{A_1}{V_{h1}}$, $\bar{A}_{h2} = \frac{A_2}{V_{h1}}$, $\bar{A}_1 = \frac{A_1}{k_{q1} S_{c4} \sqrt{S_{c3}}}$, $\bar{A}_2 = \frac{A_2}{k_{q1} S_{c4} \sqrt{S_{c3}}}$, and g_3 and g_4 are defined by

$$\begin{aligned} g_3 = \sqrt{\overline{\Delta P_1}}, \quad \overline{\Delta P_1} = \frac{1}{s_{c3}} \Delta P_1(x_3, sign(u)) \\ g_4 = k_{qc} \sqrt{\overline{\Delta P_2}}, \quad \overline{\Delta P_2} = \frac{1}{s_{c3}} \Delta P_2(x_4, sign(u)) \end{aligned} \tag{5}$$

in which $k_{qc} = \frac{k_{q2}}{k_{q1}}$.

Given the desired motion trajectory $x_{Ld}(t)$, the objective is to synthesize a control input u such that the output $y = x_1$ tracks $x_{Ld}(t)$ as closely as possible in spite of various model uncertainties.

3 ADAPTIVE ROBUST CONTROL DESIGN

Design Model and Issues to be Addressed

In this paper, we consider the parameter variations of important parameters like J_1, β_e, C_{tm}, C_{em1}, C_{em2}, and the nominal value of the disturbance d, d_n, only. Other parametric uncertainties can be dealt with in the same way if necessary. Define the unknown parameter set $\theta = [\theta_1, \theta_2, \theta_3, \theta_4, \theta_5, \theta_6]^T$ as $\theta_1 = \frac{1}{1+\frac{J_1}{J_0}}$, $\theta_2 = d_n$, $\theta_3 = \frac{\beta_e S_{c4} k_{q1}}{V_{h1} \sqrt{S_{c3}}}$, $\theta_4 = \frac{C_{tm} \beta_e}{V_{h1}}$, $\theta_5 = \frac{C_{em1} \beta_e}{V_{h1}}$, $\theta_6 = \frac{C_{em2} \beta_e}{V_{h1}}$. The state space equation (4) can be linearly parametrized in terms of θ as

$$\begin{aligned} \dot{x}_1 &= x_2 \\ \dot{x}_2 &= \frac{\theta 1}{\bar{m}_0} (x_3 - x_4 \bar{A}_c - \bar{b} x_2 - \bar{F}_{fc}) - p_q {x_2}^2 + \frac{\theta_2}{m_0} + \tilde{d}(t, x_1, x_2) \\ \dot{x}_3 &= h_1 [\theta_3 (-\bar{A}_1 x_2 + g_3 u) - \theta_4 (x_3 - x_4) - \theta_5 (x_3 - \bar{P}_r)] \\ \dot{x}_4 &= h_2 [\theta_3 (\bar{A}_2 x_2 - g_4 u) + \theta_4 (x_3 - x_4) - \theta_6 (x_4 - \bar{P}_r)] \end{aligned} \tag{6}$$

where $\tilde{d} = (d - d_n)/m_0(x_1)$. It is assumed that the range of parameter variations and the degree of uncertain nonlinearities are known, i.e.,

$$\begin{aligned} \theta \quad &\in \Omega_\theta \triangleq \{\theta : \ \theta_{min} < \theta < \theta_{max} \} \\ |\tilde{d}(t, x_1, x_2)| &\leq \delta_d(x_1, x_2, t) \end{aligned} \tag{7}$$

where $\theta_{min} = [\theta_{1min}, \ldots, \theta_{6min}]^T$, $\theta_{max} = [\theta_{1max}, \ldots, \theta_{6max}]^T$, and $\delta_d(t, x_1, x_2)$ are known. Physically, $\theta_1 > 0$ and $\theta_3 > 0$. So it is also assumed that $\theta_{1min} > 0$ and $\theta_{3min} > 0$.

At this stage, it can be seen that the major difficulties in controlling (6) are:

(1) The system dynamics are highly nonlinear, either due to the dependence of the effective mass on swing angle or the inherent nonlinearities of hydraulic dynamics such as the nonlinear flow gains (represented by $g_3(x_3, sign(u))$ and $g_4(x_4, sign(u))$) and the change of control volumes (represented by $h_1(x_1)$ and $h_2(x_1)$).

(2) The system has severe parametric uncertainties like large variation of the inertial load J_1, the change of bulk modulus due to entrapped air or temperature, etc.

(3) The system may also have large extent of lumped modeling error (represented by the uncertain nonlinearity $\tilde{d}$) including external disturbances, unmodeled friction force, etc.

(4) The model uncertainties are mismatched, i.e. both parametric uncertainties and uncertain nonlinearities appear in the dynamic equations which are not directly related to the control input $u = x_v$.

(5) As will become clear later, the "relative degree" of the system is three while the system state has a dimension of four. Thus, there exists a one-dimensional internal dynamics after an ARC controller is synthesized. It is thus necessary to consider the stability of the resulting internal dynamics, or the stability of the resulting zero error dynamics when tracking a large class of reference trajectories.

To address the above challenges, following general strategies are adopted:

a. Physical model based nonlinear analysis and synthesis will be employed to deal with the nonlinear nature of the system dynamics directly, i.e., the subsequent controller design will be based on the nonlinear physical model (6).

b. The adaptive robust control (ARC) approach [10; 12] will be used to handle the effect of both parametric uncertainties and uncertain nonlinearities effectively for high performance - certain robust feedback will be employed to attenuate the effect of various model uncertainties as much as possible while parameter adaptation will be utilized to reduce model uncertainties for an improved performance.

c. Backstepping design [15] via ARC Lyapunov function [12] will be used to overcome the design difficulties caused by unmatched model uncertainteis.

Due to space limit, only outline of the controller design is given. The detailed design procedures and stability proofs can be obtained from the authors and/or worked out using similar techniques as in [14].

ARC Controller Design

Let $\hat{\theta}$ denote the estimate of θ and $\tilde{\theta}$ the estimation error (i.e., $\tilde{\theta} = \hat{\theta} - \theta$). The following adaptation law structure will be sought

$$\dot{\hat{\theta}} = Proj_{\hat{\theta}}(\Gamma\tau) \tag{8}$$

where $\Gamma > 0$ is a diagonal matrix and τ is an adaptation function to be synthesized later, and $Proj_{\hat{\theta}}(\bullet)$ is the widely used discontinuous projection mapping [10]. For simplicity, let $\hat{\dot{x}}_2$ represent the calculable part of the $\dot{x}_2$ given by

$$\hat{\dot{x}}_2 = \frac{\hat{\theta}_1}{\bar{m}_0}(x_3 - \bar{A}_c x_4 - \bar{b}x_2 - \bar{F}_{fc}) - p_q {x_2}^2 + \frac{\hat{\theta}_2}{m_0} \tag{9}$$

The design parallels the recursive backstepping design procedure via ARC Lyapunov functions in [12] as follows.

Step 1

Define the load pressure as $P_L = x_3 - \bar{A}_c x_4$. It is seen from (6) that P_L can be thought as the input to the first two equations of (6). Furthermore, the resulting two equations have a similar form to those in [13]. Thus, the ARC design technique in [13]) can be modified to construct an ARC control function $\alpha_2(x_1, x_2, \hat{\theta}_1, \hat{\theta}_2, t)$ for the virtual control input P_L such that the output tracking error $z_1 = x_1 - x_{1d}(t)$ converges to zero or a small value with a guaranteed transient performance. The resulting control function α_2 and adaptation function $\tau_2(x_1, x_2, \hat{\theta}_1, \hat{\theta}_2, t)$ are given by

$$\begin{aligned} \alpha_2 &= \alpha_{2a} + \alpha_{2s} \\ \alpha_{2a} &= \bar{b}x_2 + \bar{F}_{fc} + \frac{\bar{m}_0}{\hat{\theta}_1}(\dot{x}_{2eq} + p_q {x_2}^2 - \frac{\hat{\theta}_2}{m_0}) \\ \alpha_{2s} &= \alpha_{2s1} + \alpha_{2s2} \\ \alpha_{2s1} &= -\frac{\bar{m}_{0max}}{\theta_{1min}}(k_{2s1} + k_2)z_2, \quad k_2 > 0 \\ \tau_2 &= w_2\phi_2 z_2 \end{aligned} \tag{10}$$

where $\bar{m}_{0max}$ represents the upper bound of $\bar{m}_0(x_1)$. The robust control function α_{2s2} is chosen to satisfy the following conditions

$$\begin{array}{ll} \text{condition i} & z_2[\frac{\theta_1}{\bar{m}_0(x_1)}\alpha_{2s2} - \phi_2^T\tilde{\theta} + \tilde{d}] \le \varepsilon_2 \\ \text{condition ii} & z_2\alpha_{2s2} \le 0 \end{array} \tag{11}$$

In (10) and (11), k_{2s1} is a positive control gain function to be specified later, k_2 is a positive scalar, ε_2 is a positive design parameter which can be arbitrarily small, w_2 is a any positive weighting factor, and

$$\begin{aligned} & x_{2eq} = \dot{x}_{1d} - k_p z_1, \qquad \dot{x}_{2eq} = \ddot{x}_{1d} - k_p \dot{z}_1 \\ & z_2 = \dot{z}_1 + k_p z_1 = x_2 - x_{2eq} \\ & \phi_2 \triangleq \left[\frac{1}{\bar{m}_0}(\alpha_{2a} - \bar{b}x_2 - \bar{F}_{fc}), \frac{1}{m_0}, 0, 0, 0, 0\right]^T \end{aligned} \tag{12}$$

Let $z_3 = P_L - \alpha_2$ denote the input discrepancy. For the positive semi-definite (p.s.d.) function V_2 defined by$V_2 = \frac{1}{2}w_2 z_2^2$, it can be shown that

$$\begin{aligned} \dot{V}_2 = & \frac{1}{\bar{m}_0(x_1)}w_2\theta_1 z_2 z_3 + w_2 z_2(\frac{1}{\bar{m}_0(x_1)}\theta_1\alpha_{2s2} - \tilde{\theta}^T\phi_2 + \tilde{d}) \\ & -w_2\frac{\theta_1}{\theta_{1min}}\frac{\bar{m}_{0max}}{\bar{m}_0(x_1)}(k_2 + k_{2s1})z_2^2 \end{aligned} \tag{13}$$

Step 2

In Step 1, as seen from (13), if $z_3 = 0$, then, output tracking can be achieved in viewing i of (11). So Step 2 is to synthesize a control input so that z_3 converges to zero or a small value with a guaranteed transient performance as follows. From (6)

$$\begin{aligned} \dot{z}_3 = \dot{P}_L - \dot{\alpha}_2 = & \theta_3[-(\bar{A}_1 h_1 + \bar{A}_2\bar{A}_c h_2)x_2 \\ & +(h_1 g_3 + \bar{A}_c h_2 g_4)u] - \theta_4(x_3 - x_4)(h_1 + \bar{A}_c h_2) \\ & -\theta_5 h_1(x_3 - \bar{P}_r) + \theta_6\bar{A}_c h_2(x_4 - \bar{P}_r) - \dot{\alpha}_{2c} - \dot{\alpha}_{2u} \end{aligned} \tag{14}$$

where

$$\begin{aligned} \dot{\alpha}_{2c} &= \frac{\partial\alpha_2}{\partial x_1}x_2 + \frac{\partial\alpha_2}{\partial x_2}\hat{\dot{x}}_2 + \frac{\partial\alpha_2}{\partial t} \\ \dot{\alpha}_{2u} &= \frac{\partial\alpha_2}{\partial x_2}[-\frac{\tilde{\theta}_1}{\bar{m}_0}(x_3 - \bar{A}_c x_4 - \bar{b}x_2 - \bar{F}_{fc}) - \frac{\tilde{\theta}_2}{m_0} + \tilde{d}] + \frac{\partial\alpha_2}{\partial\hat{\theta}}\dot{\hat{\theta}} \end{aligned} \tag{15}$$

in which $\dot{\alpha}_{2c}$ is the calculable part of $\dot{\alpha}_2$ while $\dot{\alpha}_{2u}$ is the incalculable part and has to be dealt with via certain robust feedback. Define a new virtual control input Q_L as $Q_L = (h_1 g_3 + \bar{A}_c h_2 g_4)u$. Then, it is ready to see that (14) has a similar form as those studied in the previously proposed ARC control designs [13] and the design strategy [13] can be modified to construct a virtual control law $\alpha_3(x_1, x_2, x_3, x_4, \hat{\theta}, t)$ for Q_L. The resulting control function α_3 and adaptation function $\tau_3(x_1, x_2, x_3, x_4, \hat{\theta}, t)$ are given by

$$\begin{aligned} \alpha_3 = & \alpha_{3a} + \alpha_{3s} \\ \alpha_{3a} = & -\frac{1}{\hat{\theta}_3}\left[\frac{1}{\bar{m}_0}\frac{w_2}{w_3}z_2\hat{\theta}_1 - \hat{\theta}_3(\bar{A}_1 h_1 + \bar{A}_2 h_2\bar{A}_c h_2)x_2\right. \\ & -\hat{\theta}_4(h_1 + \bar{A}_c h_2)(x_3 - x_4) - \\ & \left.-\hat{\theta}_5 h_1(x_3 - \bar{P}_r) + \hat{\theta}_6\bar{A}_c h_2(x_4 - \bar{P}_r) - \dot{\alpha}_{2c}\right] \\ \alpha_{3s} = & \alpha_{3s1} + \alpha_{3s2}, \quad \alpha_{3s1} = -\frac{1}{\theta_{3min}}(k_3 + k_{3s1})z_3 \\ \tau_3 = & \tau_2 + w_3\phi_3 z_3 \end{aligned} \tag{16}$$

where w_3 is a positive weighting factor, $k_3 > 0$ is a constant, k_{3s1} is a positive definite control gain function, and α_{3s2} is a robust control function satisfying the following two conditions

$$\begin{array}{ll} \text{condition i} & z_3\left[\theta_3\alpha_{3s2} - \phi_3^T\tilde{\theta} - \frac{\partial\alpha_2}{\partial x_2}\tilde{d}\right] \le \varepsilon_3 \\ \text{condition ii} & z_3\alpha_{3s2} \le 0 \end{array} \tag{17}$$

in which ε_3 is a positive design parameter and

$$\phi_3 = \begin{bmatrix} \frac{1}{\bar{m}_0(x_1)}(\frac{w_2}{w_3}z_2 - \frac{\partial\alpha_2}{\partial x_2}(x_3 - \bar{A}_c x_4 - \bar{b}x_2 - \bar{F}_{fc})) \\ -\frac{1}{m_0(x_1)}\frac{\partial\alpha_2}{\partial x_2} \\ -(\bar{A}_1 h_1 + \bar{A}_2\bar{A}_c h_2)x_2 + \alpha_{3a} \\ -(h_1 + \bar{A}_c h_2)(x_3 - x_4) \\ -h_1(x_3 - \bar{P}_r) \\ \bar{A}_c h_2(x_4 - \bar{P}_r) \end{bmatrix} \tag{18}$$

Once the control function α_3 is synthesized as in (16), the control input u can be backed out from the one-to-one continuous nonlinear load flow mapping, i.e., determining a $u(x_1, x_2, x_3, x_4, \hat{\theta}, t)$ such that

$$(h_1(x_1)g_3(x_3, sign(u) + \bar{A}_c h_2(x_1)g_4(x_4, sign(u))u = \alpha_3 \quad (19)$$

Theorem 1. *Let the adaptation function τ in (8) be τ_3 and the control input u being determined from (19). If the nonlinear control gains k_{2s1} and k_{3s1} in (10) and (16) are chosen such that $k_{2s1} \geq \frac{1}{2}w_2\|\Gamma\phi_2\|^2$ and $k_{3s1} \geq w_3\|\frac{\partial\alpha_2}{\partial\hat{\theta}_2}\|^2 + \frac{1}{2}w_3\|\Gamma\phi_3\|^2$ respectively, then, the control law (19) with the adaptation law (8) guarantees that*

A. *In general, tracking errors $z = [z_1, z_2, z_3]^T$ are bounded. Furthermore, the positive semi-definite function V given by $V = V_2 + \frac{1}{2}w_3 z_3^2$ is bounded above by*

$$V(t) \leq exp(-\lambda_V t)V(0) + \frac{\varepsilon_V}{\lambda_V}[1 - exp(-\lambda_V t)] \quad (20)$$

where $\lambda_V = 2min\{k_2, k_3\}$ and $\varepsilon_V = w_2\varepsilon_2 + w_3\varepsilon_3$. This implies that the tracking errors z exponentially converge to a ball, $V(\infty) \leq \frac{\varepsilon_V}{\lambda_V}$, whose size can be made arbitrarily small by increasing the controller parameters k_2 and k_3 and/or decreasing the design parameters ε_2 and ε_3 in a known form.

B. *If after a finite time t_0, $\tilde{d} = 0$, i.e., in the presence of parametric uncertainties only, then, in addition to results in A, asymptotic output tracking (or zero final tracking error) is also achieved, i.e., $z_1 \longrightarrow 0$ as $t \longrightarrow \infty$.* $\triangle$

Proof. The Theorem can be proved in the same way as in [13]. □.

Internal Dynamics and Zero Dynamics

Theorem 1 shows that, under the proposed ARC control law, the *three* transformed system state variables, $z = [z_1, z_2, z_3]^T \in R^3$, are bounded, from which one can easily prove that the position x_1, the velocity x_2, and the load pressure $P_L = x_3 - \bar{A}_c x_4$ are bounded. However, since the original system (6) has a dimension of *4*, there exists a one-dimensional internal dynamics. Although simulation and experimental results seem to verify that the two pressures are indeed bounded when tracking a non-zero speed motion trajectory, it is of both practical and theoretical interests to see if we can prove this fact, which is the focus of this subsection.

As a first step toward to the proof of the stability of the resulting internal dynamics, we will consider the stability of the zero error dynamics [16] when tracking any reference trajectory $x_{1d}(t)$. The reason for doing this is that the zero dynamics for tracking a desired trajectory $x_{1d}(t)$ is the same as the internal dynamics when $z(t) = 0, \forall t$. Thus, if the zero error dynamics is proven to be globally uniformly asymptotically stable, then, it is reasonable to expect that all internal variables are bounded when the proposed ARC law is applied, since z is guaranteed to converge to a small value very quickly by Theorem 1.

In [14], the zero error dynamics when tracking a nonzero *constant* velocity reference trajectory was considered and proved to be stable. Here, we consider the general situation of tracking any feasible time-varying reference trajectory $x_{1d}(t)$. As in [14], for simplicity, the leakage flows are neglected in the analysis since they are normally relatively much smaller than the control flows needed to track a time-varying trajectory. The results are summarized in the following Theorem.

Theorem 2. *Assume that the leakage flows can be neglected and the system is absence of uncertain nonlinearities, i.e., $\theta_4 = \theta_5 = \theta_6 = 0$ and $\tilde{d} = 0$ in (6). Let $\dot{x}_{1d}(t)$ be any feasible time-varying reference trajectory that the system can follow, i.e., there exists a bounded state equilibrium trajectory $x_e(t) = [x_{1e}(t), x_{2e}(t), x_{3e}(t), x_{4e}(t)]^T \in L^4_\infty$ and a bounded control input $u_e(t)$ satisfying the system equation (6) and having $x_{1d}(t)$ as the output (i.e., $x_{1e}(t) = x_{1d}(t), \forall t$). When tracking such a feasible time-varying trajectory $\dot{x}_{1d}$, the following results hold:*

A. *The desired load pressure trajectory $P_{Ld}(t)$ and the desired load flow trajectory $Q_{Ld}(t)$ for tracking $\dot{x}_{1d}$ perfectly are unique and given by*

$$\begin{aligned} P_{Ld}(t) &= \bar{b}\dot{x}_{1d}(t) + \bar{F}_{fc}(\dot{x}_{1d}(t)) + \tfrac{\bar{m}_0(x_{1d}(t))}{\theta_1}\Big[\ddot{x}_{1d}(t) \\ &\quad + p_q(x_{1d}(t))\dot{x}^2_{1d}(t) - \tfrac{\theta_2}{m_0(x_{1d}(t))}\Big] \\ Q_{Ld}(t) &= \tfrac{1}{\theta_3}\left[\dot{P}_{Ld}(t) + \theta_3\bar{A}_1\left(h_{1d}(t) + \bar{A}^2_c h_{2d}(t)\right)\right]\dot{x}_{1d}(t) \end{aligned} \quad (21)$$

where $h_{1d}(t) = h_1(x_{1d}(t))$ and $h_{2d}(t) = h_2(x_{1d}(t))$. In addition, if $Q_{Ld}(t)$ is zero only at discrete points, any bounded equilibrium state trajectory $x_e(t)$ and the bounded equilibrium control input $u_e(t)$ having $x_{1d}(t)$ as the output satisfy the following algebraic equations

$$\begin{aligned} x_{1e}(t) &= x_{1d}(t) \\ x_{2e}(t) &= \dot{x}_{1d}(t) \\ x_{3e}(t) - \bar{A}_c x_{4e}(t) &= P_{Ld}(t) \\ u_e(t) &= \tfrac{1}{h_{1d}(t)g_{3e}(t) + \bar{A}_c h_{2d}(t)g_{4e}(t)}Q_{Ld}(t) \end{aligned} \quad (22)$$

and differential equations

$$\begin{aligned} \dot{x}_{3e}(t) &= \tfrac{h_{1d}(t)g_{3e}(t)\dot{P}_{Ld}(t) + \theta_3\dot{x}_{1d}h_{1d}h_{2d}\bar{A}_2(\bar{A}_c g_{3e} - g_{4e})}{h_{1d}g_{3e} + \bar{A}_c h_{2d}g_{4e}} \\ \dot{x}_{4e} &= \tfrac{-h_{2d}g_{4e}\dot{P}_{Ld} + \theta_3\dot{x}_{1d}h_{1d}h_{2d}\bar{A}_1(\bar{A}_c g_{3e} - g_{4e})}{h_{1d}g_{3e} + \bar{A}_c h_{2d}g_{4e}} \end{aligned} \quad (23)$$

where $g_{3e}(t) = g_3(x_{3e}(t), sign(Q_{Ld}(t)))$ and $g_{4e}(t) = g_4(x_{4e}(t), sign(Q_{Ld}(t)))$.

B. Suppose that the reference trajectory $x_{1d}(t)$ is such that the resulting desired load pressure $P_{Ld}(t)$ in (21) is within the physical limit, i.e., $-(\bar{A}_c\bar{P}_s - \bar{P}_r) < P_{Ld}(t) < \bar{P}_s - \bar{A}_c\bar{P}_r$. Furthermore, assume that $x_{1d}(t)$ satisfies the following persistently-exciting-like condition,

$$\text{PE: } \exists T, c > 0 \text{ s. t. } \int_t^{t+T} |\, Q_{Ld}(\tau) \,| \, d\tau \geq c > 0, \ \forall t \quad (24)$$

where $Q_{Ld}(t)$ is related to $x_{1d}(t)$ by (21). Then, the resulting zero output error dynamics (i.e., the dynamics corresponding to $z_1 = 0, \forall t$) is globally [2] uniformly asymptotically stable with respect to the bounded equilibrium state trajectory $x_e(t)$ in (23). ◊

Proof: The detailed proof can be obtained from authors.

---

[2]For all pressures within physical limit

Remark 1. *Roughly speaking, the PE condition (24) is equivalent to the physical situation that the desired valve opening corresponding to the reference trajectory $x_{1d}(t)$ is not always* 0, *which includes the non-zero constant velocity reference trajectory studied in [13], and any sinusoidal reference trajectories.* ◇

Remark 2. *Result B of Theorem 2 indicates that all bounded equilibrium state trajectories $x_e(t)$ converge asymptotically to a unique trajectory depending on the reference trajectory x_{1d} only.* ◇

4 COMPARATIVE EXPERIMENTS

Experimental Setup

To test the proposed nonlinear ARC strategy and study fundamental problems associated with the control of electro-hydraulic systems, a three-link robot arm (a scaled down version of industrial backhoe loader arm) driven by three single-rod hydraulic cylinders has been set up at the Ray W. Herrick Laboratory of the School of Mechanical Engineering. The three hydraulic cylinders are controlled by two proportional directional control valves and one servovalve manufactured by Parker Hannifan company. The entire set-up is shown in Fig.3. For the purpose of this paper, experiments have been conducted on the swing motion control of the arm (or the first joint) with the other two joints fixed. The schematic of the swing circuit is shown in Fig.4. The swing circuit is driven by a single-rod cylinder (Parker D2HXTS23A) and controlled by a servovalve (Parker BD760AAAN10). The cylinder has a built-in Temposonic LDT sensor, which provides the position and velocity information of the cylinder movement. Pressure sensors (Entran's EPXH-X01-2.5KP) are installed on the each chamber of the cylinder. Due to the noises in the analog signals, the effective measurement resolution for the position is around 1 mm and for the pressure is $1.035 \times 10^5 Pa$. Since the range of the velocity provided by the Temposonic LDT sensor is too small (less than 0.09m/sec), backward difference plus filter is used to obtain the needed velocity information at high speed movement. The entire system is controlled by a dSPACE system and monitored by a host Pentium II PC. Control program is download from the host PC to the dSPACE system. Through plugged in A/D and D/A boards, all analog measurement signals (the cylinder position, velocity, forward and return chamber pressures, and the supplied pressure) are fed back to the dSPACE system and the calculated control command is then sent out to the driver of the servovalve. The supply pressure is $6.9 \times 10^6 Pa$.

System Identification

The swing inertia of the robot arm is estimated to be $J_0 = 78.69 kgm^2$. The cylinder physical parameters are $A_1 = 2.027 \times 10^{-3} m^2$, $A_2 = 1.069 \times 10^{-3} m^2$, $V_{h1} = 4.995 \times 10^{-4} m^3$, and $V_{h2} = 9.068 \times 10^{-4} m^3$. Although tests have been done to identify the friction curve, for simplicity, friction compensation is not used in the following experiments (i.e., $b = F_{fc} = 0$ in (1)) to test the robustness of the proposed algorithm to these *hard-to-model* terms. By assuming no leakage flows and moving the cylinder at different constant velocities, we can back out the static flow mappings (3) of the servovalve. The estimated flow gains are $k_{q1} = 3.590 \times 10^{-8} m^3/(\sqrt{Pa} secV)$ and $k_{q2} = 3.721 \times 10^{-8} m^3/(\sqrt{Pa} secV)$ respectively. Dynamic tests are also performed and reveal that the servovalve has a bandwidth around 10Hz. The effective bulk modulus is estimated around $2.71 \times 10^8 Pa$.

Controller Simplification

A simplification is made to the selection of the specific robust control term α_{2s} in (10) and α_{3s} in (16). Let k_{2s} and k_{3s} be any nonlinear feedback gains satisfying

$$\begin{aligned} k_{2s} &\geq \frac{\bar{m}_{0max}}{\theta_{1min}} \left[\frac{1}{2} w_2 \|\Gamma \phi_2\|^2 + k_2 + \frac{1}{2\varepsilon_2} (\|\theta_M\|^2 \|\phi_2\|^2 + \delta_d^2) \right] \\ k_{3s} &\geq \frac{1}{\theta_{3min}} \left[k_3 + w_3 \|\frac{\partial \alpha_2}{\partial \hat{\theta}}\|^2 + \frac{1}{2} w_3 \|\Gamma \phi_3\|^2 \right. \\ &\quad \left. + \frac{1}{2\varepsilon_3} (\|\theta_M\|^2 \|\phi_3\|^2 + \|\frac{\partial \alpha_2}{\partial x_2}\| \delta_d^2) \right] \end{aligned} \qquad (25)$$

in which $\theta_M = \theta_{max} - \theta_{min}$ and δ_d is defined in (7). Then, using similar arguments as in [13; 12], one can show that the robust control function of $\alpha_{2s} = -k_{2s} z_2$ satisfies (10) and (11), and $\alpha_{3s} = -k_{3s} z_3$ satisfies (16) and (17).

Knowing the above theoretical results, we may implement the needed robust control terms in the following two ways. The first method is to pick up a set of values for $k_2, k_3, w_2, w_3, \varepsilon_2$, and ε_3 to calculate the right hand side of (25). k_{2s} and k_{3s} can then be determined so that (25) is satisfied for a guaranteed global stability and a guaranteed control accuracy. This approach is rigorous and should be the formal approach to choose. However, it increases the complexity of the resulting control law considerably since it may need significant amount of computation time to calculate the lower bounds. As an alternative, a pragmatic approach is to simply choose k_{2s} and k_{3s} large enough without worrying about the specific values of $k_2, k_3, \omega_2, \omega_3, \varepsilon_2$, and ε_3. By doing so, (25) will be satisfied for certain sets of values of $k_2, k_3, \omega_2, \omega_3, \varepsilon_2$, and ε_3, at least locally around the desired trajectory to be tracked. In this paper, the second approach is used since it not only reduces the online computation time significantly, but also facilitates the gain tuning process in implementation.

Comparative Experimental Results

Three controllers are tested for comparison:

ARC1(with time varying inertial compensation) : the controller proposed in this paper with the simplification mentioned above. For simplicity, in the experiments, only two parameters, θ_1, and θ_2, are adapted; the first parameter represents the effect of external inertial load and the second parameter represents the effect of the nominal value of the lumped disturbance. Since the valve dynamics is neglected and its bandwidth is not so high (around 10Hz), not so large feedback gains are used to avoid instability; the control gains used are $k_p = k_{2s} = k_{3s} = 19$. The scaling factors of $s_{c3} = 2.8085 \times 10^6$ and $s_{c4} = 0.08588$, are used. Adaptation rates are set at $\Gamma = diag\{0.01, 0.15\}$.

ARC2 (with constant inertial compensation) : the controller that is synthesized in [14] with the assumption that the mass driven by the cylinder is constant. Scaling factors are the same as previous one. The control gains used are $k_p = k_{2s} = k_{3s} = 19$. Adaptation rates are set at $\Gamma = diag\{0.01, 0.08\}$.

Motion Controller : the state-of-the-art industrial motion controller (Parker's PMC6270ANI 2-axis motion controller) that we bought from Parker Hannifian company along with the Parker's cylinder and valves used for the experiments. The controller is essentially a PID controller with velocity and acceleration feedforward compensation. Controller gains are obtained by strictly following the gain tuning process stated in the "Servo Tuner User Guide" coming with the motion controller [17]. The tuned gains are $SGP = 20, SGI = 0.5, SGV = 22, SGVF = 100, SGAF = 0.02$, which represent the P-gain, I-gain, D-gain, the gain for velocity feedforward compensation, the gain for acceleration feedforward compensation respectively.

The three controllers are first tested for a point-to-point motion trajectory shown in Fig.3, which has a maximum velocity of $v_{max} = 0.1m/sec$ and a maximum acceleration of $a_{max} = 0.2m/sec^2$. The tracking errors are shown in Fig.4. As seen, the proposed ARC1 and ARC2 have a better performance than the Motion Controller in term of both transient and final tracking errors. Both ARC controllers reduce the final tracking error almost down to the measurement noise level. However, since ARC1 employs time varying inertia compensation which captures the nonlinear effect of the swing motion, ARC1 achieves a better tracking performance than ARC2; the tracking error of ARC1 is almost within the position measurement noise level all the time while ARC2 still has small transient error at the beginning of travel where the effective mass driven by the cylinder varies significantly with the change of swing angle as shown in Fig. 2.

To test the performance robustness of the proposed algorithms to parameter variations, a 45kg load is added at the end of the robot arm, which is equivalent to a swing inertia of $J_1 = 144kgm^2$. The tracking errors of all three controllers are shown in Fig.5. As seen, due to the fact that the adaptation algorithm of the ARC1 controller is able to pick up the change of the swing inertia, a better transient tracking performance is achieved in comparison to the constant compensation ARC2. Again, both ARC1 and ARC2 exhibit better performance than the Motion Controller.

The system is then run for a fast point to point trajectory shown in Fig.3, which has a maximum velocity of $v_{max} = 0.3m/sec$ and a maximum acceleration of $a_{max} = 1.5m/sec^2$; both are near their physical limits. The tracking errors of all three controllers are shown in Fig.6. As seen, the Motion Controller cannot handle such an aggressive movement well and a large tracking error around 15mm exhibits during the constant high-speed movement. In contrast, the tracking error of the proposed ARC1 during the entire run is kept within 3mm. Furthermore, the tracking error goes back to the measurement noise level of 1mm very quickly after the short large acceleration and deceleration periods. Again, the proposed ARC1 performs better than ARC2.

To further illustrate the effectiveness of the time-varying inertia compensation, the experiment is run for tracking a $\frac{1}{4}Hz$ sinusoidal desired trajectory($0.08sin(\pi t/2)$ with $45kg$ load for $16sec$. The errors of ARC1 and ARC2 are shown in the Fig.7. Again, since ARC1 uses time-varying inertia compensation and adapts the constant unknown payload instead, it achieves a better tracking performance than ARC2. All these results illustrate the effectiveness of the proposed time-varying inertia compensation ARC algorithm.

5 CONCLUSIONS

In this paper, a discontinuous projection based ARC controller is constructed for the precision motion control of single-rod hydraulic actuators with time-varying unknown inertia loads. The controller takes into account the particular nonlinearities associated with the time-varying inertia load and the hydraulic dynamics, and allows parametric uncertainties such as variations of robot payload and various hydraulic parameters as well as uncertain nonlinearities coming from external disturbances, uncompensated friction forces, etc. Stability of zero dynamics when tracking a kind of arbitrary trajectory is discussed and prove to be global uniformly stable. Experimental results done on the swing motion control of a hydraulic robot arm verified the effectiveness and the high-performance nature of the proposed ARC algorithm.

The paper also serves as an illustrating example on how to incorporate the particular nonlinearity of a physical system into the proposed nonlinear ARC framework for an improved performance (e.g. the nonlinear transformation between the constant swing inertia and the time-varying effective mass of the swing hydraulic cylinder in this paper).

ACKNOWLEDGEMENT

This work is supported in part by the National Science Foundation under the CAREER grant CMS-9734345 and in part by a grant from the Purdue Research Foundation

REFERENCES

[1] H. E. Merritt, *Hydraulic control systems*. New York: Wiley, 1967.

[2] T. C. Tsao and M. Tomizuka, "Robust adaptive and repetitive digital control and application to hydraulic servo for noncircular machining," *ASME J. Dynamic Systems, Measurement, and Control*, vol. 116, pp. 24–32, 1994.

[3] P. M. FitzSimons and J. J. Palazzolo, "Part i: Modeling of a one-degree-of-freedom active hydraulic mount; part ii: Control," *ASME J. Dynamic Systems, Measurement, and Control*, vol. 118, no. 4, pp. 439–448, 1996.

[4] A. R. Plummer and N. D. Vaughan, "Robust adaptive control for hydraulic servosystems," *ASME J. Dynamic System, Measurement and Control*, vol. 118, pp. 237–244, 1996.

[5] J. E. Bobrow and K. Lum, "Adaptive, high bandwidth control of a hydraulic actuator," *ASME J. Dynamic Systems, Measurement, and Control*, vol. 118, pp. 714–720, 1996.

[6] R. Vossoughi and M. Donath, "Dynamic feedback linearization for electro-hydraulically actuated control systems," *ASME J. Dynamic Systems, Measurement, and Control*, vol. 117, no. 4, pp. 468–477, 1995.

[7] L. D. Re and A. Isidori, "Performance enhancement of nonlinear drives by feedback linearization of linear-bilinear cascade models," *IEEE Trans. on Control Systems Technology*, vol. 3, no. 3, pp. 299–308, 1995.

[8] A. Alleyne and J. K. Hedrick, "Nonlinear adaptive control of active suspension," *IEEE Trans. on Control Systems Technology*, vol. 3, no. 1, pp. 94–101, 1995.

[9] B. Yao, G. T. C. Chiu, and J. T. Reedy, "Nonlinear adaptive robust control of one-dof electro-hydraulic servo systems," in

ASME International Mechanical Engineering Congress and Exposition (IMECE'97), FPST-Vol.4, (Dallas), pp. 191–197, 1997.

[10] B. Yao and M. Tomizuka, "Smooth robust adaptive sliding mode control of robot manipulators with guaranteed transient performance," in *Proc. of American Control Conference*, pp. 1176–1180, 1994. The full paper appeared in *ASME Journal of Dynamic Systems, Measurement and Control*, Vol. 118, No.4, pp764-775, 1996.

[11] B. Yao and M. Tomizuka, "Adaptive robust control of siso nonlinear systems in a semi-strict feedback form," *Automatica*, vol. 33, no. 5, pp. 893–900, 1997. (Part of the paper appeared in Proc. of 1995 American Control Conference, pp2500-2505).

[12] B. Yao, "High performance adaptive robust control of nonlinear systems: a general framework and new schemes," in *Proc. of IEEE Conference on Decision and Control*, pp. 2489–2494, 1997.

[13] B. Yao, F. Bu, and G. T. C. Chiu, "Nonlinear adaptive robust control of electro-hydraulic servo systems with discontinuous projections," in *Proc. of IEEE Conf. on Decision and Control*, pp. 2265–2270, 1998. The revised full paper '*Nonlinear Adaptive Robust Control Of Electro-Hydraulic Systems Driven by Double-rod Actuators*' is submitted to *International Journal of Control* in 1999.

[14] B. Yao, F. Bu, J. Reedy, and G. Chiu, "Adaptive robust control of single-rod hydraulic actuators: theory and experiments," in *Proc. of American Control Conference*, 1999. (to appear). The revised full paper is submitted to IEEE/ASME Trans. on Mechantronics.

[15] M. Krstic, I. Kanellakopoulos, and P. V. Kokotovic, *Nonlinear and adaptive control design*. New York: Wiley, 1995.

[16] H. K. Khalil, *Nonlinear Systems (Second Edition)*. Prentice Hall, Inc., 1996.

[17] C. D. P. H. Co., *Servo tuner user guide*. Software Mannual, Nov. 1994.

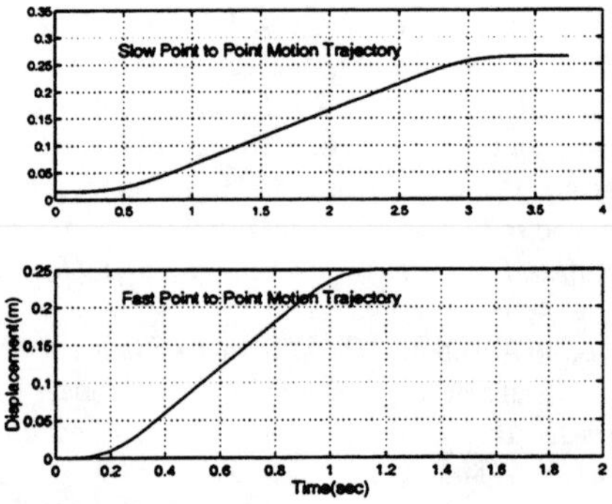

Figure 3. Point to point motion trajectory

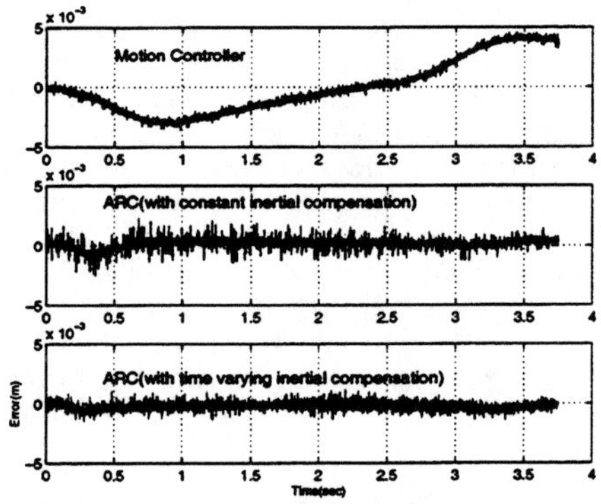

Figure 4. Tracking Errors in point to point motion without load

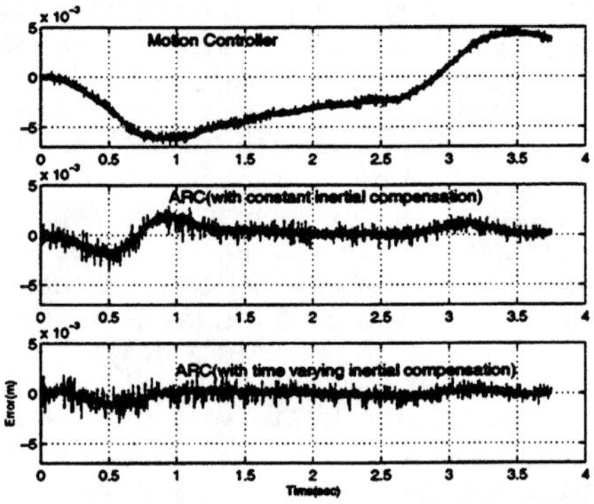

Figure 5. Tracking Errors in point to point motion with load

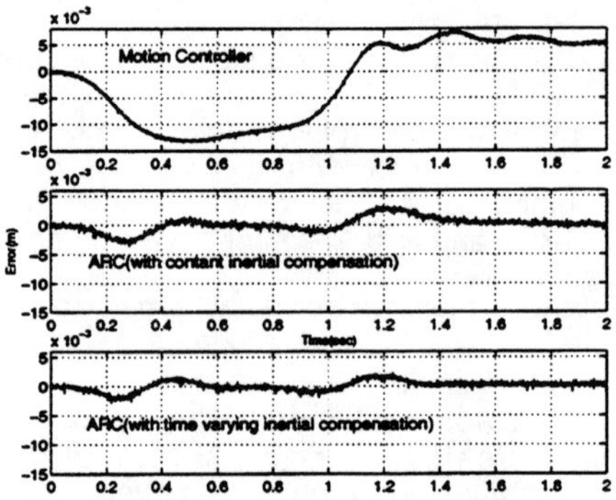

Figure 6. Tracking Errors in fast point to point motion without load

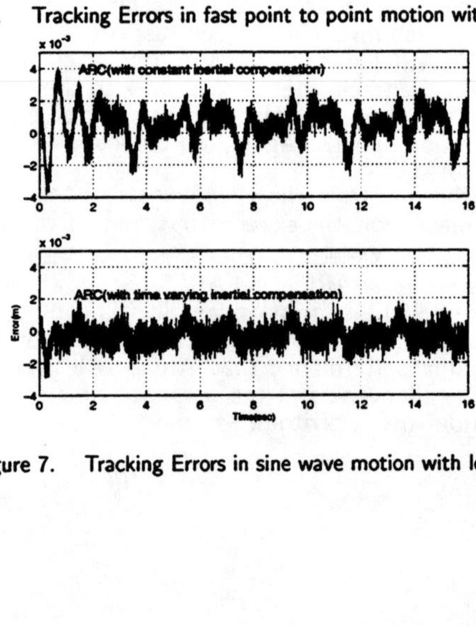

Figure 7. Tracking Errors in sine wave motion with load

FPST-Vol. 6, Fluid Power Systems and Technology — 1999
ASME 1999

MODELING AND CONTROL DESIGN OF A POWERTRAIN SIMULATION TESTBED FOR EARTHMOVING VEHICLES

Eko A. Prasetiawan, Rong Zhang, Andrew G. Alleyne, Tsu-Chin Tsao
Mechanical and Industrial Engineering Department,
University of Illinois,
Urbana, IL 61801

ABSTRACT

A nonlinear model of an Earthmoving Vehicle Powertrain Simulator is developed using both first principles as well as I/O data. The model is an interconnection of subsystem models, which consist of a prime mover, a variable-displacement pump, proportional flow valves, and fixed-displacement motor models. In addition, a typical drive or tractive load of an earthmoving vehicle is also presented and implemented in the system as one of the loads. Efforts have been directed to develop a simple model in order to cater to developing and testing powertrain controllers as well as studying the dynamic behavior of such systems. An initial control design approach utilizing the Linear Quadratic technique is also presented. The controller is aimed at regulating load speeds in presence of a step load disturbance. The simulation results of the compensated system response are presented.

1. INTRODUCTION

Power optimization of earthmoving vehicle powertrains is an important area of research for both industry and research institutions. However, there are still many issues that are open both in the modeling and control aspects of the problem. A good model that describes the overall dynamics of the system is very important for the development of an acceptable controller. In Singh [1], it is mentioned that there is a problem of limited hydraulic fluid flow, which remains unmodeled and unresolved. Oftentimes, when the powertrain provides flow to a function, fluid flow to other functions is deteriorated and their performance is sacrificed. Dasgupta and Chattopadhyay proposed a basic approach to analyze an earthmover's powertrain dynamics [2]. The approach breaks down the complex system into its individual components to be studied independently. Once the components are properly interfaced, overall understanding of all components can be derived by considering the most significant dynamic responses related to the system's expected function and performance. However, [2] does not detail the modeling aspect of their approach.

It is very important to understand the overall dynamics of these systems so as to increase design performance. In order to examine the dynamics of such systems, an Earthmoving Vehicles Powertrain Simulator (EVPS) was built at the University of Illinois as part of the Caterpillar Electromechanical Systems Laboratory. This system will be used to research the generic power optimization problem for several types of vehicle powertrains. The current paper develops a mathematical model of the system, which consists of four major types of components (a prime mover, a pump, valves and motors), for analysis and control design. A model of a typical earthmoving vehicle load is also developed.

The rest of the paper is presented as follows. In Section 2, a brief description of the EVPS is given. Then, in Section 3, individual elements of the system model will be presented. Most of the model is developed based on first principles, and it captures the system's dynamic responses in the continuous-time domain. In some cases, model components have been developed from frequency response data that is available; e.g. manufacturer's data. Efforts have been directed to achieve a reasonable trade-off between simplicity and comprehensiveness of the model. A simple model is attractive not only to facilitate the controller design but also to simplify the real-time application of possible controllers. By the same token, a model has to represent short-lived transients of the system adequately. Section 3 is broken down into several subsections and the individual model components are presented in each subsection. Additionally, a simulation is given for the overall system after the individual models are combined. In Section 4, one approach to control design is proposed for the system. The system is linearized over some arbitrary operating conditions to make use of linear control design techniques. A Linear Quadratic technique is used to determine a feedback gain matrix that minimizes an infinite horizon cost function. The objective of the controller has been primarily optimizing the overall load-speed adjustment of the system due to load variations. A conclusion then summarizes the main points of the paper.

2. SYSTEM DESCRIPTION

A schematic of the EVPS is shown in Figure 1 and the actual EVPS is shown in Figure 2. A three-phase induction motor is used as a prime mover. An electric motor is chosen because of its clean operation and its versatility to simulate other types of prime movers. Although not the focus of this paper, in future work a controller will be designed to match the dynamic responses of the motor with those of a gasoline or diesel engine, which is an actual system's prime mover. The motor drives a variable-displacement pump that provides flow to three different load-networks connected in parallel. A pressure relief valve limits the maximum value of the operating pressure to 3000 psi (207 bar). Any excess of flow triggers the valve and flow is directed back to a reservoir. A proportional flow valve regulates flow across each network. A fixed-displacement gear motor is used to convert flow to rotational energy, which is used to drive a load. To simulate a load, the gear motor is coupled with a fixed-displacement gear pump that draws fluid from a reservoir. Controlling a proportional pressure valve, which determines how much fluid can be drawn by the pump, simulates a load resistance. A separate controller will be designed to match the dynamic response of simulated load resistances with those of real world loads such as implements, drives and steering.

The experimental setup can be made analogous to an actual earthmoving vehicle powertrain configuration. Generally, the system has its prime mover in the form of a diesel engine. The engine, through some sort of torque converter, drives a pump. Similarly, the pump provides flow to load-networks configured in a similar manner. Oftentimes the configuration in an earthmoving vehicle is serial rather than parallel setup shown in Figure 1.

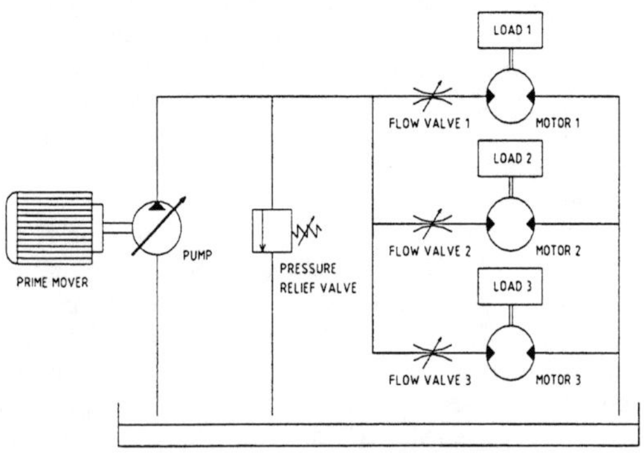

Fig. 1 EVPS schematic

Fig. 2 Actual experimental EVPS at UIUC

3. SYSTEM MODELING

The model of the system can be broken down into subsystem models that will be presented separately.

3.1 Prime Mover Model

There has been a great deal of development in the area of prime mover models for control. A paper by J. D. Powell [3] provides comprehensive survey on the subject. In the current paper, a 3.8L, spark-ignition engine model developed by Cho and Hedrick [4] is chosen. For the sake of completeness of the powertrain model development, the engine model of [4] is included in this paper.

The engine model can be broken down into five different subsystems namely throttle body, intake manifold, fuel injection, combustion and torque production, and crankshaft rotation. The throttle body describes transformation of throttle angle inputs to mass rates of air entering an intake manifold $\dot{m}_{ai}$, and is described by the equation

$$\dot{m}_{ai} = MAX_m \cdot TC \cdot PRI \tag{1}$$

MAX_m is the maximum flow rate of air through the throttle body (0.1843 kg/s), and TC is the normalized throttle characteristic as a function of throttle angles γ

$$TC = \begin{cases} 1-\cos(1.14459\cdot\gamma - 1.06^o), & \gamma \leq 79.46^o \\ 1, & \gamma > 79.46^o \end{cases} \tag{2}$$

PRI is the normalized pressure ratio influence

$$PRI = 1 - \exp(2771\cdot m_a - 9) \tag{3}$$

where m_a is the mass of air in the intake manifold. The intake manifold, which determines engine torque responses to changes in throttle angles, is modeled by the conservation of air mass rate in the manifold

$$\dot{m}_a = \dot{m}_{ai} - \dot{m}_{ao} \tag{4}$$

where $\dot{m}_{ao}$ is the mass rate of air leaving the manifold (and thus, entering the combustion chamber). The mass rate of air leaving the manifold is dependent upon engine characteristics (displacement V_e, intake manifold volume V_m and volumetric efficiency η_{vol}) and engine speeds ω_e as described by the equation

$$\dot{m}_{ao} = \left(\frac{V_e}{4\pi\cdot V_m}\right)\cdot\eta_{vol}\cdot m_a\cdot\omega_e \tag{5}$$

For this particular engine, V_e is 0.0038 m^3 and V_m is 0.0027 m^3. The volumetric efficiency η_{vol} can be determined by a non-linear relation

$$\eta_{vol} = \left(24.5\cdot\omega_e - 3.10\times10^4\right)\cdot m_a^2 + \left(-0.167\cdot\omega_e + 222\right)\cdot m_a + \left(8.10\times10^{-4}\cdot\omega_e + 0.352\right) \tag{6}$$

The fuel injection describes the relationship between fueling commands and fuel flow rates into the cylinder, which is characterized by a combination of lag and transport delays due to the discrete nature of the intake process. For a sequential-fire port fuel-injection system, the fueling model is simplified as a first-order equation in terms of the actual fuel rate entering the combustion chamber $\dot{m}_{fi}$

$$\tau_f \ddot{m}_{fi} + \dot{m}_{fi} = \dot{m}_{fc} \quad (7)$$

where the effective fueling time constant τ_f is modeled as

$$\tau_f = 0.05 + \frac{1.5 \cdot \pi}{\omega_e} \cdot \frac{\dot{m}_{fc} \cdot A/F_{des}}{MAX_m} \quad (8)$$

where $\dot{m}_{fc}$ is the commanded fuel rate and A/F_{des} is the desired air-to-fuel ratio. The combustion and torque production subsystem contains delays associated with the four combustion processes as modeled in the engine's indicated torque T_{ei} equation

$$T_{ei} = MAX_{CT} \cdot \frac{\dot{m}_{ao}(t - \Delta t_{it})}{\omega_e(t - \Delta t_{it})} \cdot AFI(t - \Delta t_{it}) \cdot SI(t - \Delta t_{st}) \quad (9)$$

where MAX_{CT} is the maximum torque capacity of an engine for a given air mass, engine speed, AFI = 1, and SI = 1 (498636 N.m/kg/s). The normalized air fuel influence AFI is described as

$$AFI = \cos\left(7.3834^o \cdot \left(\frac{\dot{m}_{ao}}{\dot{m}_{fi}} - 13.5\right)\right), \quad (10)$$

and the normalized spark influence SI is described as

$$SI = (\cos(SA - MBT))^{2.875} \quad (11)$$

where SA is the spark advance/retard from the top dead center (TDC) and MBT is the minimum-spark-advance for best torque (30°). The delays that capture the cyclic nature of a four-stroke engine are expressed as an intake to torque production delay $\Delta t_{it} \cong 5.48/\omega_e$, and a spark to torque production delay $\Delta t_{st} \cong 1.3/\omega_e$. The crankshaft rotation follows the conservation of moment about a rigid shaft

$$\dot{\omega}_e = \frac{1}{I_e}(T_{ei} - T_{ef} - T_{el}) \quad (12)$$

where I_e is the crankshaft moment of inertia (0.1454 kg.m^2), T_{ef} is the friction torque

$$T_{ef} = 0.1056 \cdot \omega_e + 15.1 \quad (13)$$

and T_{el} is the external load torque seen by the engine.

The engine model can be summarized as a non-linear model of three state dynamics. The states are m_a, $\dot{m}_{fi}$ and ω_e and the possible control inputs are γ, A/F and SA. For this paper, only γ is considered and interested readers can refer to [4] for further details.

3.2 Pump and Hose Volume Model

The displacement of the axial piston pump governs the flow being generated for a given engine speed. The displacement is determined by a position of the variable swashplate angle α. It is usually controlled by a pressure balanced hydraulic servo-system. A detailed mathematical model of the swashplate movement has been developed by Schoenau et al. [5]. However, the complexity of [5]'s model is beyond the detail level necessary for investigation. Instead, a simple model obtained from manufacturer's data is used in this paper. A block diagram below can best describe the swashplate dynamic equation

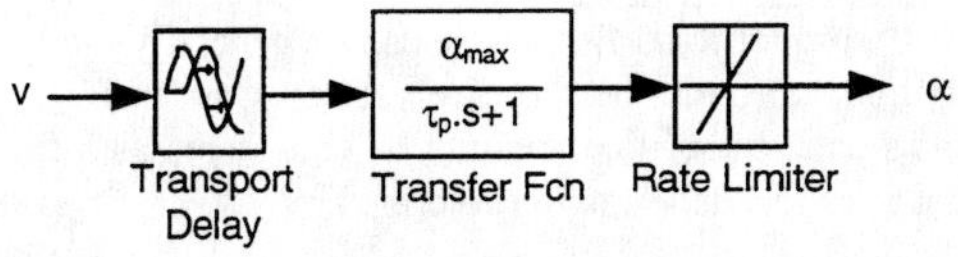

where τ_p is the pump time constant and α_{max} is the maximum swashplate angle. The normalized commanded input v is delayed by Δt_{if}, and the rise and fall rate of the swashplate angle α is limited by constants c_1 and c_2 respectively.

There is a significant volume contained in the hydraulic hose connecting the flow control valve in a manifold to the motors. This volume will contain dynamics that affect the higher frequencies in the system response. To model pressure in the hose between the pump and the valves, a lumped volume approach is taken. This uses the principle of a mass conservation in conjunction with the definition of the fluid bulk modulus [6, 7]. The differential equation governing the pressure-rise rate within the system hose can be written as

$$\dot{P}_h = \frac{\beta}{V_h} Q_h \quad (14)$$

where β is the fluid bulk modulus, V_h is the volume of the hose and Q_h is the net flow into the hose. Clearly from Figure 3, the net flow into the hose can be described as

$$Q_h = D_p \cdot \omega_p - C_l \cdot P_h - Q_{motor} \quad (15)$$

where D_p is the instantaneous pump displacement, ω_p is the pump speed, which is equal to engine speed ω_e, and C_l is the leakage coefficient. Since the pump displacement is linearly proportional to the swashplate angle for small angles, D_p can be expressed as

$$D_p = K_p \cdot \alpha \quad (16)$$

where K_p is the pump-displacement gain.

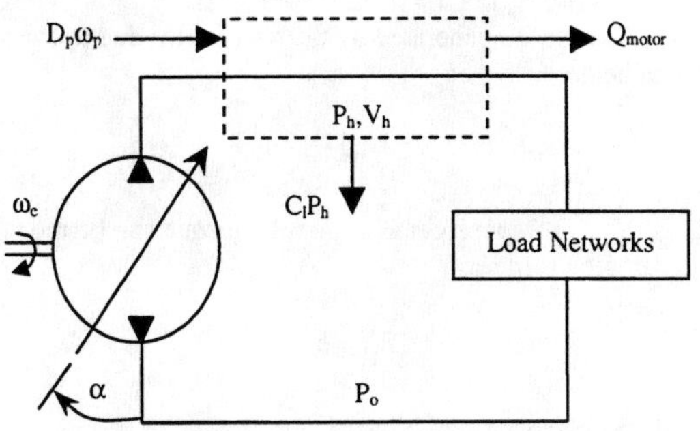

Fig. 3 Hydrostatic transmission schematic

In summary, the pump and hose pressure models have two combined states (α and P_h) and one control input v. Although the length of the hose may introduce transmission line dynamics [7], this modeling approach treats the entire volume as one lumped first-order system. Future work may be directed at adding additional model at approximate the line dynamics more accurately.

3.3 Flow Control Valve Model

The EVPS proportional flow valve can be generally modeled as a flow restriction through an orifice area. Gathered from the manufacturer's data, the orifice area can be considered as a first-order system and a transport delay Δt_{iv} as described by

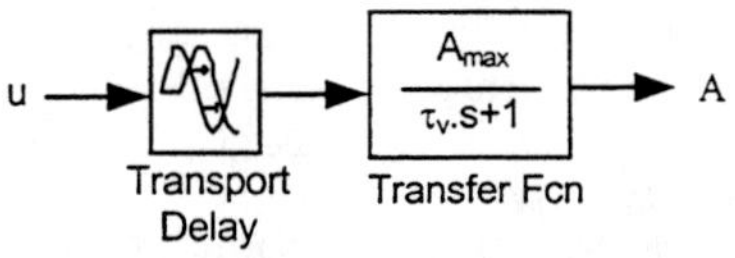

where τ_v is the valve time constant, u is the normalized commanded input voltage, A_{max} is the maximum orifice area, and A is the orifice area output. The pressure drop across the valve is inversely proportional to the square root of the orifice area according to the equation

$$\Delta P = P_h - P_i = \frac{\rho}{2 \cdot C_d^2} \cdot \left(\frac{Q_i}{A_i} \right)^2 \tag{17}$$

where ρ is the fluid density, C_d is the coefficient of discharge, P_i and Q_i are the supply pressure and flow rate to each motor on the load network i [7]. Rearranging Equation 17, the supply pressure of each motor can be determined by

$$P_i = P_h - \frac{\rho}{2 \cdot C_d^2} \cdot \left(\frac{Q_i}{A_i} \right)^2 \tag{18}$$

3.4 Hydraulic Motor Model

A torque balance relation is used to describe the motor acceleration equation

$$\dot{\omega}_i = \frac{1}{I_i} \left(T_{mi} - T_{mf} - T_{load} \right) \tag{19}$$

where I_i is the motor moment of inertia, T_{mi} is the torque generated by the supply pressure, T_{mf} is the friction torque, and T_{load} is the external torque load. The torque generated by the gear motor due to the supply pressure can be described as

$$T_{mi} = D_i (P_i - P_o) \tag{20}$$

where D_i is the motor displacement. Friction torque can be modeled as a viscous damping torque

$$T_{mf} = b_{mi} \cdot \omega_i \tag{21}$$

where b_{mi} is the viscous friction coefficient.

3.5 Load Model: Drive Load

From Figure 4 there are three different types of loads that will be considered for the EVPS. In this work, only the vehicle drive load model will be presented. This model comprises of three subsystem i.e. vehicle dynamic, tire, and longitudinal operating force models. Figure 5 shows the interrelation between each subsystem. The operating force model is not considered at this point since we assume a constant topographical condition (moving on a flat surface).

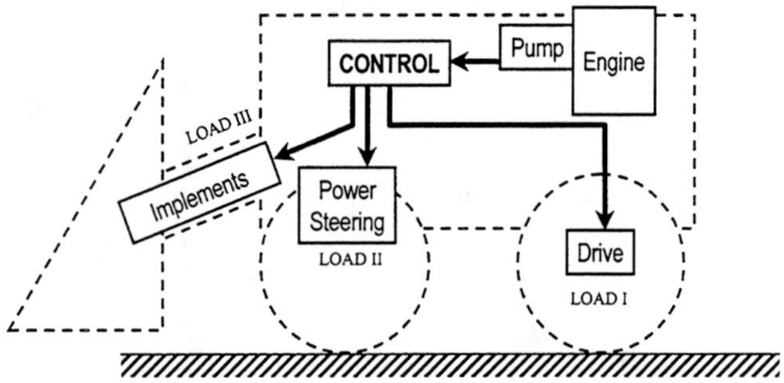

Fig. 4 Different types of load on the earthmoving vehicle powertrain

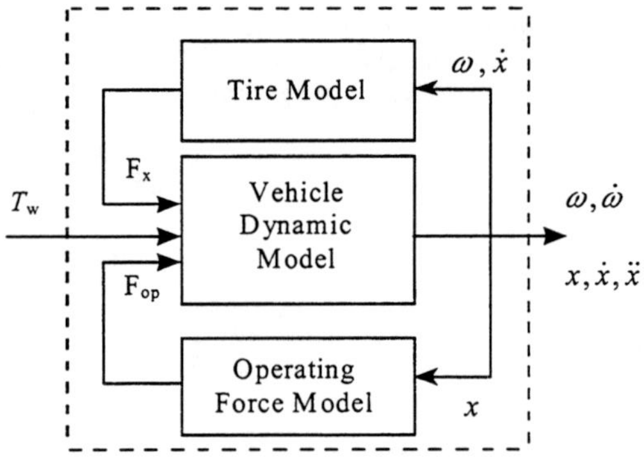

Fig. 5 Drive load model

A simplified quarter-vehicle dynamics model can be classified as rotational dynamics

$$I_w \dot{\omega} + b_w \omega = T_w - T_f - F_x r_w \tag{22}$$

and translational dynamics

$$m\ddot{x} + \frac{1}{2} \rho_{air} C_D A_v \dot{x}^2 = F_x - F_f \tag{23}$$

where:
A_v = cross sectional area of the ¼ vehicle,
b_w = rotary damping ratio of the wheel,
C_D = air drag coefficient of the vehicle,
F_f = constant rolling resistant force,
F_x = longitudinal force of the tire as a function of slip ratio λ,
I_w = wheel moment of inertia,
m = mass of the ¼ vehicle,
r_w = wheel rolling radius,
T_f = constant rolling resistant moment,
T_w = driving torque applying to the wheel,
x = vehicle linear displacement.,
$\dot{x}$ = vehicle linear velocity,
$\ddot{x}$ = vehicle linear acceleration,
ρ_{air} = air density,
ω = wheel angular velocity, and

$\dot{\omega}$ = wheel angular acceleration.

A considerable amount of research has been done in the area of tire model development. One in particular is the Magic Formula Tire Model, an empirical model developed by Pacejka and Bakker [8]. The formula is capable of producing force and moment characteristics of a tire at certain slip conditions as shown in Equation 24

$$F_x = D\cdot\sin\left(C\cdot\tan^{-1}\left(B\cdot\lambda - E\cdot\left(B\cdot\lambda-\tan^{-1}(B\cdot\lambda)\right)\right)\right) \tag{24}$$

where B is the stiffness factor (0.171), C is the shape factor (1.69), D is the peak factor (0.3), and E is the curvature factor (0.619). The coefficient values are of an on-road vehicle data. Data for an earthmoving vehicle Magic Formula tire model is not presently available, and is subject of current investigations. In this paper, we consider an acceleration case where slip ratio λ is defined as

$$\lambda = 1-\frac{\dot{x}}{\omega\cdot r_w} \quad \text{for } \omega\cdot r_w > \dot{x} \tag{25}$$

Figure 6 shows an example of simulation results for the load dynamics of a typical 150-hp, 4WD wheel loader. The angular and linear speeds behave similar to a first-order system response. The slip ratio is minimal but required to provide the longitudinal force to move the vehicle.

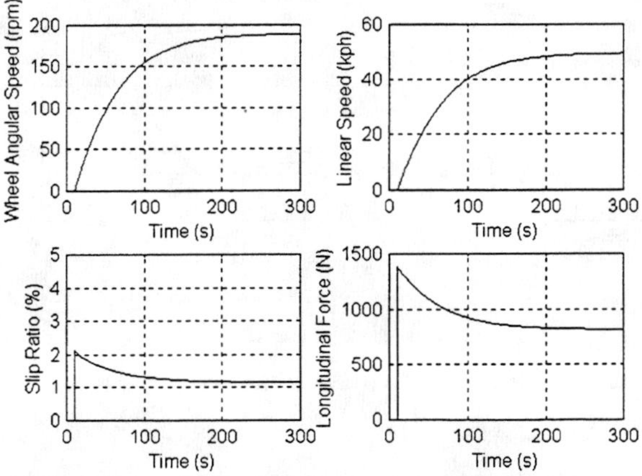

Fig. 6 Drive load dynamic simulation response

3.6 Powertrain Model Simulation Results

The overall powertrain model can be obtained by connecting the components of all individual models. Figure 7 shows a schematic of such a connection for a typical load unit. There are two connecting variables between the engine and the pump, the engine/pump speed and the pump torque load. The speed can be expressed as

$$\dot{\omega}_e = \dot{\omega}_p = \frac{1}{I_e + I_p}\left(c_3\cdot T_{ei} - T_{ef} - c_3\cdot T_{el}\right) \tag{26}$$

where I_p is the pump moment of inertia and c_3 is a scaling constant. The constant has to be applied in order to implement the 120-hp engine model in place of the 30-hp electric motor for simulation of the EVPS. The pump torque load P_{load} can be expressed as

$$P_{load} = T_{el} = D_p\cdot\left(P_h - P_o\right) \tag{27}$$

where P_o is the pressure on the low-pressure line (Figure 3). Without losing generality, P_o is assumed to be constant, which makes the hose pressure P_h a relative pressure. Between the pump and load networks, there are also two connections: the hose pressure and the summation of all flow rates at the load networks (in Figure 3, it is denoted by Q_{motor}). The motor flow rate Q_{motor} in Equation 15 can be expressed as

$$Q_{motor} = \sum_{i=1}^{3} Q_i = \sum_{i=1}^{3} D_i\cdot\omega_i \tag{28}$$

Furthermore, at each load network, there are interconnections between the valve and the motor as well as the motor and the external load. The connections between the valve and motor are the supply pressure P_i and the motor's flow rate Q_i. The connection between the motor and the load are the motor/load speed and the external torque load. For the simulation purpose, the drive load model is implemented on the second load network only. During implementation, a scaling factor has to be implemented to the model so that it is compatible with the other EVPS system models (Figure 8). The other networks are only subjected to a viscous friction load and free from other external loads.

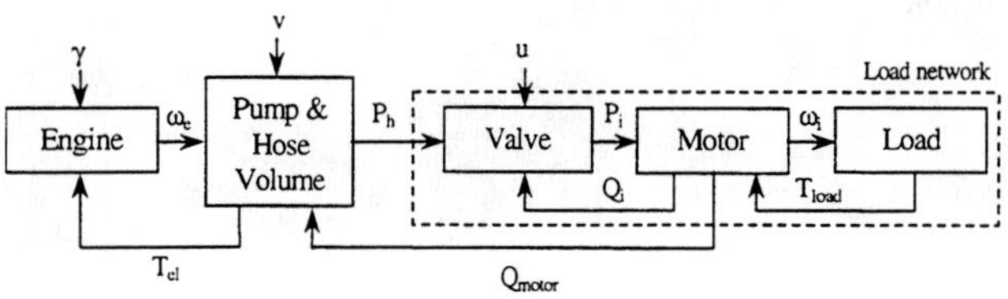

Fig. 7 Model interconnection schematic

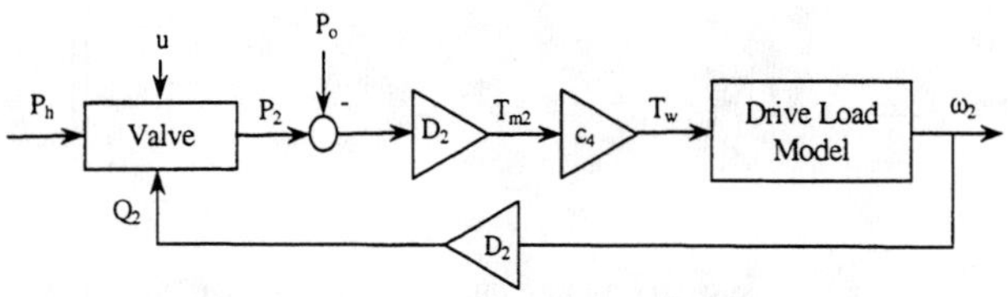

Fig. 8 Drive load model implementation

To observe open-loop dynamic response of the EVPS model, some simulation cases were carried out on MATLAB's Simulink. Parameters' values used in the simulation are given in Appendix.

The first case is when the model is subjected to a step change in throttle angle (Figure 9). A throttle angle increase of 10°, with respect to its original operating condition of 60°, is applied instantaneously at the 25th second. It is obvious that increasing throttle angle increases the power that goes into the system. Consequently, the engine speed, the hose pressure and the load speed should increase as shown by the results.

The second case is when the model is subjected to a 3.6° step input increase, with respect to its 9° operating condition, on the pump swashplate angle (Figure 10). Initially, the hose pressure rise-rate is large since the first term of Equation 15 increases rapidly before it is counterbalanced by an increase of the second term. The overall increase in the hose pressure, at steady state, produces more driving torque to the motors, and causes the motors to run faster. However, since the engine input is constant, the engine speed decreases because of a higher engine load.

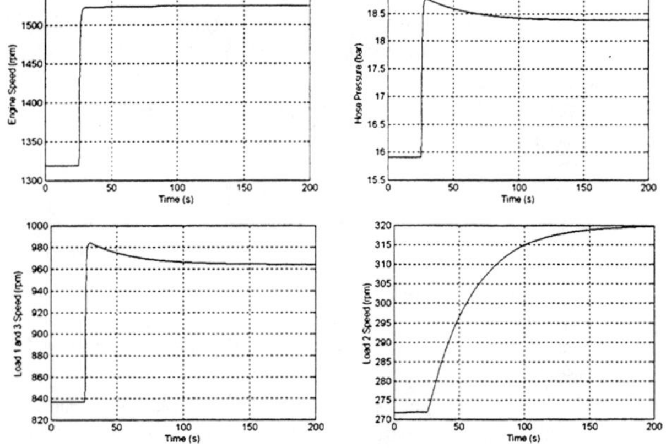

Fig. 9 Simulated system response due to a step change in the throttle angle

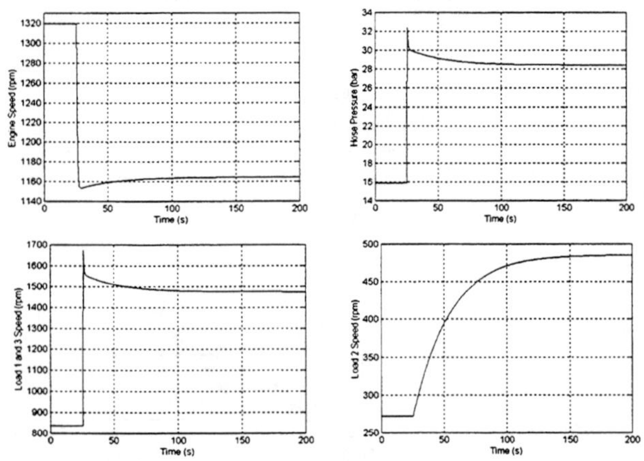

Fig. 10 Simulated system response due to a step change in the pump displacement

The third case is when the orifice area of the first load valve is reduced from full to half open (Figure 11). As a result, the load 1 speed decreases. Since the overall flow to the load network is unchanged the flow is being directed to other loads, causing them to increase in speed. The reduction of the orifice area also causes the hose pressure to increase because there is an increase in pressure drop across the valve. This increase is seen by the engine as a higher load, thus the engine speed decreases. However, the amount of changes due to the orifice area variation is quite small in magnitude so we should expect that the system would not be quite responsive to the valve orifice variation, for this range of orifice areas, as we will see later in the next section.

The last case involves the model being subjected to a step external load T_{load} of 1 Nm (Figure 12). The load is applied to the first motor. Similar to the valve orifice variation, the speed of that particular network drops whereas the speed of the other networks increase. Since the net speed (i.e. hydraulic flow) of the three motors decreases, a lower net load is seen by the pump. The pump can then produce more torque in terms of increase in the hose pressure, which is seen as a higher load by the engine. Hence, as shown in the simulation, the engine speed decreases accordingly.

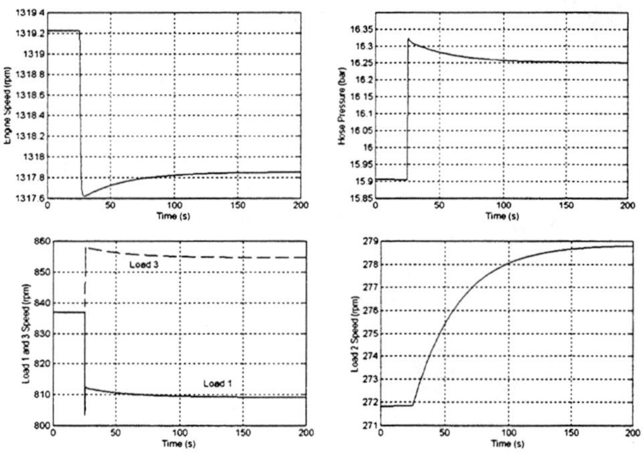

Fig. 11 Simulated system response due to a step change in the flow orifice area

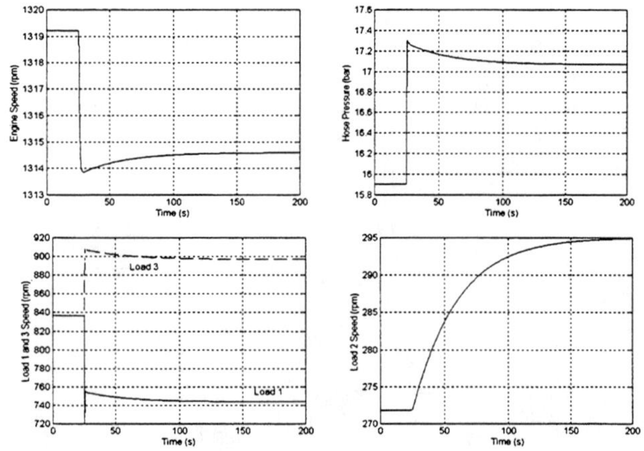

Fig. 12 Simulated system response due to a step external load on the first load network

4. LINEARIZATION AND CONTROL DESIGN

A state-feedback controller is proposed to optimize the overall load-speed adjustment of the system due to load variations. The Linear Quadratic (LQ) technique is used to obtain a control law of the form $\bar{u} = -K\bar{x}$ that minimizes a cost function

$$J = \int_0^\infty \left(\bar{x}^T Q\bar{x} + \bar{u}^T R\bar{u}\right) dt \qquad (29)$$

where the system dynamics are given by

$$\dot{\bar{x}} = A\bar{x} + B\bar{u} \qquad (30)$$

Q is a positive-semidefinite state weighting matrix, R is a positive-definite input weighting matrix, $\bar{x}$ is the state vector, and $\bar{u}$ is the input vector. In this paper, the cost function is aimed at minimizing the effect of a constant disturbance on the load speed.

The overall EVPS model consists of 11 states (m_a, $\dot{m}_{fc}$, ω_e, α, P_h, A_1, A_2, A_3, ω_1, ω_2 and ω_3) and 5 possible inputs (γ, v, u_1, u_2 and u_3). It is obvious from the governing differential equations presented in previous sections that the overall system is non-linear. To utilize the LQ technique, the system has to be linearized on some stable operating conditions, so that the equations of motion of the system can be written in the form of Equation 30.

The three-state engine model presented in Section 3 is highly non-linear. The high degree of non-linearity of the model makes it unattractive for LQ design technique. Furthermore, the existence of time delays in the torque production process will add extra poles and non-minimum phase zeros if they are approximated by Pade approximation [9]. These poles and zeros cause the order of the system model to increase, which is undesirable from control design perspective. In order to simplify the torque production model, an approximation is used instead. The process is approximated as a first-order transfer function from the throttle input γ to the engine indicated torque T_{ei}, and a transport delay as describe by a block diagram

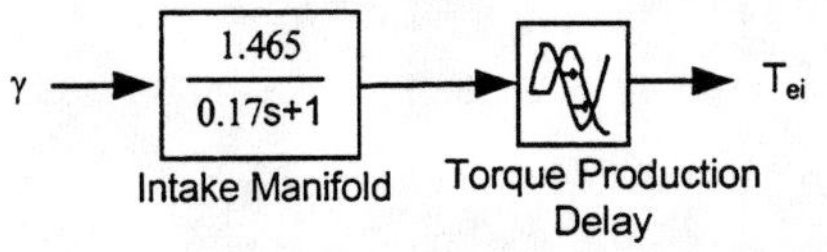

where the delay is the maximum delay of the torque production process, which is Δt_{it} = 0.1744 seconds for engine speeds greater than 300 rpm. The approximation is compared to the original model via simulation. Figure 13 shows the results of such simulation. The simplified model response is generally more sluggish in transient compared to the original model. The difference is caused by the non-linearity of the original model, which can not be predicted very well by a simple first-order model. Although there is also a steady-state error, the magnitude is very small compared to the operating condition, less than 2%. Therefore, the use of approximation in control design is justified in lieu of the complex original model within a certain range.

Transport delays that occur in the torque production, the pump and the valve model are approximated by a first-order lag

$$e^{-T_d s} \cong \frac{1}{(T_d/2)s+1} \tag{31}$$

where T_d is the delay time. The approximation introduces an extra pole to the system transfer function, but, unlike Pade's method, it does not introduce an extra zero.

As a results of approximations, the number of states grows to 15 (T_{ei}, $\dot{T}_{ei}$, ω_e, α, $\dot{\alpha}$, P_h, A_1, $\dot{A}_1$, A_2, $\dot{A}_2$, A_3, $\dot{A}_3$, ω_1, ω_2 and ω_3). Furthermore, linearization of these states over some equilibrium points allows us to write the state equation in term of Equation 30 with vector $\bar{x}$ and $\bar{u}$ are replaced by $\delta\bar{x}$ and $\delta\bar{u}$, respectively.

In achieving the performance specification, mentioned earlier, the last three states are weighted more heavily than the other states. As a result, as shown in Figure 14, a step load of 1 Nm applied on the second load is rejected much better in steady state when compared to the response of uncompensated system. The results show that the controller fails to regulate the speed of the loads, particularly 1 and 3, back to the original speed. However, this is not unusual because the LQ state-feedback controller does not have any integral action. Therefore, constant disturbances can result in steady state error. The controller tries to satisfy the flow demand by the loads through increasing the pump displacement (Figure 15), which consequently increases the hose pressure. In addition, to avoid a large drop on the engine speed due to the hose pressure increase, the controller increases the throttle angle. However, these actions of the swashplate and the throttle need more contribution from the valves, which do little in regulating the load speeds back to the original.

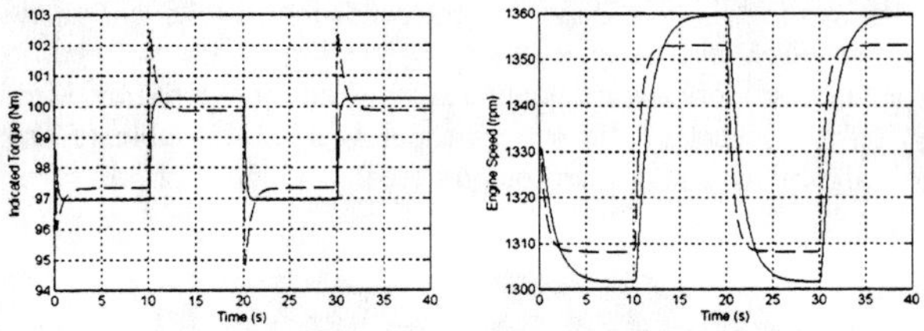

Fig. 13 Comparison of original (dashed line) and approximated engine model

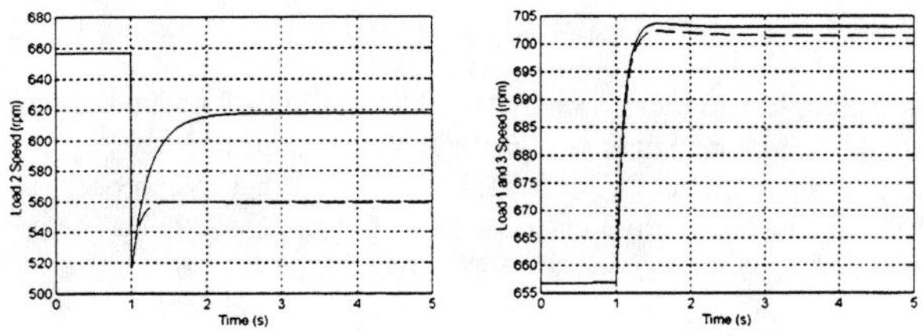

Fig. 14 Comparison of uncompensated (dashed line) and compensated responses

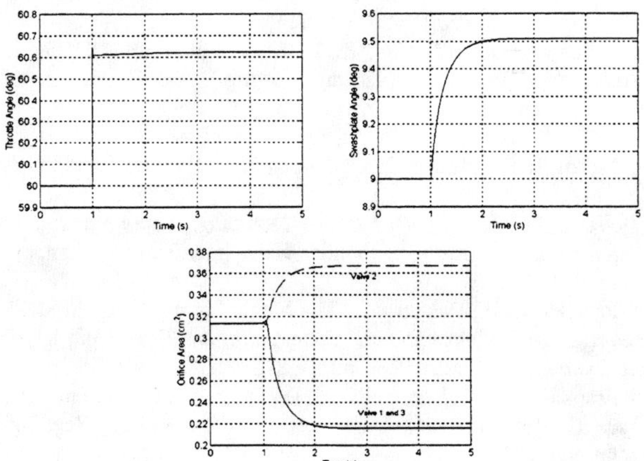

Fig. 15 Controller actions

5. CONCLUSION AND FUTURE WORK

A mathematical model describing an earthmoving vehicle's powertrain dynamics is developed. In summary, the continuous-time domain model consists of 11 states, 5 possible control inputs and 4 transport delays. Based on the simulation results, the responses of the system agree with our intuition of how the actual system would behave under certain circumstances, pending verification with experimental data. Simulation results are given both in an open loop scenario and with a full-state LQ control design. When the LQ controller is subjected to a constant disturbance it shows a steady error due to a lack of integral action.

One aspect of the future work will be focused on the matching of simulation results with experimental data taken from the EVPS test bed at UIUC. More detailed and accurate representations of the component dynamics will be developed and, if necessary, new dynamics will be incorporated. Examples of new dynamics would be the transmission line dynamics associated with the hose volume. In addition, work will be done to correlate the results to actual earthmoving vehicles. Another aspect of future work will be to develop and test more advanced and realistic control algorithms. Most likely, only a subset of the system states will be available to any control algorithm and so an output feedback approach must be investigated.

ACKNOWLEDGMENTS

The authors would like to thank Caterpillar Inc. for their generous support and encouragement.

REFERENCES

1. Singh, S., "State of the Art in Automation of Earthmoving," *Journal of Aerospace Engineering*, 1997, Vol. 10, pp. 179-88.

2. Dasgupta, S., and Chattopadhyay, A. "A basic approach by dynamic analysis of hydrostatic transmission used in a heavy earthmover," *Mining Technology*, 1991, Vol. 73, pp. 113-6.

3. Powell, J. D., "A Review of IC Engine Models for Control System Design," *Proceedings of International Federation of Automotive Control*, 1987, pp. 245-40.

4. Cho, D., and Hedrick, J. K., "Automotive Powertrain Modelling for Control," *ASME Journal of Dynamic Systems. Measurement, and Control*, 1989, Vol. 111, pp. 568-76.

5. Schoenau, G. J., Burton, R. T., Kavanagh, G. P., "Dynamic Analysis of a Variable Displacement Pump," *ASME Journal of Dynamic Systems, Measurement, and Control*, 1990, Vol. 112, pp. 122-32.

6. Merrit, H., *Hydraulic Control Systems*, Wiley, New York, NY, 1967.

7. Watton, J., *Fluid Power Systems: Modeling, simulation, analog and microcomputer control*, Prentice-Hall International, London, 1989.

8. Pacejka, H. B., Bakker, E., "The Magic Formula Tyre Model", *Proceedings of 1st International Colloquium on Tyre Models for Vehicle Dynamics Analysis*, 1991, pp. 1-18.

9. Franklin, G. F., Powell, J. D., and Emami-Naeini, A., *Feedback Control of Dynamic Systems*, Addison-Wesley, Reading, Massachusetts, 1994.

APPENDIX

A_{max}	$= 6.275\times10^{-4}$ m^2	K_p	$= 3.235\times10^{-5}$ m^3/rad^2
A_v	= 4 m^2	m	= 3250 kg
A/F_{des}	= 14.7	P_o	= 0 bar
b_{mi}	= 0.06 N.m.s	r_w	= 0.7 m
b_w	= 15 N.m.s	SA	= 30°
c_1	= 33.3	T_f	= 30 N.m
c_2	= 15.09	T_w	= 1000 N.m
c_3	= 0.196	V_h	$= 1.043\times10^{-3}$ m^3
c_4	= 245	α_{max}	= 0.314 rad
C_D	= 0.4	β	= 196.5 bar
C_d	= 0.61	Δ_{tif}	= 7.5 ms
C_l	$= 9.259\times10^{-12}$ m^3/s.Pa	Δ_{tiv}	= 24 ms
D_i	$= 3.373\times10^{-6}$ m^3/rad	ρ	= 1055 kg/m^3
F_f	= 637 N	ρ_{air}	= 1.2 kg/m^3
I_m	$= 10^{-4}$ kg.m^2	$1/\tau_p$	= 100 /s
I_p	$= 8.3\times10^{-3}$ kg.m^2	$1/\tau_v$	= 38 /s
I_w	= 20 kg.m^2		

FPST-Vol. 6, Fluid Power Systems and Technology — 1999
ASME 1999

MOTION SYNCHRONIZATION OF MULTIPLE HYDRAULIC ACTUATORS – A LINEAR MULTIVARIABLE APPROACH

Hong Sun
George T.-C. Chiu

Ray W. Herrick Laboratories
School of Mechanical Engineering
Purdue University
West Lafayette, Indiana 47907-1077
E-mail: gchiu@ecn.purdue.edu

ABSTRACT

The electro-hydraulic (EH) synchronized motion of two linear ram type hydraulic actuators is considered in this paper. It is shown that the combined system of the load and the hydraulic actuators is unstable and requires additional mechanical and/or electronic synchronization. Two EH synchronization control methods, multivariable pole placement (MPP) and Linear Quadratic Integration (LQI), are presented based on a linearized multivariable model. It is shown that the synchronization error can be penalized implicitly or explicitly using multivariable control designs. Simulation results demonstrated that decoupled control as well as flow divider techniques can not adequately address the motion synchronization issue. Both multivariable control approaches showed significant improvement compared with individual controlled cylinders.

INTRODUCTION

This paper addresses the issue of synchronizing multiple linear hydraulic actuators [3][1]. Such problem arises in hydraulic operated heavy-duty applications such as lifting equipment and rolling mills where the synchronous operation of multiple hydraulic actuators under load has important performance implications. This issue is most pronouncing in hydraulically operated multi-actuator lifting equipment. Due to uneven loading and the inherent differences in multiple hydraulic circuits and components, the lift distance among the linear actuators will be different under open-loop operation. Specifically, the overall open-loop system is unstable in the sense that the difference in lift distance will increase and will eventually results in toppling the load.

There are three approaches to address the issue of synchronizing multiple hydraulic actuators. The simplest approach is to design a flow-divider circuit that will maintain the same cylinder velocity by maintaining the same flow rate to the cylinders. The performance of the synchronization is dependent upon the performance of flow-divider as well as the compressibility of the working fluid and the consistency of the hydraulic components. Another approach is to mechanically connect the hydraulic actuators through either cabling or other linkage design. The drawback to mechanical synchronization is the added system weight and complexity as well as the limitation to the operation range of the equipment. Compared with the hydraulic and mechanical synchronization approaches, electro-hydraulic synchronization provides a flexible alternative. With EH control, synchronization control strategies can be designed to handle uneven loading as well as uncertainties and external disturbances associated with the hydraulic system. In this paper, we will investigate using multivariable control design techniques to improve the synchronization performance of multiple linear hydraulic actuators.

As the cost of electro-hydraulic component reduces, increasing amount of hydraulic systems are beginning to take the advantage of electronic control to improve performance. In 1994, Hogan and Burrow [3] looked at the issue of synchronizing unevenly loaded hydraulic cylinders. Their results indicated that to achieve synchronization of the unevenly loaded cylinder, individual control of each cylinder is required. However, with the added flexibility of individually controlled cylinders, Hogan and Burrow did not elaborate on the design of a control algorithm that will explicitly improve the synchronization performance. Xiong et al [9] proposed a model-reference adaptive control algorithm together with a cross-coupled controller [6] to improve synchronization performance as well as attempts to handle the parameter variation associated with the hydraulic systems. Chiu [2] formulated the synchronization of multiple motion axes in a geometrical framework and proposed three different approaches to explicitly address the motion synchronization issue. We will extend the linear quadratic (LQ) approach mentioned in [2] by including integral action to explicitly consider the synchronization objective as well as eliminating the steady-state synchronization error, which is defined by the difference between the two hydraulic cylinder positions.

We will consider the synchronization of two hydraulic ram type cylinders that are controlled by two individual servo valves. Figure 1 shows the hydraulic circuit diagram of this system. An obvious approach to improve synchronization is to add closed-loop position control on each of the cylinder as proposed by Hogan and Burrow [3]. However, the load coupling in the system tends to de-stabilize the system. If the feedback control does not consider the effect of load coupling, as with the individual cylinder feedback, the overall system performance will not be acceptable. To fully capture the interaction between the two cylinders, a two-input-two-output system is presented. Since the operating bandwidth of the lift is sufficiently low and majority of the operation is at constant lift velocity, a linearized multi-input multi-output (MIMO) plant model is used for designing the synchronization controller. A multivariable pole placement (MPP) approach and a LQR type design [2] are investigated. The advantage of the LQR approach is that the synchronization objective of the design can be explicitly expressed by properly selecting the performance index. To eliminate steady-state errors, the plant model is extended to include integral actions. Extensive simulation studies were done to demonstrate the effectiveness of the multivariable design. In the simulations, the linear controllers are applied to a complex nonlinear hydraulic model.

The remainder of the paper is organized as follows. The nonlinear and linearized system models are presented in the next section followed by the discussion of the pole placement and LQ control design. Simulation studies of the proposed design are presented in the fourth section. Conclusions together with discussion of on-going investigations are presented in the last section.

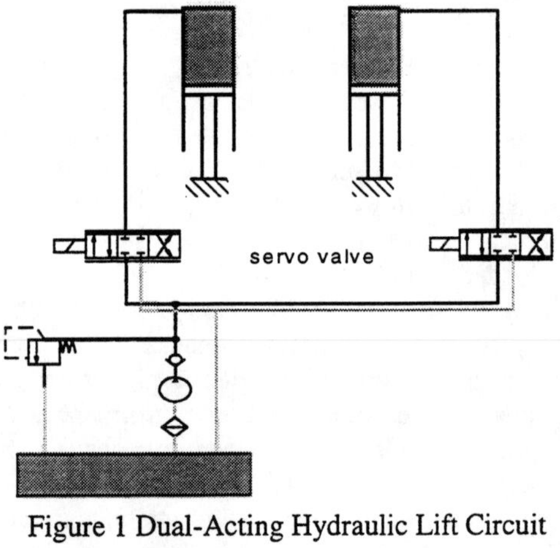

Figure 1 Dual-Acting Hydraulic Lift Circuit

SYSTEM MODELING

We will consider the synchronization of two hydraulic ram type cylinders that are coupled by four arms to lift a known load. The two cylinders are controlled by two servo-valves. The hydraulic portion of the system is shown in Figure 1 and the load coupling is shown in Figure 2. The hydraulic portion of the system can be modeled by the following differential equations [7]

$$\begin{cases} m_i\ddot{x}_i + F_{fi}(x_i,\dot{x}_i) = P_iA_i - f_i - m_ig \\ \dot{P}_i = \dfrac{\beta}{V_i(x_i)}\left(K_{qi}x_{vi}\sqrt{\left|P_S/2 + sign(x_{vi})(P_S/2 - P_i)\right|} - A_i\dot{x}_i\right) \\ \ddot{x}_{vi} + 2\xi\omega_n\dot{x}_{vi} + \omega_n^2x_{vi} = \omega_n^2u_i \end{cases} \quad (1)$$

where the subscript i =1 and 2 denote the two cylinders. In the above equations, the servo valve dynamics are modeled as second order systems with experimentally determined natural frequency ω_n and damping ratio ξ. Table 1 listed the notations used in Eq. (1).

Table 1 Notations used in Eq. (1)

m_i	respective mass of the cylinders
x_i	respective cylinder positions
P_i	pressures in the respective cylinder chambers
A_i	effective piston areas
$V_i(x_i)$	fluid volumes of the respective cylinders
β	bulk modulus of the working fluid
P_S	supply pressure
K_{qi}	valve flow coefficients
x_{vi}	valve spool displacements
u_i	valve input commands (from -10 to +10 volt)
F_{fi}	friction forces acting on the cylinder walls
F_i	effective loads on the cylinder

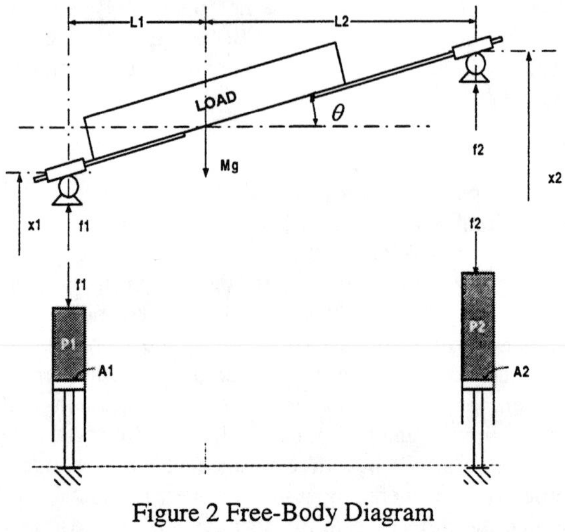

Figure 2 Free-Body Diagram

The friction forces F_{fi} are modeled using the Karnopp friction model [5] with stiction and viscous friction. The forces f_i are the coupling agents in the system which are related to the load and the cylinder motion by the following equations (see Figure 2):

$$\begin{cases} f_1 + f_2 - Mg = M\left(\dfrac{\ddot{x}_1 + \ddot{x}_2}{2}\right) \\ f_2 L_2 - f_1 L_1 = I\ddot{\theta} \end{cases} \quad (2)$$

where M is the mass of the load and the arms and I is the combined rotational inertia of the load and the arms. In Eq. (2), L_is are the moment arms from the load force f_i to the center of gravity of the system, which are assumed as unknown constants. $L = L_1 + L_2$ is the total distance between the two cylinders.

Since the design objective is to synchronize the motion of the two cylinders, it is reasonable to assume the rotating angle θ, as depicted in Figure 2, is small. Thus θ can be written as

$$\theta \simeq \tan(\theta) = \frac{x_2 - x_1}{L}$$

Substituting the above equation into the Eq. (2), we have

$$\begin{cases} f_1 + f_2 - Mg = M\left(\dfrac{\ddot{x}_1 + \ddot{x}_2}{2}\right) \\ f_2 L_2 - f_1 L_1 = I\left(\dfrac{-\ddot{x}_1 + \ddot{x}_2}{L}\right) \end{cases} \quad (3)$$

Since the servo-valve bandwidth is usually selected to be much higher than the operating bandwidth, valve dynamics are ignored in the following linearized model.

If we define the state vector

$$\mathbf{z} = \begin{bmatrix} x_1 & \dot{x}_1 & P_1 & x_2 & \dot{x}_2 & P_2 \end{bmatrix}^T = \begin{bmatrix} z_i \end{bmatrix}^T,\ i = 1, 2, ..., 6,$$

and the input vector $\mathbf{u} = [u_1 \quad u_2]^T$, combining Eqs. (1) and (3), we can obtain the following representation of the system,

$$\dot{\mathbf{z}} = \mathbf{f}(\mathbf{z},\mathbf{u}) = \begin{bmatrix} f_i(\mathbf{z},\mathbf{u}) \end{bmatrix}, \quad i = 1, 2, \cdots, 6 \quad (4)$$

where

$$f_1(\mathbf{z},\mathbf{u}) = z_2$$

$$f_2(\mathbf{z},\mathbf{u}) = \frac{1}{M_E}\begin{bmatrix} c_4\left(z_3 A_1 - F_{f1}(z_1, z_2) - m_1 g - \dfrac{L_2}{L} Mg\right) \\ -c_2\left(z_6 A_2 - F_{f2}(z_4, z_5) - m_2 g - \dfrac{L_1}{L} Mg\right) \end{bmatrix}$$

$$f_3(\mathbf{z},\mathbf{u}) = \frac{\beta}{V_1(z_1)}\left(K_{q1} u_1 \sqrt{\left|P_S/2 + sign(u_1)(P_S/2 - z_3)\right|} - A_1 z_2\right)$$

$$f_4(\mathbf{z},\mathbf{u}) = z_5$$

$$f_5(\mathbf{z},\mathbf{u}) = \frac{1}{M_E}\begin{bmatrix} -c_3\left(z_3 A_1 - F_{f1}(z_1, z_2) - m_1 g - \dfrac{L_2}{L} Mg\right) \\ +c_1\left(z_6 A_2 - F_{f2}(z_4, z_5) - m_2 g - \dfrac{L_1}{L} Mg\right) \end{bmatrix}$$

$$f_6(\mathbf{z},\mathbf{u}) = \frac{\beta}{V_2(z_4)}\left(K_{q2} u_2 \sqrt{\left|P_S/2 + sign(u_2)(P_S/2 - z_6)\right|} - A_2 z_5\right)$$

and

$$c_1 = m_1 + \frac{ML_2^2 + I}{L^2}, \quad c_4 = m_2 + \frac{ML_1^2 + I}{L^2}, \quad c_2 = c_3 = \frac{ML_1 L_2 - I}{L^2},$$

and $M_E = c_1 c_4 - c_2 c_3$.

Since the lifting operation is predominantly carried out under constant lift speed and with very low bandwidth (< 1 Hz), a linearized approach should be adequate. Linearizing the nonlinear system description, Eq. (4), at a nominal lift velocity and valve input command, a linearized model can be written as

$$\Delta\dot{\mathbf{z}} = \mathbf{A}\cdot\Delta\mathbf{z} + \mathbf{B}\cdot\Delta\mathbf{u}$$

$$= \left[\begin{array}{ccc|ccc} 0 & 1 & 0 & 0 & 0 & 0 \\ 0 & k_{22} & k_{23} & 0 & k_{c1} & k_{c2} \\ 0 & k_{32} & k_{33} & 0 & 0 & 0 \\ \hline 0 & 0 & 0 & 0 & 1 & 0 \\ 0 & k_{c3} & k_{c4} & 0 & k_{55} & k_{56} \\ 0 & 0 & 0 & 0 & k_{65} & k_{66} \end{array}\right]\cdot\Delta\mathbf{z} + \left[\begin{array}{c|c} 0 & 0 \\ 0 & 0 \\ g_1 & 0 \\ \hline 0 & 0 \\ 0 & 0 \\ 0 & g_2 \end{array}\right]\cdot\Delta\mathbf{u} \quad (5)$$

$$\mathbf{y} = \begin{bmatrix} \mathbf{y}_X \\ y_{SYNC} \end{bmatrix} = \begin{bmatrix} \mathbf{C}_X \\ \mathbf{C}_{SYNC} \end{bmatrix}\cdot\mathbf{z} = \left[\begin{array}{cccccc} 1 & 0 & 0 & 0 & 0 & 0 \\ 0 & 0 & 0 & 1 & 0 & 0 \\ \hline 1 & 0 & 0 & -1 & 0 & 0 \end{array}\right]\cdot\mathbf{z}$$

where $\Delta\mathbf{z} = \mathbf{z} - \mathbf{z}_0$ and $\Delta\mathbf{u} = \mathbf{u} - \mathbf{u}_0$. $(\mathbf{z}_0,\ \mathbf{u}_0)$ is the nominal operating point of the lift. The linearized system matrix **A** and the input matrix **B** are obtained by taking the Jacobian of the vector function **f**, i.e.

$$\mathbf{A} = \left.\frac{\partial\mathbf{f}}{\partial\mathbf{z}}\right|_{\mathbf{z}_0,\mathbf{u}_0} \quad \text{and} \quad \mathbf{B} = \left.\frac{\partial\mathbf{f}}{\partial\mathbf{u}}\right|_{\mathbf{z}_0,\mathbf{u}_0}.$$

Notice that in Eq. (5), the two cylinders' dynamics are coupled through the terms k_{ci} (i=1,2,3,4) which are given by:

$$k_{c1} = \frac{c_2 B_{p2}}{c_1 c_4 - c_2 c_3} = \frac{(ML_1 L_2 - I)B_{p2}}{m^2 L^2 + Mm(L_1^2 + L_2^2) + (M + 2m)I}$$

$$k_{c2} = \frac{-c_2 A_2}{c_1 c_4 - c_2 c_3} = \frac{-(ML_1 L_2 - I)A_2}{m^2 L^2 + Mm(L_1^2 + L_2^2) + (M + 2m)I}$$

$$k_{c3} = \frac{c_2 B_{p1}}{c_1 c_4 - c_2 c_3} = \frac{(ML_1 L_2 - I)B_{p1}}{m^2 L^2 + Mm(L_1^2 + L_2^2) + (M + 2m)I}$$

$$k_{c4} = \frac{-c_2 A_1}{c_1 c_4 - c_2 c_3} = \frac{-(ML_1 L_2 - I)A_1}{m^2 L^2 + Mm(L_1^2 + L_2^2) + (M + 2m)I}$$

with B_{pi} are the viscous friction coefficients of the respective cylinders. If the two cylinders have the identical viscous friction coefficients (B_{p1}=B_{p2}) as well as identical rod areas (A_1=A_2), then k_{c1}= k_{c3} and k_{c2}= k_{c4}, which means that the couplings between the two cylinders are the same. Since, in reality, the parameters of the cylinders are not identical, the couplings between the cylinders will depend on the cylinder characteristics as well as the loading condition (whether the load is off-centered).

The third output component, $y_{SYNC} = z_1 - z_4 = x_1 - x_2$ is the synchronization error. For lifting operations, the synchronization error is the dominating performance indicator, since the actual lift heights, y_1 and y_2, are monitored by the operators. The functional requirement of the lift is that the two cylinders will lift the load in a unisonant way ($x_1 = x_2$) while maintaining a desirable lifting speed.

The linearized system, Eq. (5), has two poles at the origin. For constant valve openings, the cylinders will reach their individual steady-state velocities. This implies that the synchronization error is unbounded under open-loop operation. Figure 3 shows the synchronization error of the nonlinear system under open-loop operation. If the system has any model uncertainty and/or subject to external disturbances, the open-loop system is unstable in the sense that the synchronization error is unbounded. Although model uncertainty and external disturbance both contribute to desynchronize the lifting operation, it is the uneven loading that has the dominating effect in desynchronize the lift.

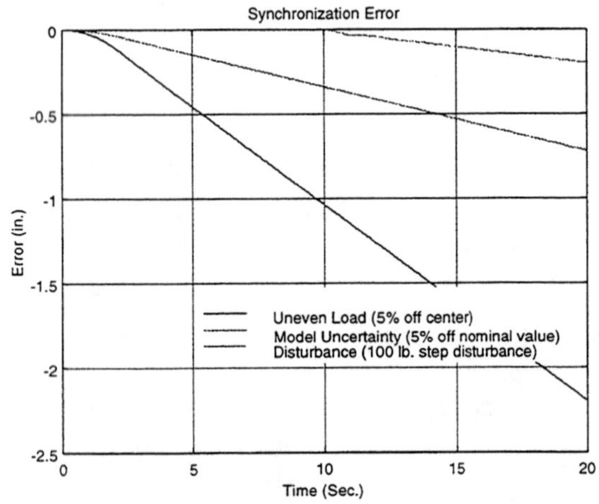

Figure 3 Open-Loop Synchronization Performance

SYNCHRONIZATION CONTROL DESIGN

Since closed-loop control is necessary to reduce the synchronization error of multiple hydraulic actuated lift systems, we will investigate two multivariable controller design approaches to address this issue. It is obvious that to reduce the synchronization error, the controller must provide special coupling effect so that the closed-loop system can reduce the effect of model uncertainties, external disturbances, and uneven loading on the synchronization error. Since the two cylinders' dynamics are already coupled through the load, see Eq. (5), a conceptually simple multivariable pole placement design should be able to improve the desired coupling. Although pole placement control can be used to assign the pole of the closed-loop system, there are no easy methods to explicitly determine the dynamics of the synchronization mode. By proper selection of the quadratic performance index, the LQ approach can be used to explicitly incorporate the synchronization objective into the controller [2].

To eliminate steady state synchronization error and reject constant external disturbances, an integral action is introduced. The integral action can be introduced by augmenting the state variable and the corresponding state equations, i.e.

$$\dot{\mathbf{z}}_E = \begin{bmatrix} \mathbf{A} & 0 \\ \mathbf{C}_X & 0 \end{bmatrix} \cdot \mathbf{z}_E + \begin{bmatrix} \mathbf{B} \\ 0 \end{bmatrix} \cdot \mathbf{u} = \mathbf{A}_E \cdot \mathbf{z}_E + \mathbf{B}_E \cdot \mathbf{u} \tag{6}$$

where $\mathbf{A}_E = \begin{bmatrix} \mathbf{A} & 0 \\ \mathbf{C}_X & 0 \end{bmatrix}, \mathbf{B}_E = \begin{bmatrix} \mathbf{B} \\ 0 \end{bmatrix}$ and $\mathbf{z}_E = \begin{bmatrix} \Delta \mathbf{z} \\ \int \mathbf{y}_X \, dt \end{bmatrix}$.

Both the multivariable pole placement and the LQ control will result in an extended state feedback control of the system described in Eq. (6). The state feedback gain can be decomposed into $\mathbf{K} = \begin{bmatrix} K & K_I \end{bmatrix}$ and the resulting incremental control law is

$$\Delta \mathbf{u} = -\mathbf{K} \cdot \mathbf{z}_E = -K \cdot \Delta \mathbf{z} - K_I \cdot \int (\mathbf{y}_X - \mathbf{y}_{Xd}) dt$$

where $\mathbf{y}_{Xd}$ is the desired cylinder position (lift height). The actual control law implemented is

$$\mathbf{u} = \mathbf{u}_0 + \Delta \mathbf{u} .$$

Figure 4 shows the implementation block diagram of the extended state feedback control.

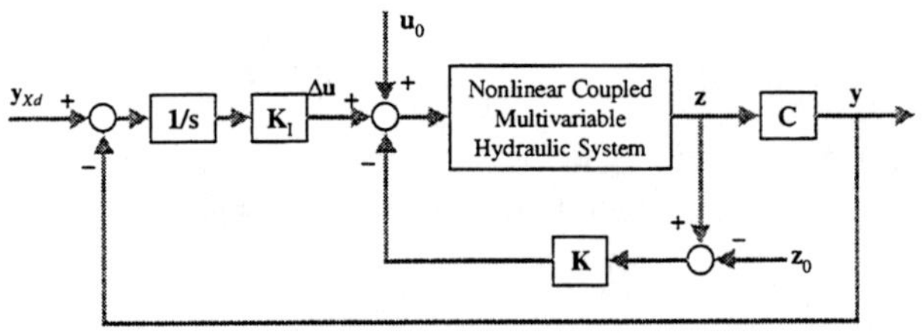

Figure 4 Synchronization Control Block Diagram

We will assume that all state information is available for feedback. When full state information is not available, a linear state estimator can be designed and the estimated state can be used in place of the actual system states.

Multivariable Pole Placement (MPP) Control

The multivariable pole placement design is based on the algorithm proposed by Kautsky and Nichols [4]. The solution of pole assignment problem for the multivariable linear system is underdetermined. To take advantage of the additional degree of freedom, a desirable property is to be considered that the poles should be insensitive to the perturbations in the coefficient matrices, i.e. $\mathbf{A}_E$ and $\mathbf{B}_E$, of the system equations. It was shown by Wilkinson [8] that the sensitivity of the closed-loop eigenvalues $\lambda_i \in S = \{\lambda_1, \lambda_2, ..., \lambda_6\}$ to the perturbations in the components of $\mathbf{A}_E$, $\mathbf{B}_E$ and the feedback gain $\mathbf{K}$ depend on the magnitude of the condition number c_j,

$$c_j = \|y_j\|_2 \|x_j\|_2 / \|y_j^T x_j\|_2, j = 1, 2, \ldots, 6,$$

where x_j and y_j are the left and right eigenvectors of the closed loop system $\mathbf{A_c} = \mathbf{A_E} - \mathbf{B_E K}$, respectively. Wilkinson [8] also showed that

$$\max c_j \le k_2(\mathbf{X}) = \|\mathbf{X}\|_2 \|\mathbf{X}^{-1}\|_2, \ \mathbf{X} = [x_1, x_2, \ldots, x_6].$$

Therefore, the MPP design problem can be formulated as:

Given the system as in Eq. (6) with $\mathbf{A_E}$, $\mathbf{B_E}$ *and the desired closed-loop eigenvalues* $\Lambda = diag(\lambda_1, \lambda_2, \cdots, \lambda_6)$, *find a real feedback gain matrix* $\mathbf{K}$ *satisfying*

$$(\mathbf{A_E} - \mathbf{B_E K})\mathbf{X} = \mathbf{X}\Lambda \quad (7)$$

and minimizing the robustness measurement condition number c_j.

The condition of the existence of such a **K,** which satisfies Eq. (7) is given in [4]:

Given eigenvalues Λ *and non-singular* $\mathbf{X}$, *then there exist a* $\mathbf{K}$ *satisfying* $(\mathbf{A_E} - \mathbf{B_E K})\mathbf{X} = \mathbf{X}\Lambda$, *if and only if*

$$U_1^T(\mathbf{A_E X} - \mathbf{X}\Lambda) = 0 \quad (8)$$

where $\mathbf{B_E} = [U_0, U_1]\begin{bmatrix} Z \\ 0 \end{bmatrix}$ *with* $U = [U_0, U_1]$ *orthogonal and Z non-singular. In addition the state feedback gain* $\mathbf{K}$ *is given by*

$$\mathbf{K} = Z^{-1}U_0^T(\mathbf{A_E} - \mathbf{X}\Lambda\mathbf{X}^{-1}) \quad (9)$$

The feedback gain matrix **K** can be obtained through following steps:

Step 1. Choose an initial eigenvector **X** which satisfies the Eq. (8).

Step 2. Calculate the condition number c_j.

Step 3. Adjust the eigenvector **X** according to the method 0 in [4] to find a new eigenvector, and recalculate c_j.

Step 4. Repeat step 3 until the condition number is minimized to some desirable value.

Step 5. Find the $\mathbf{A_c} = \mathbf{A_E} - \mathbf{B_E K}$ by solving $\mathbf{A_c X} = \mathbf{X}\Lambda$ and compute the gain matrix **K** using Eq. (9)

Linear Quadratic (LQ) Control

To explicitly incorporate the synchronization objective into the controller structure, the synchronization error should be explicitly included in the LQ performance index. For the extended system described in Eq. (6), the following performance index is used

$$J = \int_0^\infty \left\{ \begin{bmatrix} \rho_1(\Delta z_1^2 + \Delta z_4^2) + \rho_2(\Delta z_1 - \Delta z_4)^2 + \\ \rho_3(\Delta z_2^2 + \Delta z_5^2) + \rho_4(\Delta z_2 - \Delta z_5)^2 + \rho_5(y_{1f}^2 + y_{2f}^2) \end{bmatrix} \\ + \Delta\mathbf{u}^T \cdot \mathbf{R} \cdot \Delta\mathbf{u} \right\} dt$$

$$= \int_0^\infty \left\{ \mathbf{z}_E^T \cdot \begin{bmatrix} \mathbf{Q}_X & 0 \\ 0 & \mathbf{Q}_I \end{bmatrix} \cdot \mathbf{z}_E + \Delta\mathbf{u}^T \cdot \mathbf{R} \cdot \Delta\mathbf{u} \right\} dt$$

$$= \int_0^\infty \left(\mathbf{z}_E^T \cdot \mathbf{Q}_E \cdot \mathbf{z}_E + \Delta\mathbf{u}^T \cdot \mathbf{R} \cdot \Delta\mathbf{u} \right) dt \quad (10)$$

where $y_{if} = \int \mathbf{y}_X(i)dt, (i = 1,2)$. **R** is a positive definite matrix and $\mathbf{Q}_X$ and $\mathbf{Q}_I$ are positive semi-definite symmetric matrices. Notice that in Eq. (10), the weighting coefficient ρ_2 and ρ_4 corresponds to the penalty weightings on the position synchronization and velocity synchronization error, respectively.

In synchronization control, the synchronization error should be emphasized more than the tracking error terms by selecting $\rho_2 > \rho_1$ and $\rho_4 > \rho_3$. The steady-state feedback control gain that minimize the performance index J described by Eq. (10) is given by [1]

$$\mathbf{K} = [K \ \ K_I] = \mathbf{R}^{-1}\mathbf{B}_E^T\mathbf{H}$$

where the matrix **H** satisfies the algebraic Riccati equation

$$\mathbf{HA}_E + \mathbf{A}_E^T\mathbf{H} - \mathbf{HB}_E\mathbf{R}^{-1}\mathbf{B}_E^T\mathbf{H} + \mathbf{Q}_E = \mathbf{0}$$

SIMULATION RESULT

Realistic simulations were performed to evaluate the performance of the proposed synchronizing control algorithms. The linear controllers are simulated with the nonlinear coupled hydraulic system model, Eqs. (1) and (3). To evaluate the performance of the proposed controller, the following scenarios are studied:

1) Model uncertainty: One of the EH valve's flow coefficient is subjected to 5% variation from its nominal value.
2) Uneven load: The weight center of the loading is 5% off middle point between two cylinders (L/2 point).
3) Disturbance: A 100 [lb.] step friction force acts on one cylinder at time t=10 second.

As a baseline comparison, two identical SISO controllers were designed to individually control each cylinder motion, i.e. no electronic coupling between the two cylinders. Figure 5 shows the synchronization response of the SISO individual control under uneven loading when the dominating pole of the closed-loop system is set to –0.1. The result is not very impressive with a maximum synchronization error of a little over 0.1 [in]. To further increase the individual closed-loop bandwidth, another SISO control law is designed to place the dominating closed-loop pole at –1. Each individual cylinder control system is stable and has acceptable position tracking performance. However, when the two cylinders are coupled to lift an off-centered load, the load coupling effect induced oscillatory response, see Figure 6.

A multivariable pole placement controller is designed by placing a pair of dominating closed-loop poles at $-2\pm 2j$. The load coupling effect has been taken into consideration when designing the multivariable controller. When the system is subjected to the similar uneven loading as with the case shown in Figure 6, the multivariable system showed significant improvement in synchronization performance. Comparing Figure 5 with Figure 7, we see that there is much less control input chattering at the initial transient. Also the control inputs in Figure 7 reach their own steady state values, while the control inputs shown in Figure 6 are oscillating. The multivariable pole placement controller eliminated the steady-state synchronization error as well as limited the maximum synchronization error to a little over 6×10^{-3} [in].

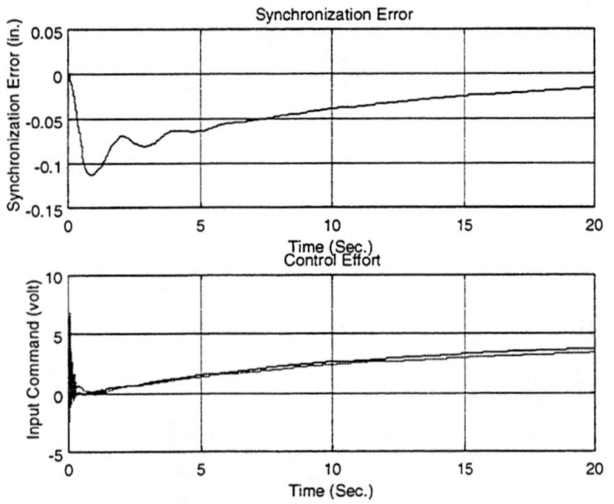

Figure 5 Individual Control with Dominating Poles at –0.1

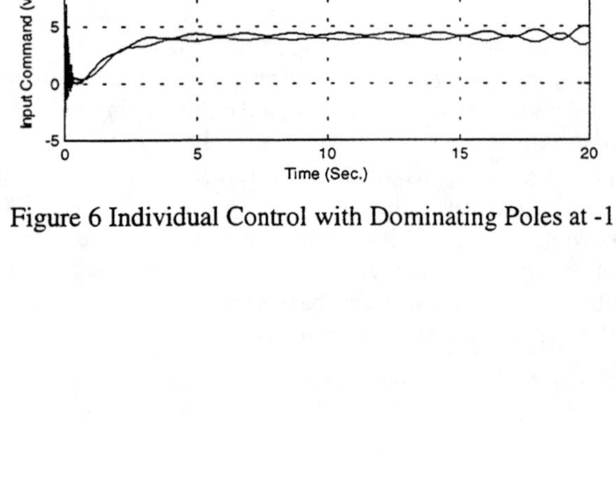

Figure 6 Individual Control with Dominating Poles at -1

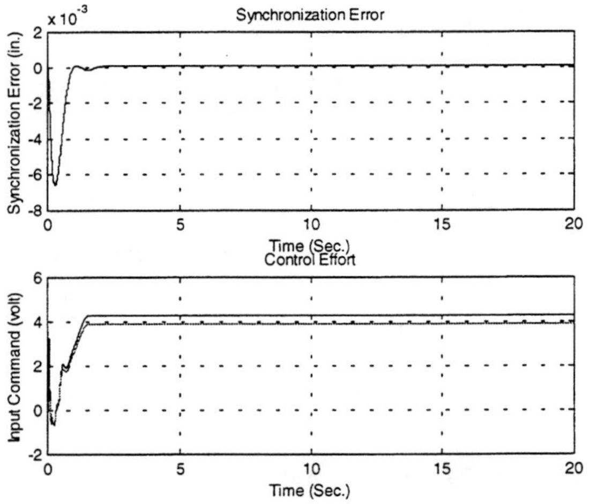

Figure 7 MPP Control with Integral Action

Another synchronizing controller is designed using the LQ approach by selecting the proper weighting coefficients to penalize the synchronization errors. The resulting dominating pole position is at –1.3. The dominating pole for the LQ approach is a bit slower than that of the pole placement design. However, the LQ synchronization performance under uneven load is about twice better than that of the multivariable pole placement control (compare Figure 8 with Figure 7).

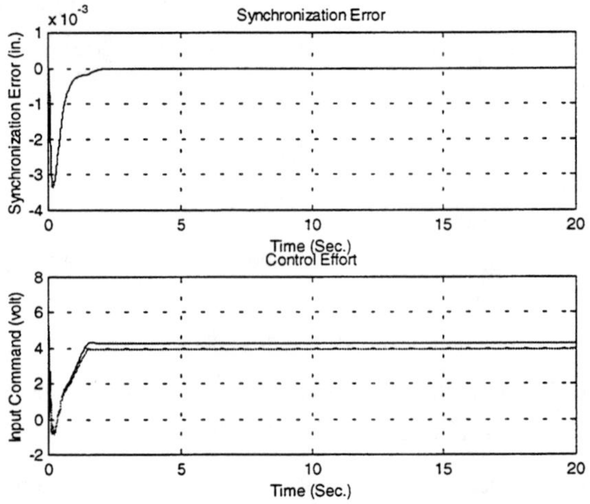

Figure 8 LQ Control with Integral Action

Under nominal operating condition with identical hydraulic components, both the pole placement and LQ synchronization controller showed significant improvement over the individually controlled system. The LQ design, which explicit consideration in minimizing the synchronization error, performs better than the comparably designed multivariable pole placement control, which is consistent with what we expected.

Simulation studies under more realistic conditions were performed. Figure 9 shows the comparison of the synchronization error among the three controllers when a step external disturbance of magnitude 100 [lb.] is applied to one of the cylinders. Both the MPP and LQ control showed significant improvement over the individually controlled system. The LQ control is still about twice better than the MPP control. Figure 10 shows the synchronization error among the three controllers under uneven loading. Figure 11 shows the synchronization error among the three controllers when one of the servo valve flow coefficients is off by 5% of the nominal value. Uncertainty in the valve flow coefficient has relatively little impact on synchronization performance. However, both the MPP and LQ show superior robustness compared with the individual control. The LQ controller performs better than the MPP controller. Figure 12 shows the result under both uneven loading and external disturbance. Similar performance improvement can also be observed.

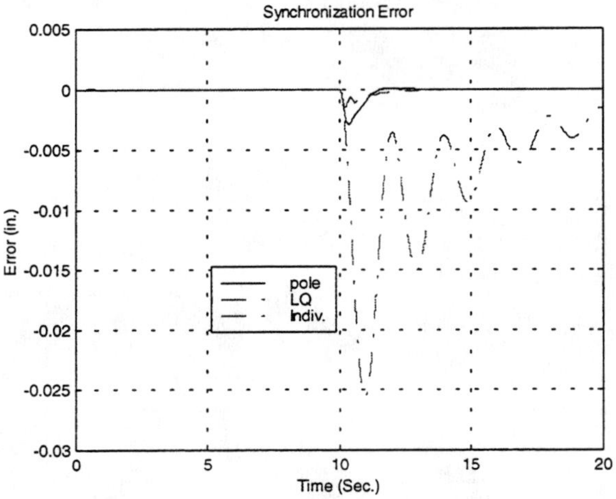

Figure 9 Synchronization Error under External Disturbance

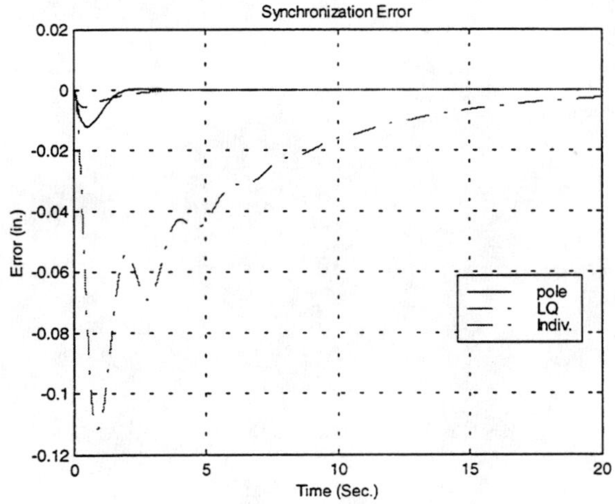

Figure 10 Synchronization Error under Uneven Loading

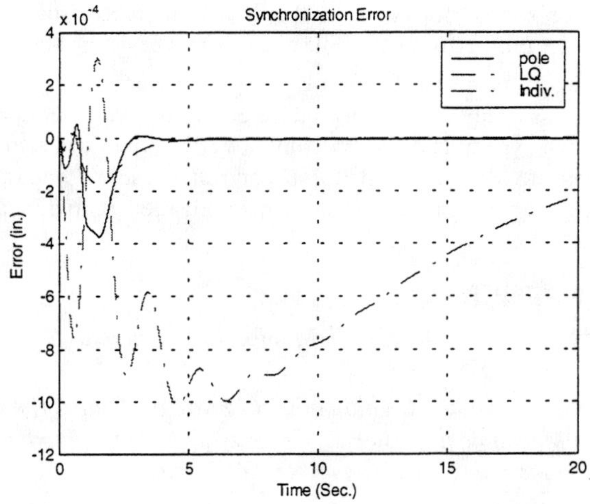

Figure 11 Synchronization Error under 5% Model Uncertainty

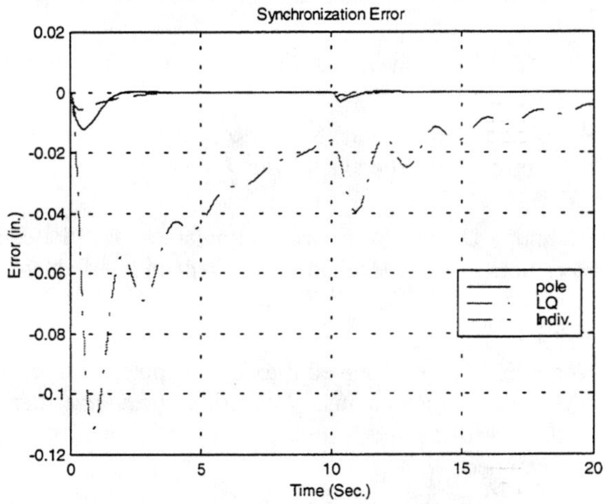

Figure 12 Synchronization Performance

CONCLUSION

The electro-hydraulic synchronized motion of two linear ram type hydraulic actuators is considered. It is shown that the combined system of the load and the hydraulic actuators is open-loop unstable in the sense that the synchronization error grow unbounded for constant velocity references. Additional mechanical and/or electronic synchronization control is needed for the system to be operational. A nonlinear multivariable model of the system is developed as well as the linearized model. The multivariable modeling approach to synchronization control is necessary because of the load coupling between the two cylinders. Two linear multivariable controllers are designed based on simple pole placement as well as LQ optimal control. Simulation results demonstrated that the multivariable approach to synchronization control could significantly improve the system performance as well as robustness. In addition, by explicitly consider the synchronization objective in the LQ formulation, the LQ controller out-performs the pole placement types controller.

Experimental verification is being performed. Analytical robustness analysis of using linear controller with nonlinear hydraulic system is also under investigation. Work is also in progress to design a nonlinear controller using Lyapunov synthesis approach for motion synchronization of redundant hydraulic actuators.

REFERENCE

[1] Anderson, B.D.O. and Moore, J.B., *Optimal Control,* Prince Hall, 1990

[2] Chiu, G.T-C, *Coordination Control of Multiple Axes Mechanical Systems: Theory and Applications to Machining and Computer Disk File Systems*, Ph.D. Dissertation, University of California at Berkeley, 1994.

[3] Hogan, P. and Burrows, C.R., "Synchronizing Unevenly-Loaded Hydraulic Cylinders," *Proceedings of the ASME Fluid Power and Systems Technology Division,* Vol. 1, pp 75-80, 1994.

[4] Kautsky, J. and Nichols, N.K., "Robust Pole Assignment in Linear State Feedback," *Int. J. Control,* Vol. 41, pp. 1129-1155, 1985.

[5] Karnoop, Dean, "Computer Simulation of Stick-Slip Friction in Mechanical Dynamic Systems," *Transactions of the ASME,* Vol. 107, pp100-107, March 1985.

[6] Koren, Y., "Cross-Coupled Biaxial Computer Control for Manufacturing Systems," *Journal of Dynamic Systems, Measurement, and Control,* Vol. 102, pp 265-272, December 1980.

[7] Merritt, H.E., Hydraulic *Control Systems,* Wiley, New York, 1967.

[8] Wilkinson, J.H., The Algebraic Eigenvalue Problem, Oxford University Press, 1965.

[9] Xiong, Y.F., Lequoc, S. and Cheng, R.M.H., "Adaptive Control of a Synchronizing Servo-System," *SAE Technical Paper Series,* p 1-5, 921621, 1992

[10] Yeaple, F., *Fluid Power Handbook,* Dekker, New York, 1984.

AUTHOR INDEX

FPST-Vol. 6
Fluid Power Systems and Technology — 1999

Book Number: G01141